GEOGRAPHY

FOR CCEA GCSE LEVEL

Tim Manson

ISBN: 978-1-78073-032-5

First Edition
Second Impression, 2014

Layout and design: April Sky Design
Printed by: W&G Baird Ltd, Antrim

Page 281 constitutes an extension of this copyright page.

Colourpoint Educational
An imprint of Colourpoint Creative Ltd
Colourpoint House
Jubilee Business Park
Jubilee Road
Newtownards
County Down
Northern Ireland
BT23 4YH

Tel: 028 9182 6339
Fax: 028 9182 1900
E-mail: info@colourpoint.co.uk
Web site: www.colourpoint.co.uk

The Author

Tim Manson learned to love Geography from an early age. He is a graduate of QUB, the University of Ulster and the Open University. He started the Geography department at Slemish College in 1997 and teaches GCSE and A Level Geography, and Travel and Tourism. He is a Senior Examiner for an awarding body in Geography and is a keen advocate for creative uses of ICT in learning and teaching. He has a highly successful website: www.thinkgeography.org.uk

Acknowledgements

I'd always wanted to write a book but it took a lot more time and effort than I ever thought possible. Thanks to Colourpoint for giving me the opportunity to realise a dream. Big thanks to my editor Rachel Irwin who kept me on my toes throughout, to Wesley Johnston for the great diagrams and to Margaret McMullan for her advice and critical eye! Thanks also to my colleagues Michael Bennett and Gillian Johnston and the Slemish College Geography department – two of the best geography teachers that I have come across and who continue to inspire me to be better, every day! Thanks to all the students who have to put up with me and in particular to the Manson Year 12 class of 2012 – some of their ideas appear here – be proud! Thanks to the family for the holidays, craic and babysitting! Finally, to my wife, Helen and kids who have had to put up with a lot of absences in the last year – thanks for helping make all the 'field trips' special!

For Erin and Isaac

How to use this book

This book is designed to be as engaging as possible. It is not enough to read through any book, learn the content and think that you are ready for the exam. To that end I have tried to include as many questions as possible throughout this book to give you practice at answering the many types of question that could be asked in the final examination.

The text covers the entire CCEA GCSE Geography course and has been written to follow the specification in a logical pattern. It begins by addressing the three Physical Geography themes that make up the Unit 1 examination paper, followed by the three Human Geography themes that make up the Unit 2 examination paper. Each theme is broken down into three to four parts, which follow the content of the specification and address the learning outcomes. The final section of the book looks at Unit 3, the Controlled Assessment, and provides some advice on how to write up the fieldwork task, worth 25% of the final GCSE mark.

The text is illustrated throughout by diagrams, photographs, maps, graphs, numerous examples and case studies to reinforce learning. It also uses several different types of exercises to develop understanding:

- *Key questions*: These are examples of the types of questions you would expect to find on a GCSE exam paper. They can be used at the beginning of each part to see how much students already know about the topic. Once the topic has been studied, students can answer the questions again, this time in as much detail as possible, to see how much they have learned.
- *Key words:* These are lists of the important terms connected to each topic and they appear at the beginning of each part. Exam questions often ask students to 'complete the definitions of the key words listed'. This exercise encourages students to start writing their own definitions using the text in the book.
- *Test yourself:* These questions are designed to test understanding and ideal for teachers to set as homework. Each question is followed by the number of marks (in brackets) that would be awarded if it was an exam question. This gives students an indication of the detail needed for each answer.
- *Structured notes:* These are extended activities which will provide detailed notes to aid revision preparation.
- *Organise your revision:* These provide tips, advice and specific revision guidance to help break the subject content down into digestible chunks.
- *Skills questions*: These questions are designed to test and practice specific geographic skills. They focus on reading Ordnance Survey maps, aerial photographs and GIS maps.
- *Exam practice questions:* These are a selection of past paper questions from the most recent GCSE examination papers. These are designed to test students' knowledge and understanding, either in class or set as homework. As with the test yourself questions, each question is followed by the number of marks (in brackets) to indicate the detail needed for each answer and how long to spend on writing it. Mark schemes for the different questions can be found on CCEA's Geography microsite (http://www.ccea.org.uk/geography/)

The GCSE Geography course expects both Foundation and Higher Tier students to learn the same geographical information. With this in mind, the book has been written to appeal to all abilities, with the language designed to engage those studying for the Foundation Tier but containing sufficient material to challenge those studying for the Higher Tier. Past paper questions are also provided to enable both sets of students to practice their exam answering skills.

Introduction

Welcome to *Geography for CCEA GCSE Level*. If you are reading this it probably means that you have chosen to study Geography for GCSE. Well done and congratulations on choosing one of the most exciting, vibrant and up-to-date subjects! The thing that makes Geography different from many other subjects is that it is always changing, and amazingly we still do not know everything about the world that we live on. Scientists are constantly analysing and changing what we know to be true about our planet, so I have included, where possible, facts and figures for some of the most recent global examples.

Using this book will not guarantee an A* in GCSE Geography – but it will give students a really good start! The text is written to cover the entire CCEA GCSE Geography course, with Units 1 and 2 addressing the exam part of the GCSE, and Unit 3 offering some guidance on how to write up the Controlled Assessment. The book mostly follows the structure of the specification, with a little reordering of topics in Unit 1 where I felt it benefited the teaching order. However, the clearly unitised structure is designed to be flexible and allow teachers to organise their teaching in a way that suits them and their students.

The only way for students to get the grade they want is to make sure that they work hard and think their way through the whole course. It's one thing to know all the geographical facts, but it's something else to be able to use these facts to argue a point in a GCSE answer. With this in mind, this book is full of exam style (test yourself) and past paper questions (examination practice questions) to give plenty of practice at structuring exam answers to achieve the best marks possible.

All the best in your studies.

Tim Manson

The author has produced a number of other resources that could add to your learning experience:

- Online Learning – www.thinkgeography.org.uk
- GCSE Revision Podcasts – www.thinkgeography.org.uk (can also be found on Apple iTunes)

Contents

Contents...

Unit 1:
Understanding Our Natural World

1A The Dynamic Landscape

1B Our Changing Weather and Climate

1C The Restless Earth

1A The Dynamic Landscape

"A man of wisdom delights in water"
Confucius

One of the most fascinating features of geography is the interaction between the various components in our natural world. Our world is always changing. There are numerous processes at work that are constantly altering our landscape. Sometimes these processes move at a fast rate, making sudden and dramatic differences to our landscape, whilst at other times the changes can be so slow that they go unnoticed, even over many years.

This theme takes a close look at how rivers and coasts help to shape the land and then looks at some of the ways that we try to manage these changes.

Theme 1A is divided up into five parts:

1. **The Drainage Basin**
2. **River Processes and Features**
3. **Sustainable Management of Rivers**
4. **Coastal Processes and Features**
5. **Sustainable Management of Coasts**

PART 1: THE DRAINAGE BASIN

(a)

The components of the drainage basin cycle
Inputs: *precipitation*
Stores: *interception by vegetation*
Transfers: *surface runoff/overland flow, infiltration, throughflow, percolation and groundwater flow*
Outputs: *river discharge*

The characteristics of a drainage basin
Watershed
Source
Tributary
Confluence
River mouth

Key words

Water Cycle
Drainage Basin
Watershed
Source
Tributary
Confluence
Mouth
Discharge
Interception
Groundwater
Surface runoff
Overland flow
Infiltration
Throughflow
Percolation
Groundwater flow

Key questions

By the end of this section you will be able to answer the following questions:

1. State the meaning of a drainage basin.
2. State the meaning of a watershed.
3. What affects the amount of interception?
4. What is the difference between infiltration and percolation?
5. What affects the rate of infiltration?
6. Name three transfers/flows in the drainage basin.
7. Study the OS map of a river on page 22. Give the six figure grid reference for the confluence of a river and its source.
8. Study the OS map on page 22. Do you think the Glenarm river would have a large or small discharge. Give one reason for your answer?

The components of the drainage basin cycle

A **drainage basin** is described as the area of land that is drained by a river and its tributaries. When a droplet of water falls onto the land **(precipitation)** the force of gravity pulls the water downhill and back towards the sea. The drainage basin is a very important part of the **water cycle.**

The water cycle

The water or hydrological cycle is a natural system where water is in constant movement above, on or below the surface of the earth, and is changing state from water vapour (gas), to liquid and into ice (solid).

Figure 1
The water cycle

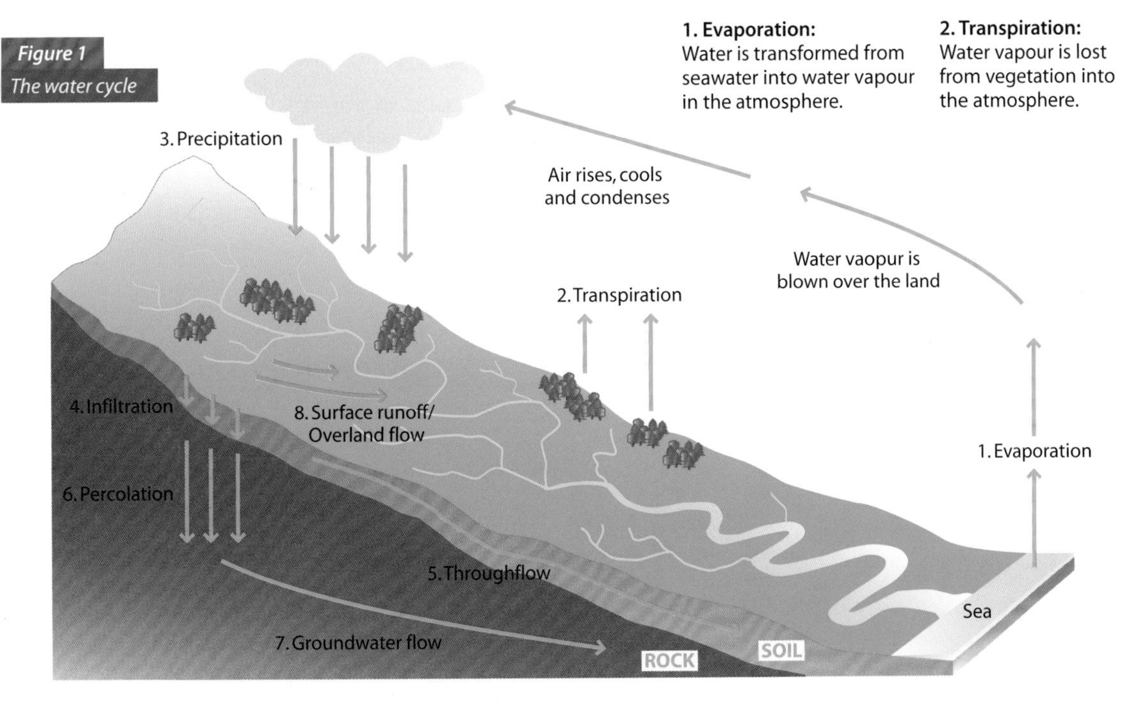

1. Evaporation: Water is transformed from seawater into water vapour in the atmosphere.

2. Transpiration: Water vapour is lost from vegetation into the atmosphere.

3. Precipitation: Water vapour condenses into drizzle, rain, sleet, snow and hail, and this falls towards the surface of the land.

4. Infiltration: Water soaks (filters) into the soil.

5. Throughflow: Water moves downhill through the soil.

6. Percolation: Water moves from the soil and into the rock.

7. Groundwater flow: Water moves slowly through the soil and rocks back into the sea.

8. Surface runoff/ Overland flow: Water moves across the surface of the earth, becoming a stream, tributary or river.

The drainage basin system

A major component of the water cycle is the drainage basin system. This system contains inputs, stores, transfers and outputs, examples of which are listed below.

INPUTS	STORES	TRANSFERS	OUTPUTS
precipitation	interception (from vegetation)	surface runoff/overland flow	river discharge
	soil moisture	infiltration	
	groundwater	throughflow	
	surface storage	percolation	
		groundwater flow	

Figure 2
Components of the drainage basin system

Figure 3
The drainage basin system

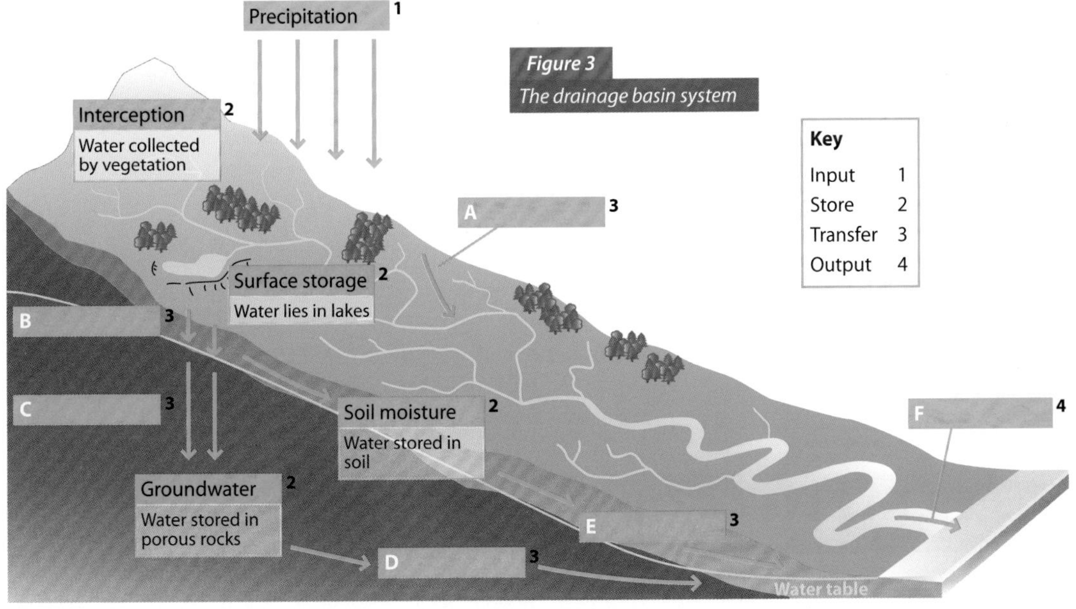

1. Inputs: are when water is introduced or put into the system.

2. Stores: occur when water is kept within the system and not moved through it.

3. Transfers: are processes or flows within the system, where water is moved from one place to another.

4. Outputs: occur in the river system when the water is carried through the river and back into the sea.

Structured notes

1. Make sure that you know and understand the definitions of all of the parts of the water cycle and drainage basin system.
2. Make a copy of Figure 3, the drainage basin system, and replace boxes A, B, C, D, E and F with labels that identify the five transfers and one output that make up this system.

The characteristics of the drainage basin

The amount of water on the earth is approximately 1.4 billion cubic kilometres. 97% of this is stored in the earth's seas and oceans, 2% is stored as icecaps and glaciers, and about 0.3% is freshwater on the surface of the earth as part of the drainage basin system.

Water arrives on the land and makes its way back to the sea. The force of the water on the land is an important factor in shaping our dynamic and ever-changing landscape.

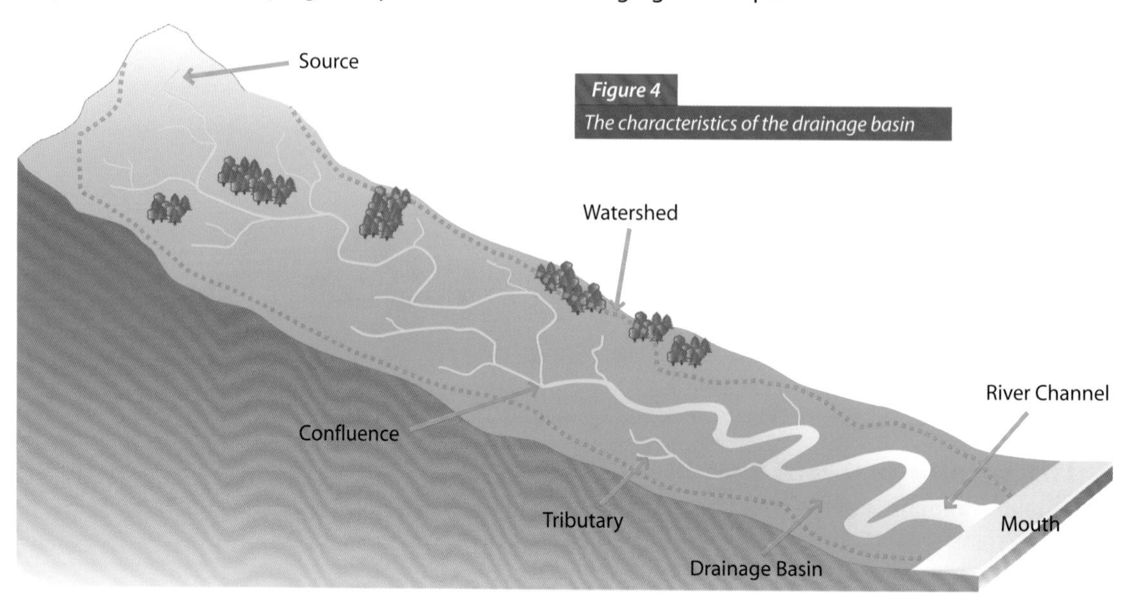

Figure 4
The characteristics of the drainage basin

Structured notes

1. Make a copy of Figure 4, the components of the drainage basin.
2. Match up the following parts of the drainage basin with the correct definition.

Source	A small river or stream.
Confluence	Where drops of water join to start a river.
Watershed	The place where the river flows into the sea.
River channel	Where two rivers meet and join.
Mouth	The main body of water, flowing downhill.
Tributary	The dividing line between one drainage basin and another.

Check your learning

Now that you have studied Part 1: The Drainage Basin, return to page 8 and answer the Key Questions for this section.

Test yourself

1. What is the name given to the key input of the river basin system? (1)
2. Name all of the transfers/flows in the system. (5)
3. What affects the amount of interception? (1)
4. Is river discharge an input, store or output of the system? (1)
5. Define the following terms:
   • Infiltration • Percolation • Throughflow • Surface Runoff (4)

PART 2: RIVER PROCESSES AND FEATURES

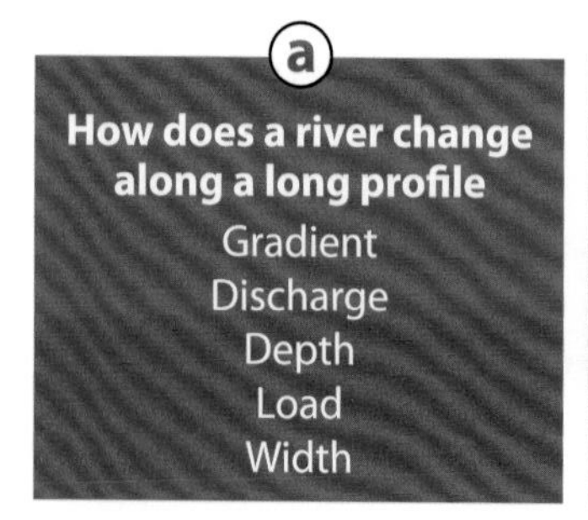

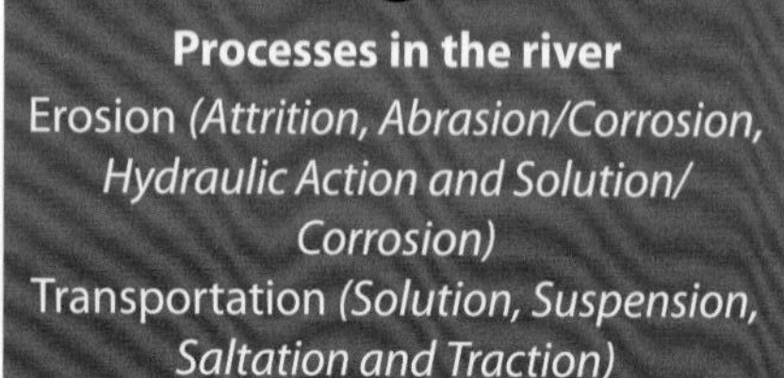

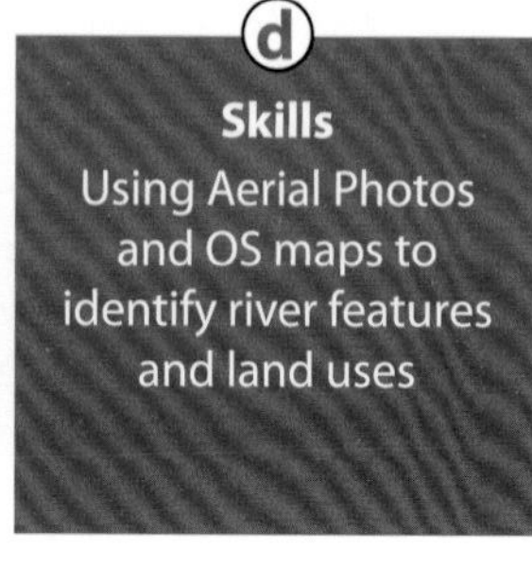

Key questions

By the end of this section you will be able to answer the following questions:

1. State the meaning of 'discharge'.
2. Describe and explain how a river's discharge changes as you move downstream. Why might it alter throughout the year?
3. The load is smallest near the mouth of the river. State fully one reason why this is so.
4. What form of transportation is most common in a river?
5. Describe the process of corrosion (solution).
6. A river becomes wider and deeper as you move downstream. Explain the processes that cause this.
7. Using labelled diagrams, explain how a waterfall forms.
8. Draw a cross section through a meander and add labels.
9. Why does deposition occur on the inside of a meander?
10. Draw a cross section through a floodplain.
11. On a floodplain, why is the finest material deposited furthest from the river?
12. What are the main features of a river that you would expect to be able to recognise easily on an OS map?

Key words

Erosion
Attrition
Abrasion (Corrasion)
Hydraulic Action
Corrosion (Solution)
Transportation
Solution
Suspension
Saltation
Traction
Deposition

How does a river change along a profile?

As we have already seen, rivers form a very important part of the drainage basin system. As they transfer water from one place to another they are constantly changing the land over which they flow. However, the journey that any river makes on its way from mountains to the sea is never easy and the characteristics of the river change as the river descends.

Rivers can be divided up into three distinct stages: the upper course, the middle course and the lower course. In each course a different set of influences are working on the river, which change the features and processes in each stage.

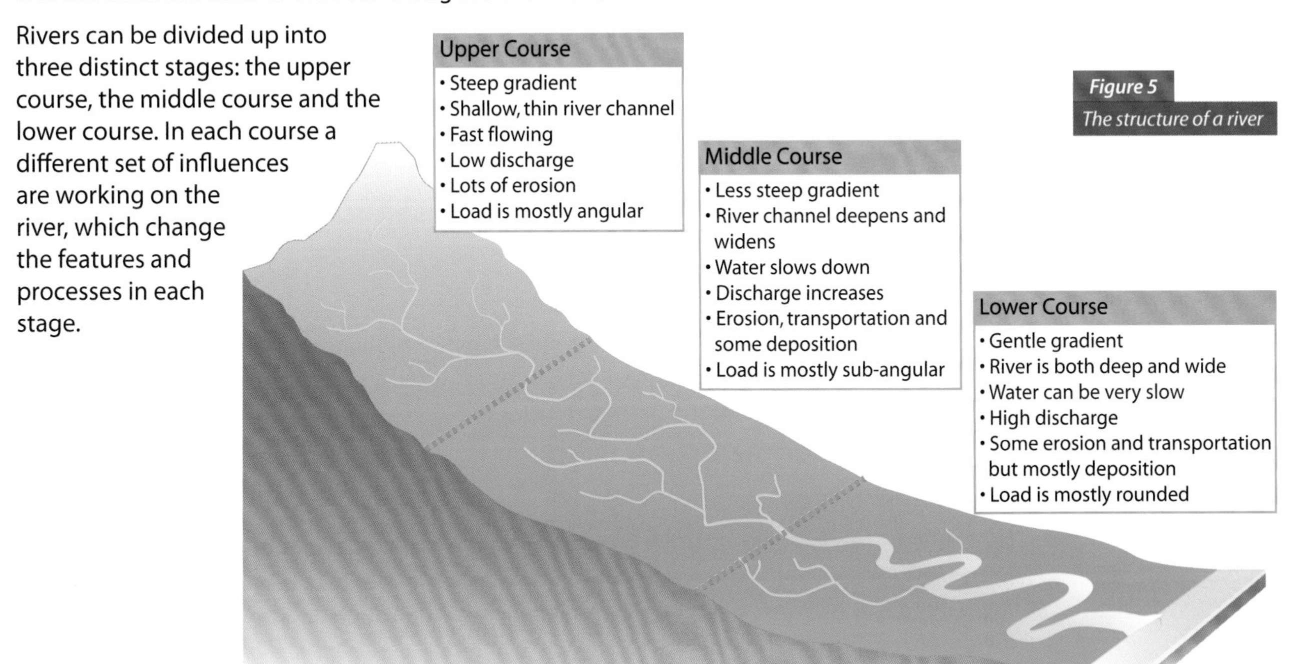

Figure 5
The structure of a river

What causes changes in the river?

The **gradient** in a river changes considerably through the different courses. In the upper course, in the hills and mountains, the gradient will be very steep and gravity pulls the small amounts of water downhill at a relatively fast rate. As the gradient starts to level off in the middle course, the force of gravity on the water and the speed of the water both decrease. In the lower course, the gradient becomes gentle to almost flat by the time that the river reaches the sea at its mouth.

In the upper course of the river, the **depth** and **width** of the river are small. This is because the river is only beginning its journey to the sea. However, as tributaries and streams join up with the main channel, more and more water will start to make its way downstream and cause the river channel to deepen and widen.

Upper Course	Middle Course	Lower Course
The river uses weathering and erosion to erode downwards in the hills to form a v-shaped valley. The river channel is thin and shallow.	The river starts to meander, cutting from side to side to erode the valley. There is some erosion at river cliffs. A floodplain starts to form. The river channel is deeper and wider.	The river meanders over the wide flood plain in s-shaped curves. There is deposition of alluvium/silt across the floodplain. The floodplain is flat and wide. The river channel is deep and wide.

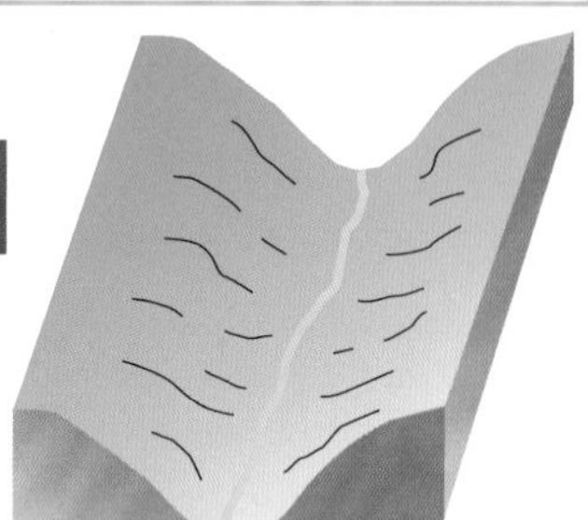

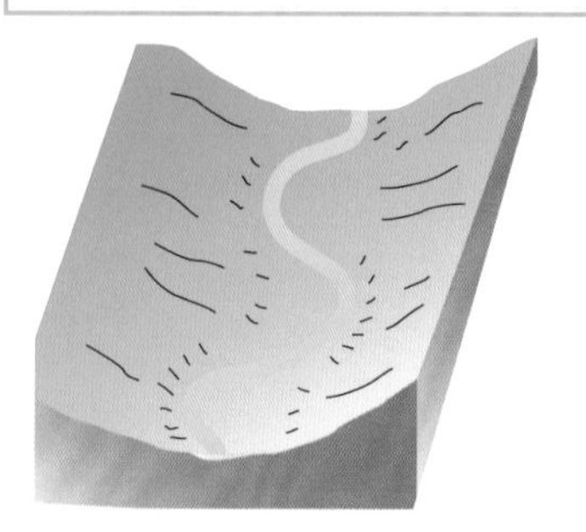

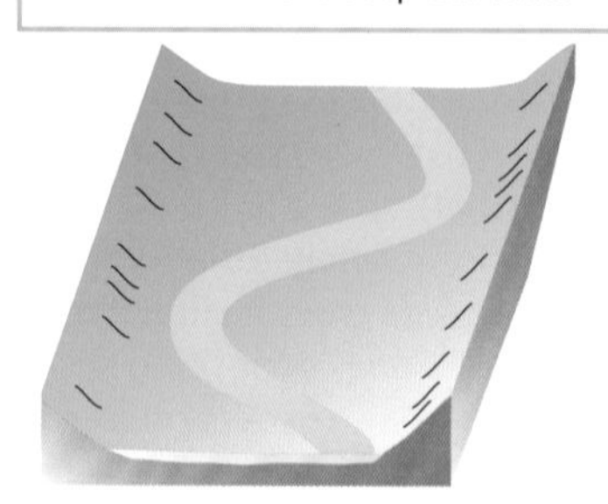

Figure 6
Changes to the river channel downstream

The **discharge** in a river is the amount of water that passes a particular point in a river at a particular time. This is usually measured as cubic metres of water per second (cumecs). Discharge is calculated by multiplying the cross sectional area of the river channel at a point along the river by the speed of the water (or velocity).

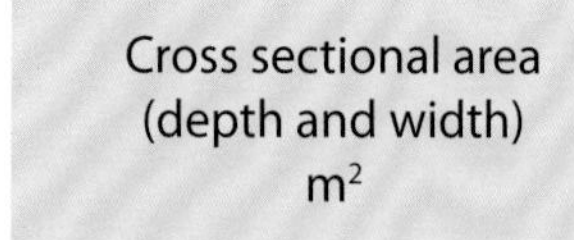

×

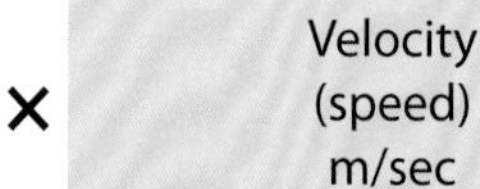

=

The **load** of a river is the material that the river is carrying. This is mostly material that has been eroded from the river bed and the river banks. The size of the load can range from tiny pieces of sediment up to huge boulders. The faster the velocity (and discharge) of a particular river, the more load that can be carried. In river flood conditions, huge lumps of material can be transported through the river system.

Structured notes

1. Use Figure 5 to make notes about the main differences between each of the sections of the course in the river.
2. Figure 6 indicates some of the key changes that happen to the river as it moves from one course to the next. Draw a diagram for each of the courses and see if you can add three or four labels to explain what is happening in each case.

Test yourself

1. Look at Resource A and Resource B below and list the main differences that you notice between the characteristics of each river. (4)
2. Resource A was photographed in the upper course of the river and Resource B in the lower course. Copy and complete the table in Resource C in as much detail as you can and attempt to describe the river. (12)
3. Resource D shows the Bradshaw model of downstream change in a river. What are the main things that it summarises? (5)

Resource A

Resource B

Resource C

	Upper course	Lower course
Gradient		
Depth of river		
Width of river		
Av speed of water		
Discharge		
Load size		

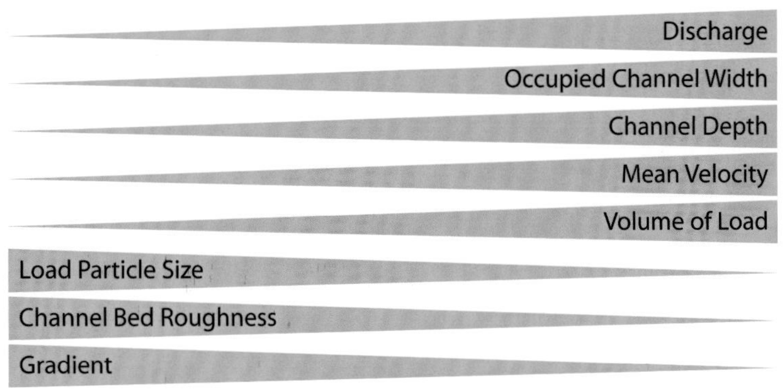

Resource D

Discharge
Occupied Channel Width
Channel Depth
Mean Velocity
Volume of Load
Load Particle Size
Channel Bed Roughness
Gradient

Processes in the river

A river process is simply a description of something that happens in the river. These processes work together to shape the river. The amount of water in the river system is always changing so this constantly modifies the impact of the processes at different times and at different sections along the course of the river.

Erosion processes

Erosion happens in a river when parts of the river bed and/or river bank get eroded and are removed from the landscape.

There are four types of erosion in the river:

Attrition

Attrition takes place when stones that are being carried downstream knock against each other and start to wear each other down. This knocks the edges of the stones and results in smaller, rounder sized stones further downstream.

Abrasion/corrasion

The force of the moving water in the river throws the stones and other eroded particles that it is carrying against the bed and banks of the river, and this dislodges more material. It works like a piece of sandpaper grinding and eroding the rocks.

Hydraulic action

This is when the force of the water pounds into the river bed and banks and dislodges more material. It works a bit like a power hose.

Solution/corrosion

This is when weak acid (chemicals) in the water react with the rock and dissolves soluble minerals. It happens often in limestone areas.

Structured notes

1. When you have taken some notes on the four different types of erosion in the river, see if you can draw a diagram or cartoon that will help you remember how these different types of erosion can take place.
2. Why do you think more erosion might take place when the river is experiencing flood conditions?

Test yourself

Read the following passage to see if you can identify the type of erosion being described at numbers 1–4 below.

Canoeing down the Six Mile Water

By Mark McNeilly

"The river was just about safe enough for us to canoe from close to the Dunadry hotel down into Antrim and to our cars in the Loughshore car park. The river was nearly in full spate. As I paddled down over a small weir I noticed that the speed of the water was removing some of the rock at the bottom of the river bed.1 Further downstream, at a big bend, some rocks were being lifted by the river and were banging against the soil and rock and causing it to break off.2 When I was getting closer to Antrim the speed of the water increased and I noticed that some stones were being carried in the river and were banging into each other.3 At the end of my trip I was quite thirsty and I went to scoop up some of the water in my hands to drink but noticed that there were tiny black and white specks floating in the water.4"

ORGANISE YOUR REVISION

Some people like to revise using lists and can 'bullet point' the main facts behind each type of erosion. For those of you who like to see 'the big picture' and how what you learn connects to other things, you might like to try a mind map.

Resource A

How to draw a mind map: Erosion in the river

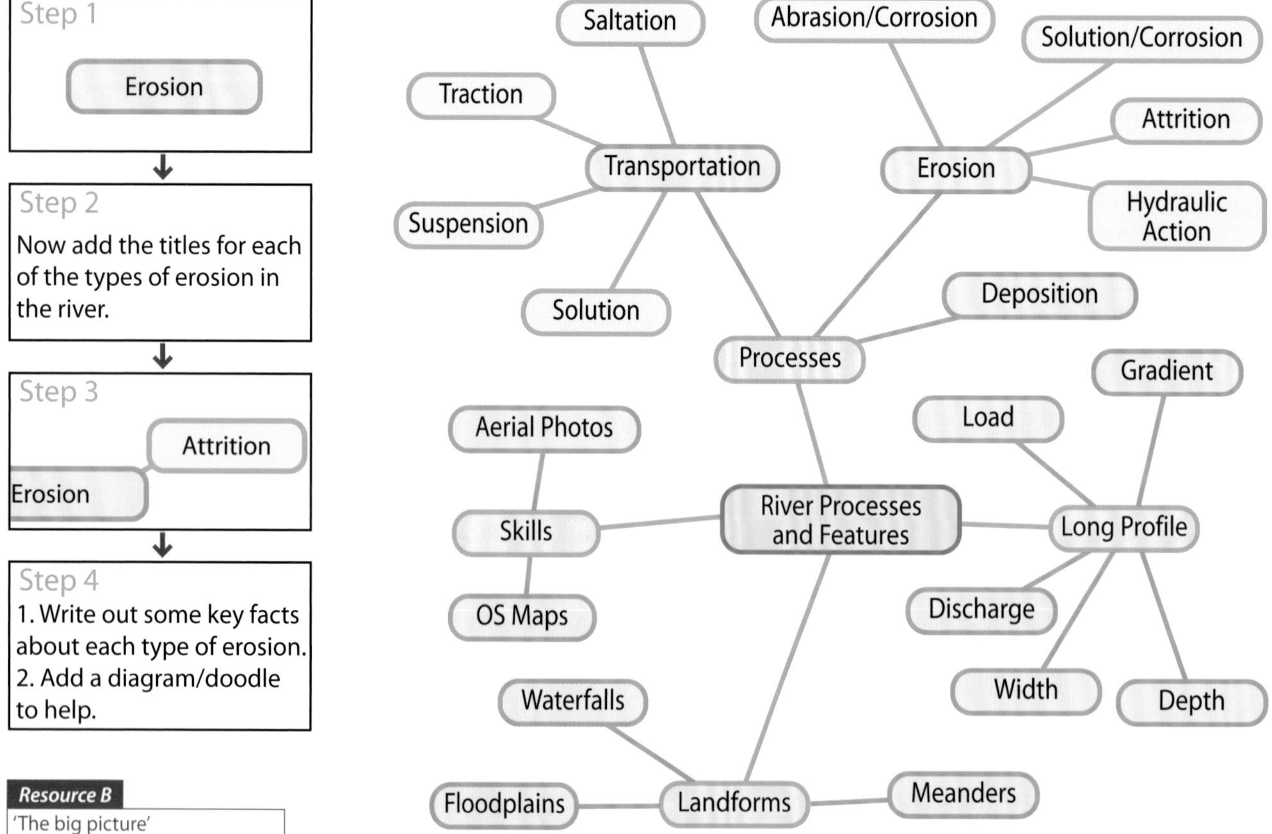

Resource B

'The big picture'
An example of a mind map for river processes and features.

Transportation processes

Transportation is when the eroded material in the river is carried from one place to another through the river system. There are four types of transportation in the river:

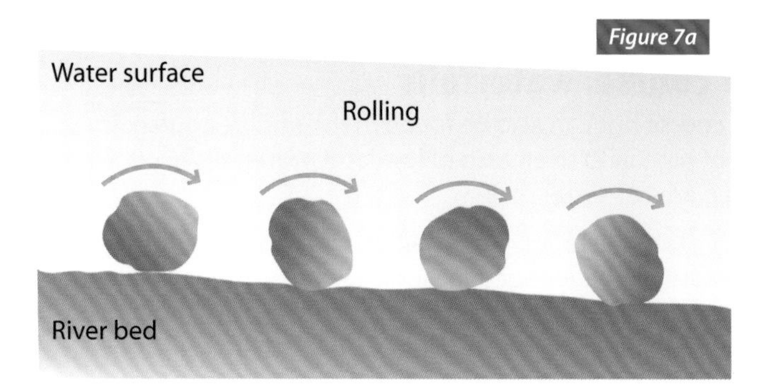

Traction

This is when the heaviest particles of eroded material are **rolled** along the river bed. Usually these stones and boulders can only be moved when the river has a large volume of water in it.

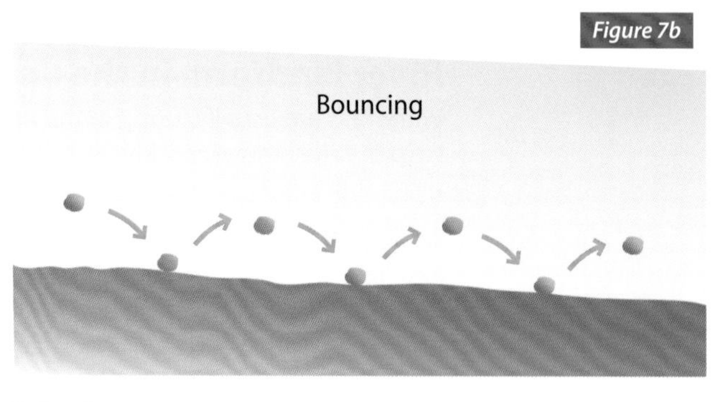

Saltation

This is when some of the heavier particles are not held up in the flow of the river all of the time. Instead they may be **bounced** along the river bed.

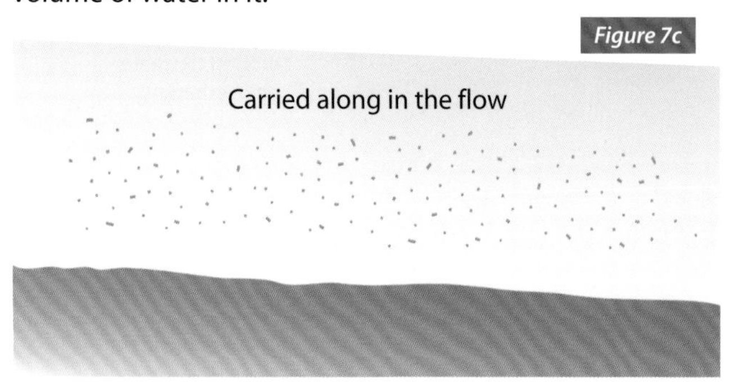

Suspension

As the speed of the water increases, the river is able to pick up larger particles and stones in its flow. When particles are **carried along in the flow** of the water and do not make contact with the river bed, they are suspended within the water.

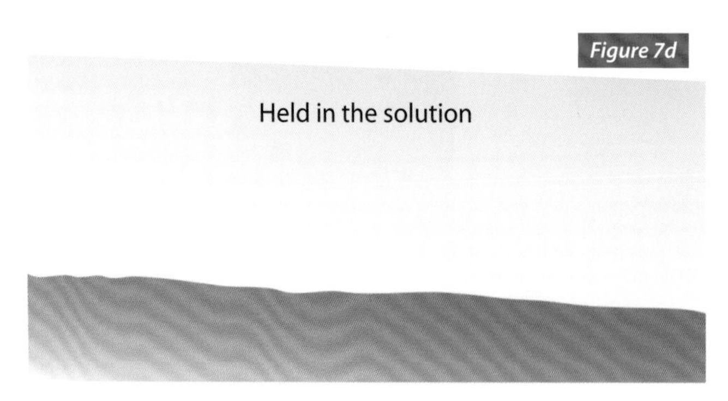

Solution

This happens when some minerals (like limestone) dissolve easily in water and the microscopic particles are **held up in the solution of the water.**

Deposition processes

Deposition is where the river load becomes too heavy for the river to carry and is dumped or deposited along the course of the river.

Test yourself

1. What evidence would you look for to find erosion in a river? (2)
2. Which type of transportation in the river do you think is the most effective? (3)
3. Resource A shows an annotated diagram of a cross section through a river bend. Make a copy of the diagram and complete the labels.

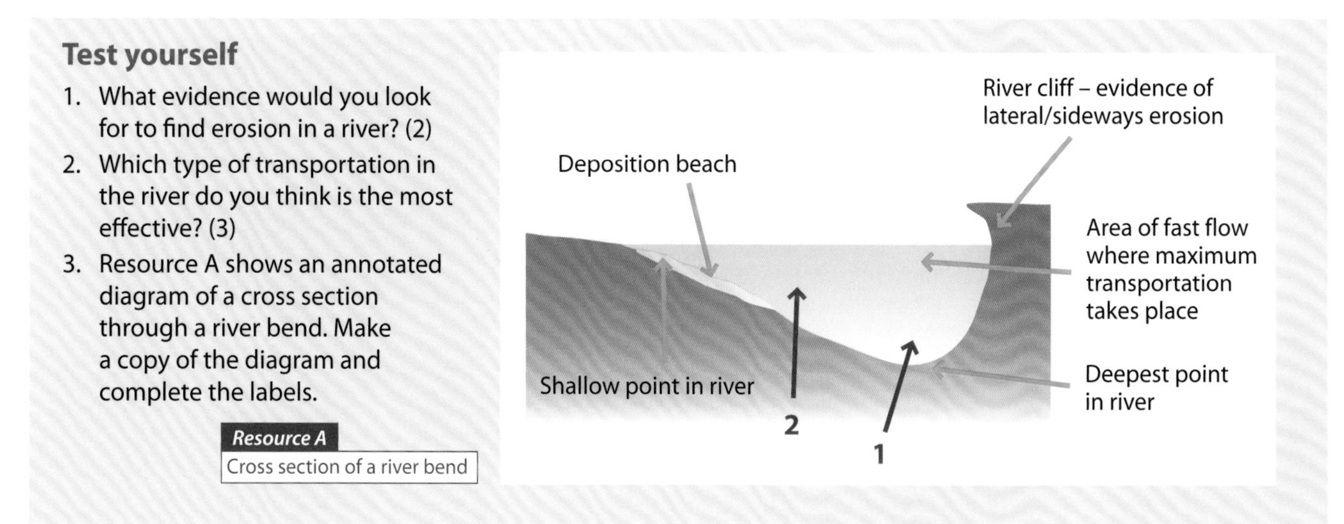

Resource A
Cross section of a river bend

River landforms

As the water (surface runoff) moves across the surface of the earth, the three distinct processes of erosion, transportation and deposition work together to shape the land further. The processes create specific types of landform in each of the different courses.

River landform in the upper course: waterfalls

Waterfalls are usually found in the upper course but can also be found in the middle course. They help the river move quickly from an area of high land to an area of low land. Waterfalls occur when rivers move from an area of hard rock to an area of softer rock.

As the water moves over the hard rock (for example, Granite or Basalt), it will be able to erode the exposed and less resistant soft rock more easily. Hydraulic action, abrasion and attrition all work to erode the rock, which is then transported further downstream by the river. However, the main process at work here is erosion. It defines the nature of this landscape, causing the edge of the waterfall to recede and leaving a gorge in its wake.

Figure 8
The formation of a waterfall

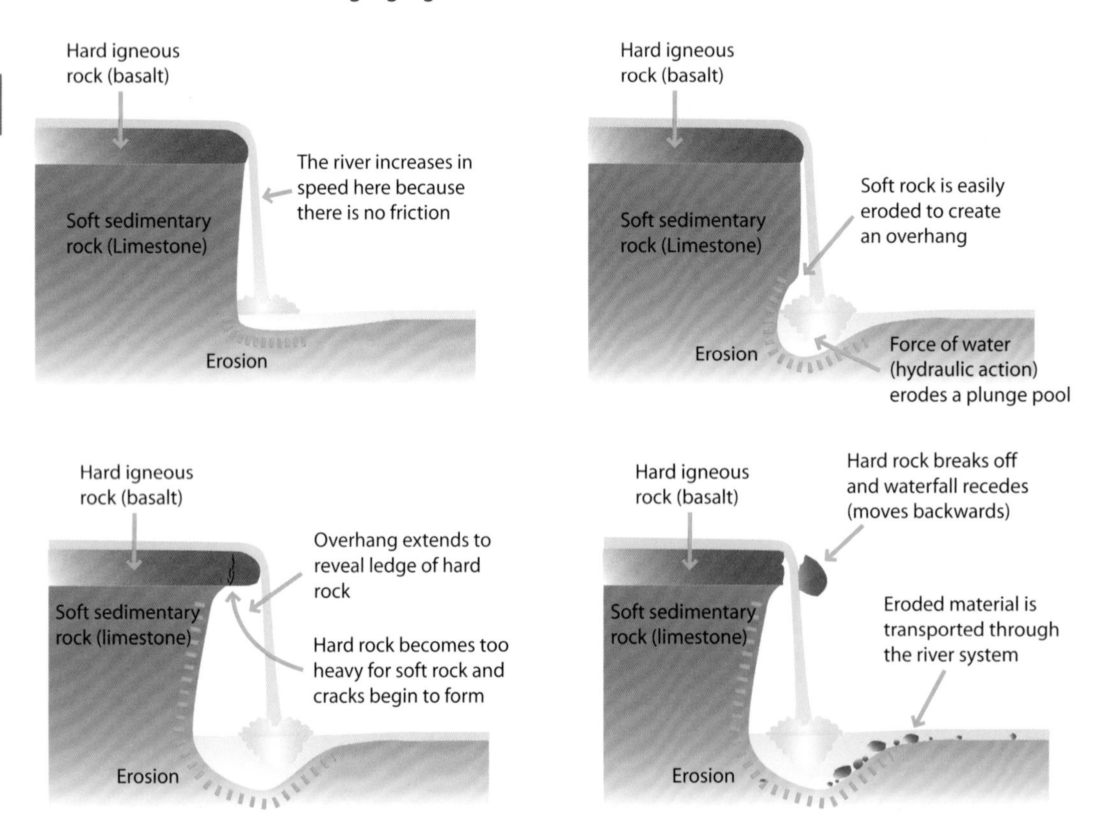

Figure 9
The formation of a Gorge

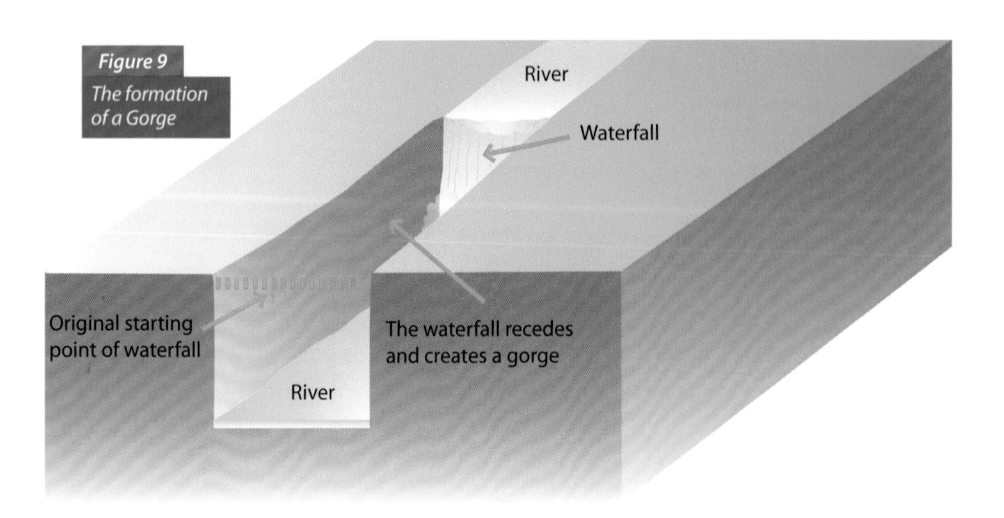

A famous waterfall where this occurs is on the border between the USA and Canada, on the Niagara River. The waterfalls which form here have been called the Niagara Falls and there is evidence that these are receding at a rate of about 1 metre each year.

Figure 10
Niagara Falls

The island of Iceland has many different types and shapes of waterfall as the meltwater from glaciers is taken from the high land of the basalt rocks towards the sea. In each case the force of the water has carved a new channel through the rock.

Figure 11
Erosion of the soft rock at the base of the cliff at the Gulfoss waterfall, Iceland

Figure 12
Erosion features at the Skogarfoss waterfall, Iceland

Erosion cliff
Deposition beach
Erosion cliff (erosion due to the speed of the water hitting the river bank on the **outside** of the bend)
Deposition beach (build-up of material on the **inside** of the bend)
Erosion cliff
Deposition beach
Deposition beach
Erosion cliff
Arrow indicates place of fastest flow within the river.
X Y

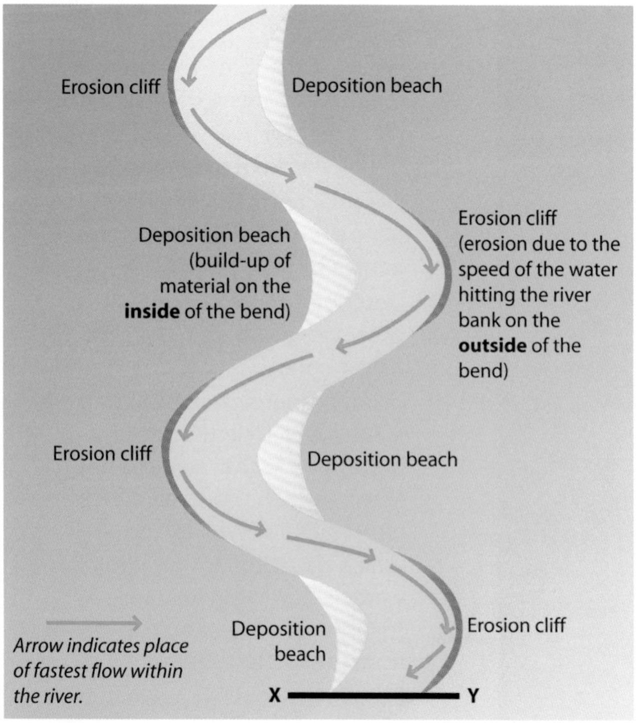

Figure 13
Erosion and deposition in the river channel

River landform in the middle course: meanders

Meanders are the bends in the river channel that usually start to occur in the middle course of the river.

Rivers rarely flow in straight lines. Geographers are still unsure as to why meanders form but erosion and deposition have a big part to play. As the river makes its way down the river valley, the volume and velocity of the water increase (the discharge increases). When the water moves fast, this increases the opportunity for erosion to take place. However, at some points in the river, the water will move slowly and deposition will take place. Within the middle and lower course of the river erosion can happen in close proximity to deposition.

As the water moves down through the river meanders, the fastest

Figure 14

Meanders and ox bow lakes on a river in Northern Alberta, Canada

flow tends to be towards the outside of the river bend. This faster flow causes more erosion (hydraulic action, abrasion), which deepens the river bed and can widen the river channel by eroding a river cliff at the side. The slowest flow is on the inside of the river bend. Here, the river does not have enough energy to continue to transport the eroded material, so it starts to lay this material down (deposit) on the inside of the river bend. As more and more material is deposited, a build up of silt/alluvium is created. This is called a deposition beach.

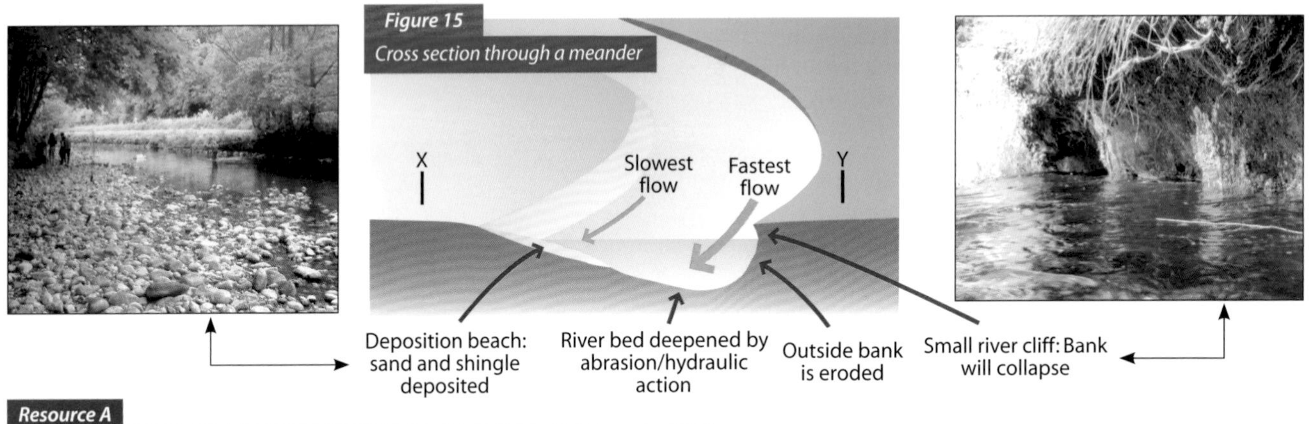

Figure 15

Cross section through a meander

Resource A

The Mighty Mississippi (near the border between Arkansas and Mississippi, USA)

Source: NASA Goddard Space Flight Center

Test yourself

Study Resource A which shows meanders formed along the course of the Mississippi river in the USA. Answer the questions below in as much detail as you can.

1. What are the three river processes that help to shape the river into a meander? (3)
2. Define the term meander. (2)
3. Describe and explain what happens on the outside of a river bend. (3)
4. Describe and explain the processes at work on the inside of a river bend. (3)
5. Using Resource A, explain one feature which demonstrates deposition in the river and one feature which shows evidence of erosion in the river. (4)
6. Which direction do you think the water is flowing in Resource A? Justify your answer with evidence from the satellite image. (3)
7. Explain in detail two ways that erosion is shaping the channel at location A in Resource A. (4)

The formation of ox bow lakes

Ox bow lakes are river features that usually occur in the lower course when erosion and deposition can modify the path that a river takes.

Figure 16
The formation process of ox bow lakes

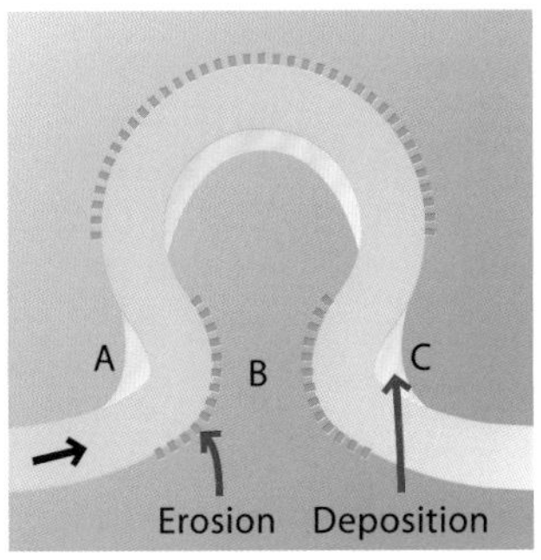

Step 1:
The river is meandering as usual, with deposition taking place on the inside of the bend (A and C) and erosion occurring on the outside of the bend to create river cliffs (B).

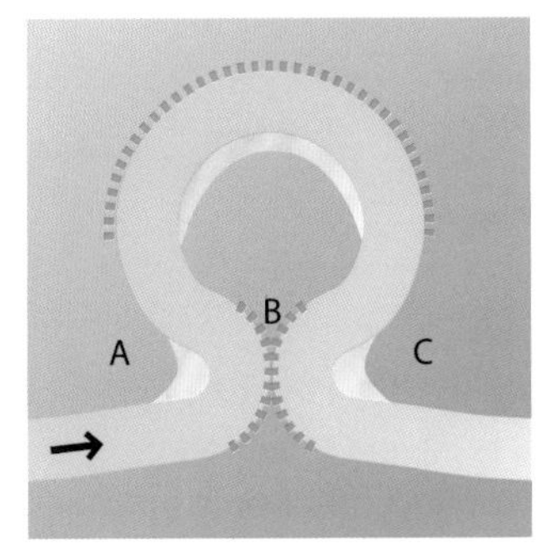

Step 2:
Erosion continues at B, on the outside of the bend. During a river flood, when the velocity and discharge of the river are much higher, there is more erosion and the river cuts through the neck.

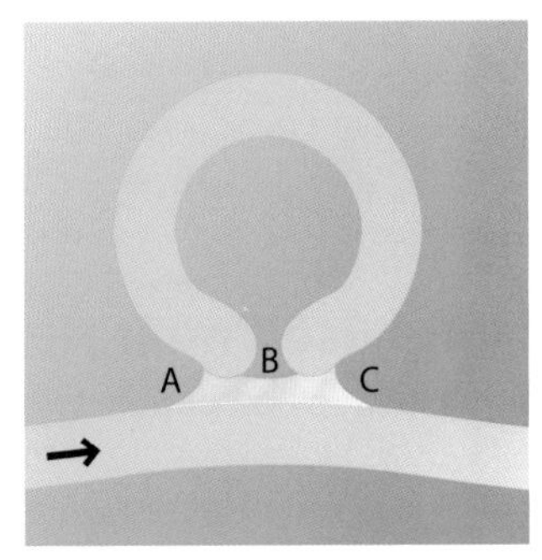

Step 3:
Deposition takes place and starts to block up the neck of the river (A, B and C). The meander is permanently cut off from the new, straight river channel.

Step 4:
The former river channel now forms an 'ox bow lake', where the water will gradually infiltrate into the soil and evaporate, leaving the river bed exposed and likely to be populated by surrounding plant species over time.

River landform in the lower course: floodplains

Figure 17
The floodplain

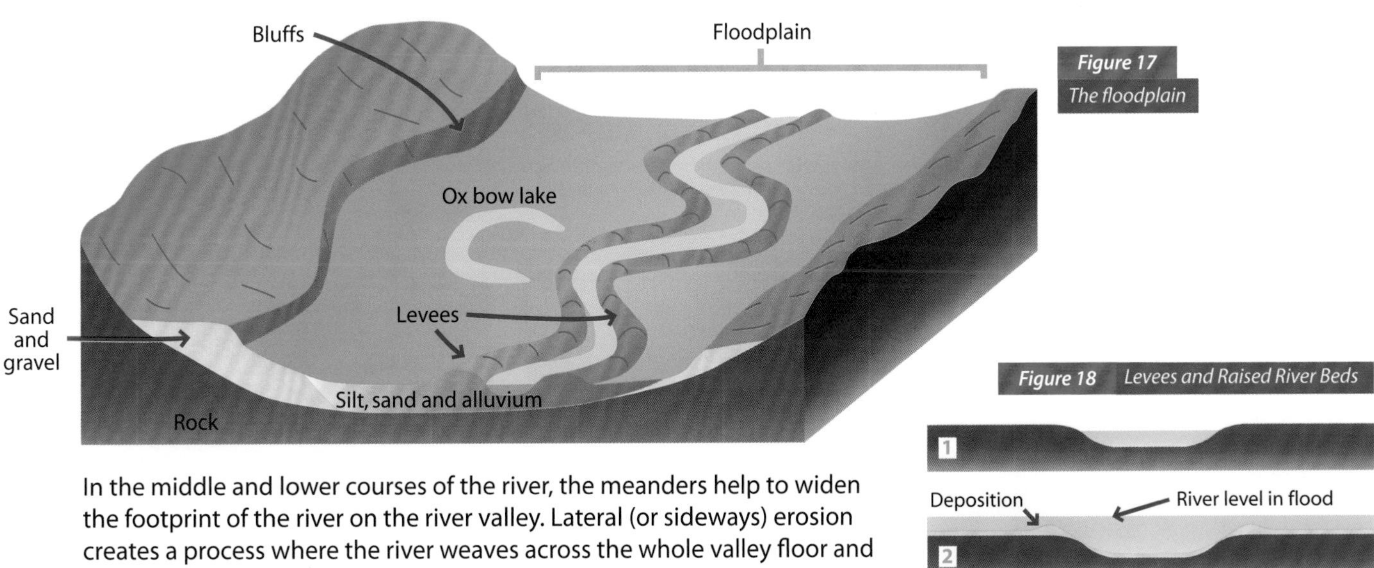

In the middle and lower courses of the river, the meanders help to widen the footprint of the river on the river valley. Lateral (or sideways) erosion creates a process where the river weaves across the whole valley floor and is always trying to change the shape or path of where it goes. At times of flood, the erosion and deposition processes in the river can radically change the river's position.

The floodplain is the area of land that has been covered with the silt deposited by the changing course of the river. It is essentially the area of land over which the river is likely and able to flood.

Every time that the river does flood, it will deposit a layer of fine sand/silt, called alluvium on the valley floor. This helps to maintain the flat river floodplain.

Figure 18 *Levees and Raised River Beds*

1

Deposition River level in flood

2

When the river floods, it bursts its banks and deposits coarse material close to the river bank.

This process continues year upon year, over a long period of time and gradually more material is deposited at the river banks and along the river bed.

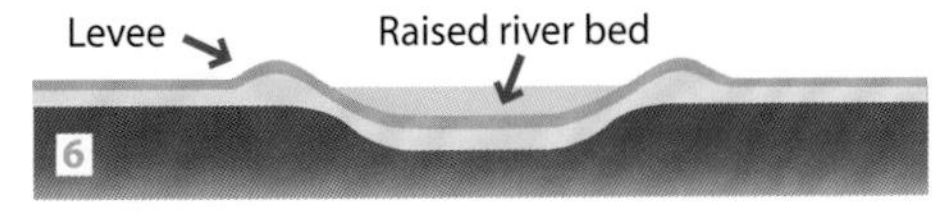

Over time, the river has built up raised banks called levees and a flood plain consisting of fine material.

The formation of levees

As the river floods and continues to deposit material across the valley floor, the largest, coarsest material will be dumped first, close to the river bank. This forms a raised bed called a levee. The levees are formed by the repeated flooding of the river. These can build up over a long period of time (even centuries).

Structured notes

Use the information about ox bow lakes, flood plains and levees to makes notes on the following key points:

Ox bow lakes	**Flood plains**	**Levees**
What are the main processes that cause ox bow lakes to happen?	How do flood plains get built up over time?	How and why do levees form at the side of rivers?

Figure 19

Some examples of land uses near rivers

How is land used near rivers?

Throughout history, when people were trying to decide where to build a house or a settlement, they often favoured locations close to rivers. Rivers provided people with a defensive position, drinking water, a source of power (watermills), water for irrigation and silt to add nutrients to the soil. Rivers also allowed easy transport from one place to another, which promoted trade.

Photograph A shows the town of Enniskillen, Northern Ireland, which is built on an island within the River Erne.

Source: Adfern

Photograph B shows how a watermill was used to power simple machinery over 200 years ago in Annalong.

Source: John Holm

Photograph C shows how water from the rivers in Nebraska is used to irrigate the surrounding arid landscape.

Photograph D shows how commerce and business has built up around the River Thames in London.

Source: Jim McDougall

Photograph E shows how the river is used to transport goods and materials inland on the River Rhine.

Photograph F shows how the river can attract people for recreation and tourism on the River Cam in Cambridge.

Test yourself

1. Photograph A shows the town of Enniskillen, Northern Ireland which is built around some islands in the River Erne. The building marked with an X is Enniskillen Castle. Describe why this might have been seen as a good place to build a defensive site. (3)
2. Take a look at Photograph B. Explain why factory owners might have wanted to locate their factory close to a river. (3)
3. Photograph C shows irrigation being used from river water in Nebraska, USA. Give two reasons why you think that local farmers might value land close to the river. (4)
4. Photograph D shows an example of how many major cities across the world have built up surrounding major rivers. What might the disadvantages for this be? Describe and Explain one disadvantage of a city being so close to the river. (4)
5. Photograph E shows how rivers can be used for business and commerce. Describe one positive issue of this process. (3)
6. Many people use rivers for recreation (Photograph F). Think of one use and explain in detail one negative issue with this use. (3)

Skills: Using aerial photos and OS maps to identify river features and land uses

Using aerial photographs

Photographs can be an essential primary and secondary source of information for geographers. They help to visualise an area, providing a better idea of the different physical and human geographical influences that are at work, and showing how the theory in geography can happen in the 'real world'.

Resource A

The Hudson and East Rivers surrounding Manhattan Island, New York City

Source: Doc Searls

Test yourself

1. Identify how the land is being used at location A. (2)
2. Draw a field sketch of this photograph. Start by drawing the frame for the photo and then sketch the course of the river through the frame.
   See if you can label the following on your sketch map:
   - Manhattan Island (the long thin island between the two rivers).
   - Governors Island (south of where the Hudson and East rivers meet).
   - Central Park (a long rectangular park on Manhattan island).
   - Hudson river (to the West).
   - East river (to the East).
   - The Bridges which cross the East River are called (from South to North)
     - Brooklyn Bridge
     - Manhattan bridge
     - Williamsburg bridge
     - 59th Street Bridge
   - The island in the middle of the East river is called Roosevelt Island.
   - The Bridge at the top of the picture (on the Hudson) is called the George Washington bridge. (11)
3. Name two features on the aerial photo which give evidence that deposition takes place at this part of the river. (2)
4. Give two examples which show evidence that erosion has helped to shape this landscape. (2)

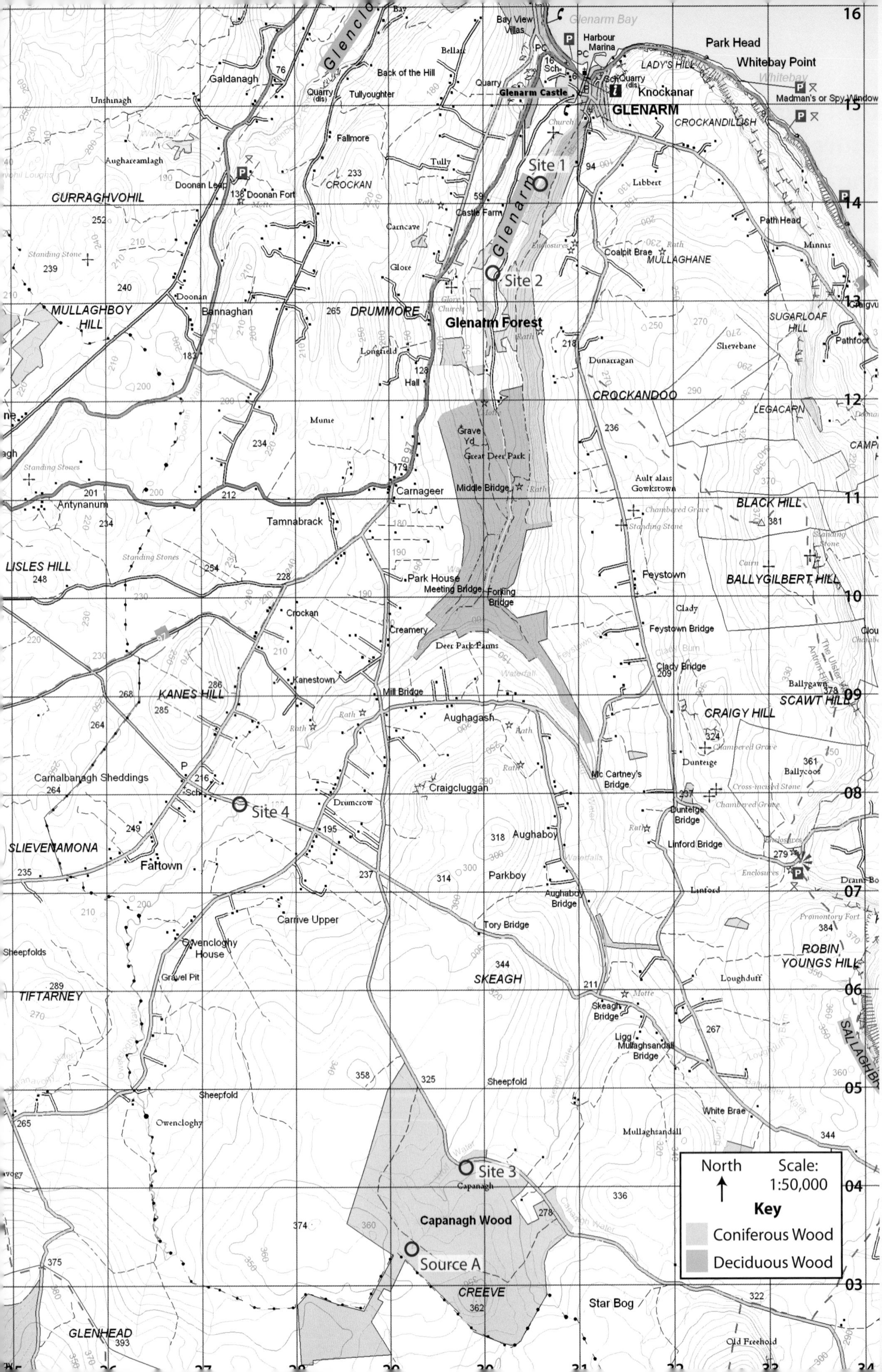

Resource B

Map of the Glenarm river basin

North ↑

Scale: 1:50,000

Key

Coniferous Wood

Deciduous Wood

Using OS maps to identify river features

Rivers are physical features that really stand out on maps. It can be quite easy to see the journey of a river in the upper and middle course on a map, as this takes place on high ground. However, many rivers then flow through urban areas, where identifying the river features can be more difficult.

Study Resource B carefully and see if you can answer the questions which follow.

Test yourself

Basic map skills

1. What is the name of the forest in grid square 3012? (1)
2. What is found at the following six figure grid references? (3)
   a. 265104
   b. 320093
   c. 304093
3. The mouth of the Glenarm river is beside a church with tower in Glenarm village. What is its six figure grid reference? (2)
4. What height is the top of Tiftarney in grid square 2505? (1)

Understanding rivers on maps

5. What is the name of the river feature found at 300100? (2)
6. The tributaries of the Glenarm river have many starting positions. Source A is one such source. Give six figure grid references for two other sources of this river. (4)
7. Site 3 is found at grid reference 298042 on Wolf Water.
   a. Which direction will the water be flowing? (1)
   b. What impact do you think the forest in this area might have on the volume of the water in this place? (3)
   c. Name two river features that you notice affecting Wolf Water in Capanagh Wood. (2)
8. Calculate the straight line distance between the bridge over the river at 310151 and the feature at 300100. (3)
9. Calculate the distance the river covers between the bridge over the river at 310151 and the feature at 300100. (3)
10. In the section of the river labelled 'Linford Water' there are some waterfalls. What are their six figure grid references? (2)
11. How would you describe the shape of the river valley along the Linford Water near McCartney's Bridge? (2)
12. Make a copy of the table below. Identify three different land uses along the path of the Glenarm river and complete the table. (9)

Resource C LAND USES ALONG THE GLENARM RIVER

TYPE OF LAND-USE	SIX FIGURE GRID REFERENCE	BRIEF DESCRIPTION
1.		
2.		
3.		

Check your learning

Now that you have studied Part 2: River Processes and Features, return to page 11 and answer the Key Questions for this section.

PART 3: SUSTAINABLE MANAGEMENT OF RIVERS

a

The causes of flooding (physical and human)
Case Study: *The Boscastle flood, 2004 (a case study from the British Isles)*

b

The impacts of flooding
On people: *loss of life, property and insurance cover*
On the environment: *pollution and wildlife*

c

River management strategies
Hard engineering: *dams, levees/embankments, flood walls, straightening and deepening the river, storage areas*
Soft engineering: *washlands, land-use zoning and afforestation*

d

Case study:
the Yangtze river, China *(a case study from outside the British Isles)*
Evaluate the river management strategies used

Key words

Flooding
Hard engineering
Levees
Embankments
Soft engineering
Washlands
Afforestation
Land-use zoning
Boscastle
Yangtze

Key questions

By the end of this section you will be able to answer the following questions:

1. Why is flooding common in a drainage basin that has experienced steady rain over several days followed by a short period of intense rainfall?
2. How does urbanisation affect the likelihood of flooding?
3. Name two physical and two human causes of the flood in Boscastle.
4. When did the Boscastle flood occur? What was its impact on people?
5. What is the difference between hard and soft engineering?
6. Name two different management strategies that have been used to control flooding on the Yangtze.
7. How might river straightening and embankment enlargement make flooding worse?
8. How might land-use zoning help to reduce the impact of flooding?
9. What are washlands and how effective are they?
10. What advantages are there of allowing a river to flood naturally?

The causes of flooding (physical and human)

Rivers are nature's pipes. They are there to carry water as part of the water cycle from one place to another and in the process they will shape the land over which they are running. Yet, rivers are meant to flood. They are meant to be efficient systems that will drain excess water away quickly.

Floods happen when there is a temporary, extra amount of water in the river system, which causes the water in the river channel to overflow, covering areas of land and the surrounding flood plains that are usually dry.

If flooding is a natural part of the drainage basin cycle, why do we then become so concerned when rivers do burst their banks? This is because, over time, many people have decided to live close to rivers, putting them at risk of the river flooding and causing damage to their property.

So what is it that causes these rivers to flood in the first place? To answer this question, we need to look at each of the ways that water can enter the river system. We will see that things happen here which cause the amount of water (the discharge) to vary.

Physical causes

1. Precipitation

Flooding occurs more often when there are periods of heavy rainfall over a few days. The general rule is that the more water that falls onto a drainage basin over a short period of time, the more rapidly the water will fill up the air spaces in the soil, and the quicker the water will be forced to run off the surface and into the river. Sometimes the worst flooding is associated with short but intensive thunderstorms. Heavy snowfalls can also cause flooding, as water is melted by a rise in temperature. If this rise in temperature is accompanied by rainfall, there can be serious flooding.

However, flooding can also happen following periods of dry weather and drought. In the summer, the soil can become baked by the sun, making it hard and the rainwater can struggle to infiltrate into the soil. The water runs across the surface and creates a flash flood, which is when the water levels in the river rise very quickly following a rain event.

2. Soil and underlying rock

Different types of soil will influence how quickly or how slowly any rainwater will infiltrate through the soil. Sandy soils are permeable and allow water to pass through quickly. Clay soils are less permeable and will stop the water from passing through, which can cause flooding at the surface and increased surface runoff at a much faster rate.

Rock type also influences the passage of water through. Some rocks are permeable (porous) and allow the water to pass through with relative ease (for example, limestone, chalk and sandstone). However, some rocks have very small pores (non-porous) and are said to be impermeable, as water does not pass through them (for example, basalt, slate and granite).

3. Land use/vegetation

Another important factor that can influence flooding is how the land is used within the drainage basin. Any drainage basin that has very little vegetation will be more likely to experience flooding than areas where there is much vegetation and forests. This is because trees intercept water through their leaves and root system, and any water stored in this way will reduce the amount of water reaching the river system.

4. Steepness of drainage basin

The size and shape of the drainage basin can also dictate if a river is more or less likely to flood. A large drainage basin means that the water will take a long time to reach the river and is less likely to flood quickly. A small drainage basin means that the water will get to the river channel quickly and has a higher chance of flooding. Sometimes drainage basins can be small and the slopes within the area quite steep. This means that gravity will move the water more quickly towards the river channel and again, this is more likely to cause a flash flood.

Human causes

1. Deforestation

As noted above, if the presence of vegetation helps to slow down the return of water to the river channel, then the removal of this vegetation is going to cause an increase in the surface runoff, and therefore the chances for flooding. Removal of trees also means that the roots that would have helped to hold the soil structure together will not be able to do this any more. This causes an increase in soil erosion, which can also lead to a rapid increase in flooding.

Figure 20
Human action can accelerate flooding

2. Urban growth

The world is urbanising at a fast rate, and more and more farmland is being converted into concrete jungle instead. This change of land-use means that less water is held in storage within the drainage basin system. It also means that any water that falls onto buildings and roads gets channelled into drainage pipes very quickly (less infiltration and throughflow), which rapidly returns the excess water to rivers (increased surface runoff). Urban growth can cause a big increase in flooding along a river.

3. River management

Sometimes the actions that local governments take to try to change the shape of a river channel can actually make a flood more likely. For example, the 1953 flood of the River Lynn in Lynmouth was amplified by the decision to try to move the river channel, to allow a hotel to be built in a particular position. Changing the river channel, by making it more narrow or building a bridge at an inappropriate position, can cause the river capacity to be reduced and make flooding more likely.

4. Global warming

As more water is released from ice stores (glaciers and polar ice caps) due to the process of global warming, more water becomes available to fall on the drainage basin. This means that the rivers have to be able to cope with more water than they have had to experience in the past.

CASE STUDY

Causes of the Boscastle flood, 16 August 2004

(a case study from the British Isles)

Introduction

Boscastle is a small fishing/tourist village on the north Cornwall coastline. It is part of an Area of Outstanding Natural Beauty, with much of the land in and around Boscastle being owned and managed by the National Trust. In 2004 it experienced an unprecedented river flood, which washed cars downstream and reshaped some of the town.

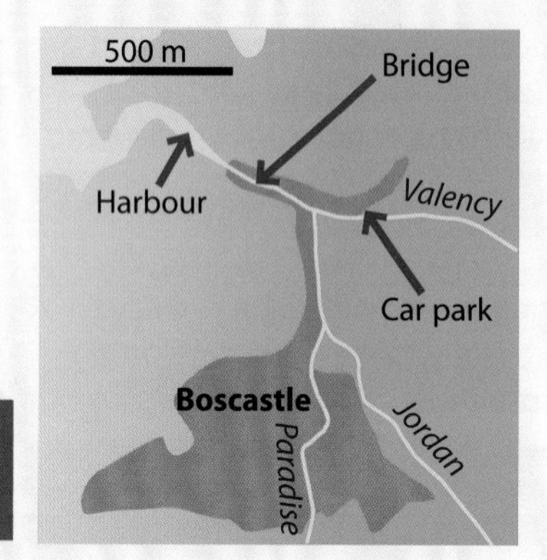

Figure 21
Map showing the location of Boscastle and the nearby rivers

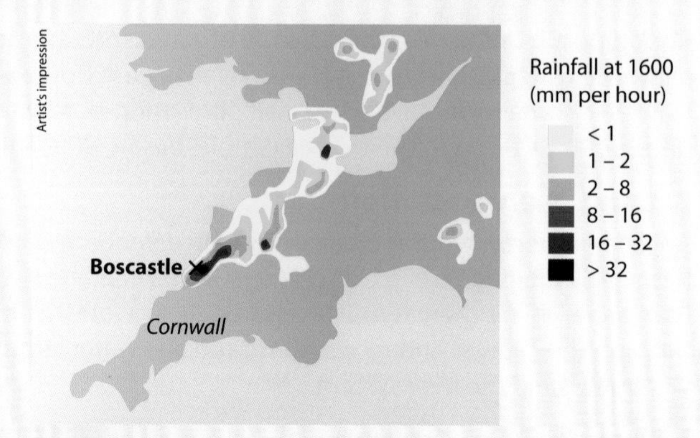

Figure 22
Rainfall Radar map for 1600 on 16 August

Causes of the flooding

Physical factors

1. Prolonged rainfall	The whole month of August was very wet. It had rained for 12 out of 14 days.
2. Heavy rainfall	There was a large depression (low pressure area) which produced thunderstorms around Boscastle on 16 August. 200 mm of rain was recorded at Otterham and Lesnewth in a 24 hour period, though most of the rain fell in a five hour period. Peak intensities were in excess of 300 mm/hr. The storm was classified as 'extreme' by the Environment Agency.
3. Steep valley sides	The valley sides are very steep in the surrounding area. The river rises to more than 300 m in only 6 km. This meant that the water flowed very quickly in and through the river system.
4. Underlying rock	The impermeable slate rock did not allow groundwater to infiltrate and water became surface runoff almost immediately.
5. Thin, impermeable soils	The surrounding impermeable peat soils became saturated and water could not percolate through the soil, meaning that the water very quickly started to flow across the surface.
6. Small drainage basin	The drainage basin area of the rivers that flow through Boscastle is very small. The catchment area is noted by the Environment Agency as being 20 km^2 or 7.7 $miles^2$.
7. Little interception	There are few trees within the drainage basin due to deforestation in recent years. This meant that nearly all of the water falling onto the drainage basin was flowing towards the river channel.

Human factors

1. Deforestation	Many trees in the locality had been cut down to allow farming to take place. Few trees were available to intercept rainwater and this meant that the water was moving into the river channel faster.
2. Urbanisation	The more buildings that are built in a drainage basin, the more tarmac and concrete, and impermeable surfaces there will be. This meant that there was very little infiltration and the water was directed into the river channel faster.
3. The Bridge Factor	Some of the bridges over the River Valency and River Jordan were too small and they got quickly blocked up with flood debris, which trapped trees. This blockage acted like a small river dam, which caused a surge of water to move down the river when the force of the water broke through the temporary barrier.

Structured notes

Case studies are a very important part of what you need to know for your GCSE Geography exam. It is important that you learn the key facts and figures from the case study carefully and then practice answering questions using the information.
Use the following guidance to help you make notes about the physical and human causes of flooding in Boscastle in 2004:

1. Describe the location of Boscastle.
2. What is the name of the main river involved?
3. What were the three main causes of the flood that happened on 16 August 2004? (These are physical causes.)
4. What additional physical causes played a part in the intensity of the flood event?
5. What human factors played a part in making the flood a major disaster?

Further research

There are some excellent sources available online which can help you to understand the causes of the flood in Boscastle. Look up some of them.

The impacts of flooding

Impact on people

Flooding can have both a positive and negative impact on people. However, it is common to dwell on the negative. Flooding can cause huge amounts of damage to property, causing water damage and in exceptional cases actually pulling buildings down. Floods can lift objects such as cars and caravans, and transport them far downstream. Floods can also kill people. The Lynmouth flood in 1952 killed 35 people. However, some of the world's deadliest floods have occurred in China. The floods in 1931 are estimated to have left over three million people dead and around 50 million people homeless for a time.

Crops grown across flood plains are often lost in flash flood conditions. Animals may be drowned and livelihoods erased. Floods can also cause public health issues, as floodwater mixes with sewage, causing problems with drinking water and an increase of waterborne diseases.

Many people in Less Ecomomically Developed Countries (LEDCs) do not have enough money for home insurance and as a result they often find it difficult to rebuild their homes and to replace their possessions following a flood. Aid is usually needed from the local government or from international organisations in order to support poor people who have been affected by flooding. In More Ecomomically Developed Countries (MEDCs) there is another issue – some people who live close to rivers find that insurance companies will not allow them to take out a policy. They have to live in the knowledge that if the nearby river floods, their property will not be insured for damage.

However, flooding can sometimes bring advantages. An excess of water can help to replace drinking water supplies and many farmers who live across flood plains use the floodwater to help irrigate their crops. Farmers in Egypt who had land close to the River Nile were most annoyed when flooding was controlled by the building of the Aswan dam in 1970. Previously, they had relied on the floodwaters to help their crops grow and survive in a desert climate. They also enjoyed the silt (alluvium) that was laid down in a flood event, as this provided a natural fertiliser for their crops.

Impact on the environment

Floods are a necessary part of life for the environment. Some species of fish need flood conditions to be able to breed and to be able to swim upstream to their breeding grounds. Floods also bring relief from drought conditions, providing animals with access to drinking water.

Unfortunately flooding can also wash chemicals, waste and sewage into rivers, and cause pollution, which can harm water supplies and kill wildlife. Industrial pollution is a common reason for fish-kills in rivers. Animals can also drown or be washed away by flood events. Most river wildlife lives in a very delicate balance with their surroundings, so even the slightest change in the temperature, speed or volume of water can have a big impact on the survival of the animals.

Structured notes

1. Make a copy of the table below. Use the information on page 27 to develop the advantages and disadvantages of flooding listed in the table.

IMPACTS OF FLOODING ON PEOPLE AND THE ENVIRONMENT		
IMPACTS ON	ADVANTAGES OF FLOODING	DISADVANTAGES OF FLOODING
People	1. Drinking water 2. Irrigation 3. Silt (alluvium)	1. Damage to property 2. Death, injury and homelessness 3. Damage to crops 4. Public Health issues 5. Insurance issues
The environment	1. Fish 2. Drought relief	1. Chemicals 2. Crop/animal losses

2. What was the most deadly flood in the world? Carry out some research to out and put together some information on the flood. When did it happen? What caused it? What impact did it have on the people that it affected? Also, see if you can find out what the worst flood event within the UK in recent years was.

CASE STUDY

Impacts of the Boscastle flood, 2004

Figure 23

The flood waters caused huge amounts of damage to the town of Boscastle. The River Valency picked up cars and carried them down towards the mouth of the river.

At first, no-one in Boscastle realised that the heavy August rains were going to do anything more than raise the water levels a little. However, it quickly became clear that the amount of rainwater being moved through Boscastle was starting to have some unexpected impacts. The water levels in the Rivers Valency and Paradise continued to rise throughout the afternoon, with the peak flow in the River Valency about 140 m^3/sec (tonnes) between 5 pm and 6 pm, and the water speed in excess of 4 m/sec (10 mph), which is more than sufficient to cause structural damage. Two million tonnes (440 million gallons) of water flowed through Boscastle that day.

Impact on people

1. **Death, injury and homelessness**

   The flood affected around 1,000 residents and visitors. They witnessed and were involved in the largest peacetime rescue in the history of Great Britain. There were no fatalities and only one casualty was reported – a broken thumb. Seven RAF helicopters airlifted 1,000 people to safety. People had to stay in the town hall, in caravans and with friends for a few weeks while homes were being repaired.

2. **Damage to property**

   Buildings: 98 properties were flooded and damaged, four properties were demolished by the force of the water and four footbridges across the River Valency were washed away.

Cars: 115 cars were washed into the river. About 84 of these were recovered from Boscastle's harbour while the remainder are still unaccounted for. Oil and petrol from the cars caused pollution.

3. **Impact on the economy**
   The cost of rebuilding the town and completing work to make sure a flood like this could not happen again was estimated at £50 million. The town has also lost business in the aftermath of the flood, as tourists are afraid of being 'trapped' in Boscastle as they were on this day. This has affected jobs and incomes within the town.

Impact on the environment

The flood had minimal impact on the natural environment. However, pollution was caused by the cars in the river while the increased amount of water flooded the sewage system so that dirty water floated through the town.

Test yourself

Read the following two eyewitness accounts of the floods in Boscastle and answer the questions which follow.

3.45 pm Car Park Rescue

Holidaymakers sat sheltering from the rain in their cars in Boscastle car park when it started to flood. Among them were Rachelle Strauss and her husband Richard, on holiday from Gloucestershire with their three-year-old daughter.

"Cars started to move and I watched the water rising. In less than a minute it was up to the bottom of the car doors," said Rachelle. They managed to drive up the hill and Richard went back to help people who were by now stranded in their cars.

"I heard people screaming and realised that they were trapped in their cars unable to get out because of the force of the water against the doors," Rachelle said. "People were calling out of their sunroofs for help, some frozen by fear and unwilling to get out of their cars; they had to be physically dragged out." Her husband and two other men linked arms and waded through deep water to rescue people trapped in their cars. One driver slipped and fell but Richard managed to grab her before she was swept away.

4.00 pm Trapped in the Visitor centre

Two families were trapped in the Visitor Centre by rapidly-rising water. Centre manager Rebecca David ushered the five adults and six children up a stepladder into the attic as the Centre took the brunt of the floodwater pouring down the Valency.

Then a tree smashed into the building, demolishing two-thirds of it. They scrambled out of a skylight on to the remaining roof as floodwaters reached the guttering.

The visitors and staff had been trapped for an hour when a Royal Navy helicopter from Culdrose winched them up one by one in the largest single rescue of the day.

4.30 pm Rescue in Valency Row

Emily Maughan was knocked off her feet when the floods smashed down her front door in Valency Row. "The door just flew towards me and a tidal wave of water came over the top," said Emily, who was swept out of her house by the current. She clung to a drainpipe in chest-deep water and screamed for help. She was heard by neighbour John McLaughlin who opened his front door – letting in floodwater three feet deep – and hauled Emily inside to safety.

Source: Living with the Risk, the Environment Agency, 2005

1. How did some of the people mentioned in the eyewitness accounts come to realise that the bad weather was turning into a dangerous flood? (3)
2. List the different ways that the river caused damage in the eyewitness accounts. (4)
3. Name two ways that people were rescued from the eyewitness accounts. (2)
4. John Maughan, a local builder, made the following statement following the incident:
   "When buildings started collapsing, I was thinking of all these people waving out of their upstairs windows, that all their buildings would collapse. I remember thinking 'I bet you there's 50 people drowned tonight.'"
   A similar flood in Lynmouth in 1952 (20 miles to the north of Boscastle) left 35 dead. Why do you think no-one died as a result of this flood? (5)
5. See if you can find some pictures of the damage in Boscastle. You might even be able to find some interesting pieces of video of the flood event.

ORGANISE YOUR REVISION: PRACTICE ESSAYS

In the Higher Tier paper you need to be able to write structured short essays to get up to 9 marks. This is a skill that you need to practice as part of your revision.

Write an essay on 'The causes and effects of the Boscastle flood'.

Part 1: What were the main causes (Physical and Human)? (6)
Part 2: What were the short term impacts (on the day of the flood)? (4)
Part 3: What were the long term impacts? (5)

Try to include at least two photos and one map to back up what you are writing and annotate (label) them appropriately. (5)

River management strategies

It is very difficult to minimise the impact of the physical causes of flooding and especially difficult to ensure that flooding never happens again in an area such as Boscastle. However, measures can be taken to make the river more efficient at moving the water through the drainage basin, releasing the excess water into the sea instead of amongst urban areas. This will reduce the damage to the surrounding areas.

In an attempt to prevent and control river flooding, humans have tried to manage rivers. Two main strategies are used:

Hard engineering strategies: These require major alterations and changes to the river to try and stop the river from flooding. Generally these involve big building projects, where machinery is used to change the river and new walls or banks are built. These measures are not sustainable in the long term.

Soft engineering strategies: This is where limited alterations take place and flooding is more managed than prevented. These are generally more sustainable, as they use more natural processes and do not damage the environment to the same extent.

Evaluation of river management strategies

Hard engineering strategies

1. **Channel enlargement (deepening and widening)**
   Flooding can be reduced by increasing the size of the river channel so that it is capable of carrying more water. This means that in times of very heavy rainfall the river is less likely to flood its banks than before the enlargement, as it will be able to transport the water quickly and effectively through the river channel. The advantages of this strategy are that it helps to protect both sides of the river, the look of the river is not changed dramatically and it usually continues to protect the natural habitat of animals.

Figure 24
Channel enlargement

©iStockphoto.com

However, there are a number of disadvantages. Firstly, work on channel enlargement needs to happen regularly, as the bottom of the river often silts up rapidly. Secondly, heavy machinery is used to dig out the bottom and sides, which can be expensive and can have a negative impact on the local ecosystem. Thirdly, some scientists argue that the natural process of flooding helps to slow the river down. They suggest that removing the bottom of the river bed and the banks will cause a high river velocity, which will transport the water very quickly through the river to areas further downstream, which might not have been protected.

2. Bridge widening
Bridges can be widened to stop debris such as trees and boulders from becoming trapped behind them and creating a dam, as this would cause a large wall of water to build up.

Source: what'sthatpicture / James Morley

Source: David Dickenson

Figure 25 *The bridge in Lynmouth before and after the 1952 flood*

However, this replacement process can be very expensive and since the new bridges are modern constructions, they are often not as attractive as the former structures (for example, old arched bridges).

3. **River straightening**

Rivers that meander across the river flood plain can be straightened to help increase the velocity of the water. This reduces flooding as the water drains away from the flood plain more quickly. Some parties want rivers straightened for reasons other than reducing flood risk. For example, property developers find it easier to build alongside straight rivers than on bends and farmers prefer more regular field shapes and sizes, as it allows their machinery to reach the corners of fields, so there is little land wastage.

Figure 26
A stretch of the straightened river Kissimmee in Florida

Source: US Army Corps of Engineers Digital Visual Library

However, the problem with straightening a river is that it does not always work. Rivers can often return to their original, natural course. There are also environmental impacts. For example, fish like to lay eggs along the shallow, inside of bends, where the water is slower. Straightening the river removes these bends.

4. **Levees and embankments (flood walls)**

Levees and embankment walls can be built on either side of the river. The walls mean that even if the river floods above the river bank, the water cannot spread over the flood plain and cause damage.

The construction of levees and embankments can be expensive but can be carried out with little impact on the river itself. Some people argue that walls can spoil the look of the river. Others note that embankments can increase the speed of water in the river and if the water does break through the embankments, the flood can be more destructive because the water is travelling so fast. When a river floods naturally it usually deposits silt (alluvium) on the flood plain but this is prevented with the building of levees. These levees also prevent water from draining back into the river quickly following a flood event.

Source: US Army Corps of Engineers Digital Visual Library

Figure 27
A river levee near Hamburg, Iowa is breached by a flood, June 2011

5. **Dams/reservoirs**

In many ways dams across the upper course of the river are an obvious way of controlling the amount of water that can travel through the river system. They ensure that the risk of flooding is almost completely reduced and can be used to generate hydro-electric power, which is a cheap and renewable way to produce electricity. The lake/reservoir behind the dam can also be used for water-based recreational activities.

Source: Antoine Taveneaux

Figure 28
Hoover Dam in Nevada

However, dams are extremely expensive and they have a major impact on the natural environment. They completely change the natural landscape and ecosystem of an area beyond all recognition.

6. **Storage areas**
Water can be pumped out of the river and stored in temporary lakes. This reduces the amount of water flow in the river at peak times and the impact of flooding in more sensitive areas. When it is needed it can be pumped back into the river. These storage areas are effective but they take up a lot of space. A good example of this has happened along the Mississippi river, where the flood water has been diverted into Lake Pontchartrain.

Source: US Army Corps of Engineers

Figure 29
A spillway diverts water from the Mississippi into Lake Pontchartrain in May 2011

Soft engineering strategies

1. Land-use zoning

Land-use zoning involves dividing the flood plain up into areas which experience different degrees of flood risk. For example:

Red zone = places with a high chance of experiencing a flood
Amber zone = flooding possible but unlikely
Green zone = flooding very unlikely

Once the flood plain has been divided up into the different zones, an appropriate land-use can be chosen for each zone. For example:

Figure 30
Land-use zoning

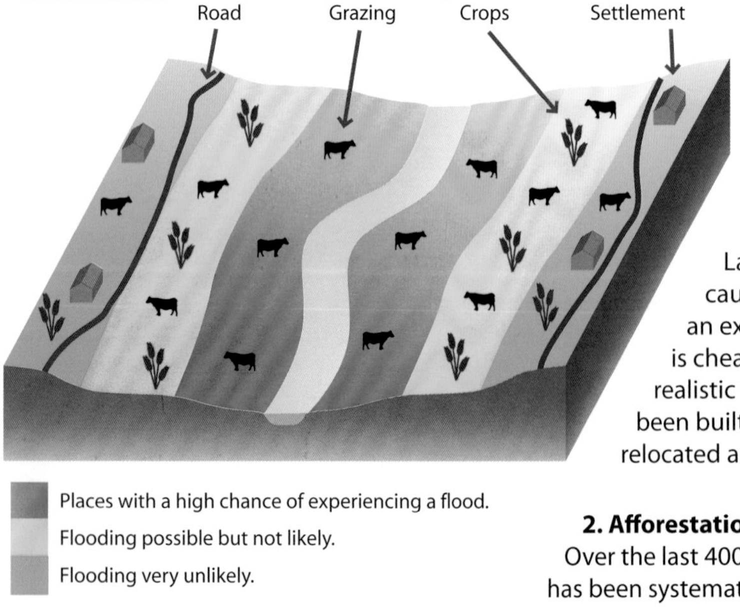

Road
Grazing
Crops
Settlement

Places with a high chance of experiencing a flood.
Flooding possible but not likely.
Flooding very unlikely.

Red zone = non-residential land, such as golf courses, parks and farmland
Amber zone = car parks, sports facilities, some houses (but these should have flood protection considered in their design)
Green zone = residential housing (flood protection does not need to be considered)

Land-use zoning allows the river to flood naturally and causes minimal disruption and damage to people. It is an extremely effective method of managing floods as it is cheap and easy to carry out. Unfortunately it is not very realistic to apply land-use zoning to areas that have already been built up beside a river, as people would have to be relocated and businesses moved. This would be unpopular.

2. Afforestation

Over the last 400 years much of the natural tree coverage of the UK has been systematically removed and replaced with farming. Planting trees in the upper course of the drainage basin helps to reduce flooding as trees intercept and store the water. This will help stop some of the water entering the river channel and can reduce the risk of flooding.

Afforestation cannot prevent flooding but it can help to reduce its likelihood. These new woodlands can provide better natural habitats for animals and timber can be harvested for additional income. However, it is only possible where there is a lot of spare land available for planting and all land owners and users must agree.

Figure 31 Afforestation in Nepal

Source: Nepal YMCA

3. Washlands

Washlands are areas of land that water can wash into during a flood. They are usually found in the lower course of a river and act a little like storage areas, with the added benefit of helping to increase friction and slow the river down.

Washlands are created by leaving a large area of land next to a river empty to receive any floodwaters. It can sometimes be difficult to find large tracts of land to complete this process, especially close to cities.

Figure 32 *Washlands on the River Ouse in Cambridgeshire (the washes were flooded in July as water was pumped from the river to cover the farmland)*

Source: Hugh Venables

Sustainable management strategies

The big question is: Are these measures sustainable? The United Nations defined sustainability in 1987 as "development that meets the needs of the present without compromising the ability of future generations to meet their own needs." In 2005 the World Summit noted that this required the reconciliation of environmental, social and economic demands – the 'three pillars of sustainability'. Resource use needs to be able to meet human needs while preserving the natural environment. Therefore we need to be able to identify whether these strategies fit the sustainable bill. The activity below will help you with this.

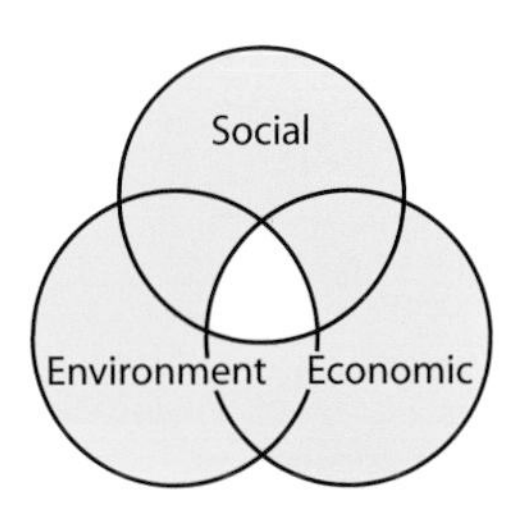

Figure 33 *Sustainable development*

Structured notes

1. Make a copy of Resource A below and complete the table for each of the river management strategies. Resource B will help you.

Resource A

		Score 5 = very good and 1 = poor			Out of 15
River management technique	*Hard or soft*	*Environmental sustainability*	*Social sustainability*	*Economic sustainability*	*Total score*
Channel enlargement					
Bridge widening					
River straightening					
Levees and embankments					
Dams/reservoirs					
Storage areas					
Land-use zoning					
Afforestation					
Washlands					

Resource B

Environmental sustainability	Social sustainability	Economic sustainability
Is the strategy good or damaging to the environment? Does it help protect wildlife and animal habitats?	Is the strategy something which will add to people's lives? Will it allow them to live close to each other without risk of death or injury?	Is the strategy expensive to build and maintain or is it just a one off cost? Will the strategy last for a long time?

2. Once you have completed the table, decide which strategies you think are the most sustainable river management techniques and which are the least sustainable. Once you have decided, list the three most sustainable, followed by the three least sustainable.

CASE STUDY

The Yangtze river, China
(a case study from outside the British Isles)

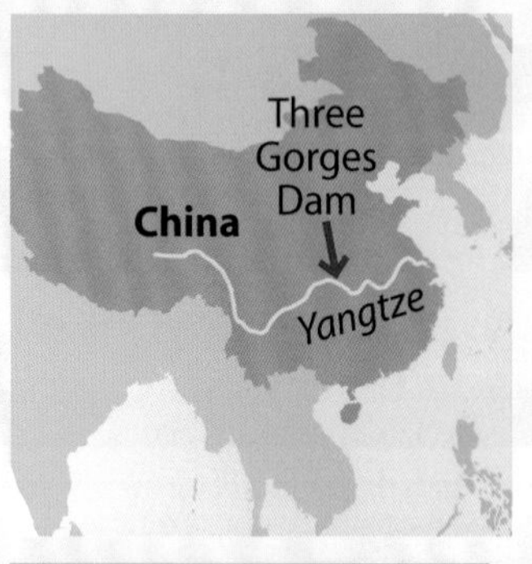

Figure 34 Yangtze river location

The Yangtze (Changjiang) river is 6,300 km long and is the largest river in China and Asia. Its name in Mandarin means 'long river'.

The need for a river management scheme

Like most rivers, the Yangtze river floods regularly. However, due to the population explosion in China from the 1950s onwards, the amount of human activity surrounding this area has grown rapidly and this has further amplified the impacts of any flood event. The Chinese government has been trying a number of flood protection and management measures for the last 50 years, with varying degrees of success.

Yangtze river fact file

- It is the third largest river in the world (only the Nile and Amazon are longer and have bigger drainage areas).
- Its source is the snow-capped Geladandong peak of Mt Tanggula in SW China.
- It includes more than 700 tributaries over a drainage basin area of 1.8 million km^2. (Northern Ireland is nearly 14,000 km^2 so this river basin is about 128 times bigger than Northern Ireland.)
- The main rainy season, which generates between 70–90% of the annual rainfall, is from May to October.
- The average annual rainfall over the basin is 1,100 mm.
- The population in the area was estimated in 2000 at 427 million and the area is identified as one of the most economically developed regions within China.
- The Yangtze is very prone to floods induced by rainstorms.
- Since 1877, 25 floods have exceeded the channel capacity.
- One of the worst floods was in 1931, over some of the middle and lower courses of the river. An area of 130,000 km^2 of land was flooded. 28.5 million people were affected, the death toll was 145,000 and economic losses were estimated at 1.5 billion dollars.

River flooding has been a huge problem for the Chinese people over the last 200 years. In a list of the ten deadliest floods worldwide, six Chinese floods events are recorded, which killed an estimated 7.3 million people (in 1887, 1911, 1931, 1935, 1938 and 1975).

Figure 35 The Three Gorges Project and Reservoir

The river management strategies used

Strategy No 1

Hard engineering: The Three Gorges Dam Project (commonly referred to as TGP)

Three Gorges Project factfile

- ☐ The TGP's role in a flood situation should be to:
  - control a drainage area of 1 million km^2
  - provide flood storage for 22.15 billion m^3
- ☐ As well as providing a solution to flooding problems, the Chinese government has planned that the TGP will also help solve power supply and river transport issues along the Yangtze.
- ☐ The project built a dam across the Yangtze river. Work started in 1994 and the estimated cost was £50 billion.

- ☐ The completed dam includes 32 hydro-electric power (HEP) generators, which are capable of producing 10% of China's energy needs.
- ☐ The building of the dam and the increase in water depth along the course of the river allows 10,000 tonne ships to travel further upstream than before and river safety is much improved.
- ☐ Over 100 million tonnes of earth and rock were dug up and replaced with 27 million tonnes of concrete.
- ☐ The dam itself is 2335 m long and 185 m high. The lake area is 1,000 km^2 and flooded 140 towns, resulting in the forced evacuation of 1.2 million people.
- ☐ The water level in the dam rises to about 175 m above the original Yangtze river bed.

Advantages

- The building of the dam is designed to protect 100 million people downstream.
- The dam is designed to regulate the amount of water in the river system more easily, which should be able to stop flooding and allow better access upstream for shipping.
- The TGP creates 'green energy' through the HEP plant. Scientists expect that emissions to the atmosphere of 120 million tonnes of CO_2, more than 2 million tonnes of SO_2 and large quantities of dust will be avoided, thus helping China to reduce their greenhouse gas emissions. Less air pollution should also bring health benefits.
- New settlements have provided people with better houses and more amenities.

Disadvantages

- Most floods in recent years have come from the tributaries that join the Yangtze below the TGP.
- The land often experiences earthquakes and landslides, which may put the dam at risk of causing a catastrophic flood event.
- It is possible that the port might become silted up due to deposition.
- Over 1.2 million people had to move, sometimes long distances from where they used to live. Much of the land made available for resettlement is above 800 m and has poor soils, making farming very difficult.
- It took nearly 20 years to complete the project.
- With more people now forced to live in cities, pollution and sewage from the cities increased. Pollution also came from the construction work – making the river very dirty.
- The dam interferes with the local aquatic ecosystem. Local animal species such as the Yangtze river dolphin (Baiji) are under threat and some already believe the dolphin to be close to extinction.
- The area surrounding the TGP needs more trees to be planted but pressure on the land for farming and for urban living space makes this very difficult.
- Many archaeological treasures and the limestone scenery of the Three Gorges were drowned.
- The cost of £50 billion put the country in debt and did not directly benefit many of the local poor.

Figure 36

The changes to the Yangtze river from 1997 to 2004 caused by the Three Gorges project

Source: United States Department of the Interior, US Geological Survey

Figure 37

Raised levee to protect the city of Badong on the Yangtze river

Strategy No 2

Hard engineering: levees

- Over 3,600 km of levees have been built up along the main channel of the river, with a further 30,000 km of secondary levees.
- They have been built to help protect an estimated 80 million people.
- Levees were used along much of the area where the TGP now exists. The area from Yichang and Wuhan was protected by a 180 km long, 15 m high levee.
- The levees are designed to be able to handle a 10 to 30 year frequency flood.
- Levees have a lot less impact on the natural environment than dams, as animal habitats are not totally destroyed and homes are not flooded.
- In 1998 the Yangtze still flooded over the top of the levees, killing 3,000 people and leaving 30 million homeless.

Structured notes

Using the information above and your own knowledge of levees, see if you can create a list of the advantages and disadvantages of the levees and embankments along the Yangtze.

Strategy No 3

Soft engineering: washlands (detention basins)

Figure 38

Detention Basin in the Bayanbulak area of Xinjiang province

- Detention basins are often natural lakes or low-lying areas that have been designated for temporary flood storage.
- They have been in use for about 70 years on the Yangtze, with 40 major flood detention basins and storage potential for 50 billion m^3 of water but have limited impact.
- The most important is called 'Jingjiang basin'
  - It is a 2000 km^2 gated basin.
  - It has a storage capacity of 5.4 billion m^3.
  - Water depths can reach between 8 and 10 m.
  - If used, flood water would affect the homes of ½ million people.
  - It was last used in 1954.

Advantages

- Detention Basins are easier to construct than dams and reservoirs.
- The land can be used for other purposes. For example, farming and residential property can exist in the basin and water will only be allowed into the areas when there is a major flood.
- Following the completion of the TGP there has been a renewed effort to restore flood plains and increase the flood retention capacity in the hope that this will help to improve biodiversity, conservation and local ecosystems.
- Floods are now easier to manage and the environment is better able to cope with the consequences if a flood does occur.

Disadvantages

- The Dongting Lake is a natural detention area but its capacity has declined in recent years because of siltation and land reclamation for farming. The threat of flooding has been made worse recently due to the loss of these storage systems.
- One big issue for flood detention basins is that the safety of people cannot be guaranteed. There is no flood-proofing of buildings within the basin and there is a lack of proper main roads to evacuate people in the event of a flood. Consequently, during the 1998 flood the Jingjiang detention basin could not be used, as the safety of people living here could not be assured. When it was finally decided that the basins had to be used, 300,000 people had to be moved at a cost of over £2.4 million.
- These basins are sometimes called 'sacrificial' areas, as they will be sacrificed in order to save more valuable industrial and urban areas further downstream. However, many of the artificial detention areas still contain high-value farmland and other assets.

Strategy No 4

Soft engineering: Flood warning systems

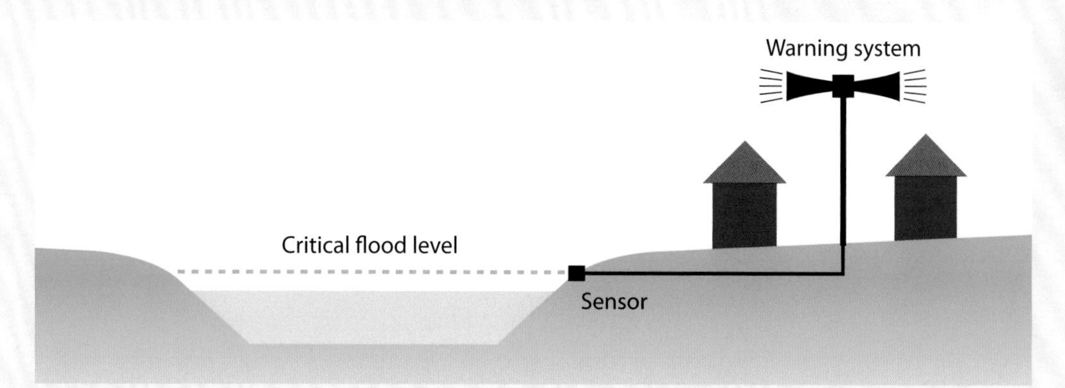

Figure 39
How a flood warning system can save lives

These help provide time for people to move their possessions and evacuate areas. A new series of 180 flood control monitoring stations have been built along the Yangtze. The systems rely on telephone and satellite communications. Information from the river, plus rainfall predictions can then be fed into more than 100 computer models of possible flood scenarios most likely to occur. The system can suggest how long a flood will last, where it is likely to expand or shrink and offer options of controlling it, such as diverting water into the detention basins. The project for the Yangtze cost about £9 million and it is hoped that it will be instrumental in saving lives.

Structured notes

Using the information above and your own knowledge of Flood Warning systems, create a list of the advantages and disadvantages of this type of flood management along the Yangtze.

The sustainability of these projects

In China, the extreme nature and frequency of the floods during the twentieth century meant that the Chinese government had little opportunity to plan for the future.

The social, environmental and economic impacts of the management strategies for these areas

Below are some of the key questions that you need to think about when trying to decide how sustainable a particular management project is.

What is a social issue?

How do the management strategies impact people? Does it change their lives? Does it affect how they live, where they live and their routine? Does it affect food supply? Does it affect travel? Do you think the management strategies are largely positive or negative?

What is an environmental issue?

How do these management strategies impact the environment? Do they cause pollution? Does it affect drinking water? Where does all the sewage go? Does it affect animals and their natural habitats? Do you think the management strategies are largely positive or negative?

What is an economic issue?

How do these management strategies impact people and money? Has the building of new management strategies brought prosperity or poverty to local people? Does it affect jobs? Have the strategies changed the traditional way of life in the area? Do you think the management strategies are largely positive or negative?

Structured notes

Make a copy of the table below.

STRATEGY			SUSTAINABILITY		
Strategy	Advantages	Disadvantages	Social issues	Environmental issues	Economic issues
1. The Three Gorges Dam Project	• • •	• • •			
2. Levees	• • •	• • •			
3. Detention basins	• • •	• • •			
4. Flood warning system	• • •	• • •			

1. Complete the table by identifying some of the main **advantages** and **disadvantages** of the management strategies used along the Yangtze. Try to come up with three of each.

2. Use some of the questions on page 37 about social, environment and economic issues to help you complete the three columns on sustainability.

Test yourself

1. Which of the strategies do you think is the **most** sustainable and **why?** (4)
2. **Describe** and **explain** which of the strategies you think is the **least** sustainable? (5)

Resource A

A new era?

In a presentation in 2011, the Changjiang Water Resources Commission (CWRC) have noted their basic philosophy in a new era:

"Ensuring the health of the Yangtze, Promoting the harmony between human and nature"

They go on to note that the basic requirement of flood protection and disaster mitigation of the Yangtze in this new era is the "harmonious relationship between human and water".

3. Read Resource A. Does this sound like the CWRC is gradually thinking about more sustainable measures? Give reasons to support your answer. (4)
4. What evidence would you like to see, to ensure that the CWRC are taking this message seriously? (4)
5. What other things do you think the CWRC could consider to make flood management of the Yangtze more sustainable? (4)

Check your learning

Now that you have studied Part 3: Sustainable Management of Rivers, return to page 24 and answer the Key Questions for this section.

EXAM PRACTICE QUESTIONS

Some of these questions are from previous CCEA GCSE examination papers and others have been written in the same format to give you practice at answering 'exam style' questions.

Try to answer the questions with as much detail as possible. Also consider the number of marks that each question receives, as this will give you a good indication of the amount of depth that your answer needs.

Resource A

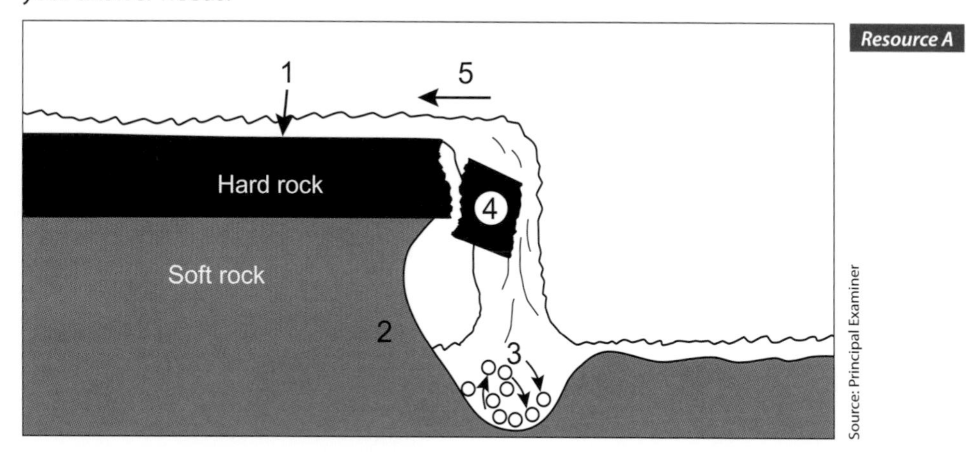

Source: Principal Examiner

Resource B

STATEMENT	NUMBER RESOURCE A
Erosion of softer rock causes undercutting	
The undercut rock collapses	
The river flows over a layer of hard rock	
Erosion leads to the formation of a plunge pool	
The position of the waterfall moves backwards	5

Resource C

SITUATION	MORE SURFACE RUNOFF	LESS SURFACE RUNOFF
Light rain falling on dry ground		✓
Lots of tarmac surfaces		
Heavy rain falling onto wet ground		
Lots of trees		

Resource D

	SITE 1	SITE 2	SITE 3	SITE 4	SITE 5
Distance from Source (km)	0.8	1.35	3.5	6.0	8.9
Average length of rock in cm (long axis)	12.90	11.25	10.88	9.55	7.90
Average roundness	Angular	Sub angular	Sub rounded	Rounded	Well Rounded

Measuring the long axis of a rock collected from the river.

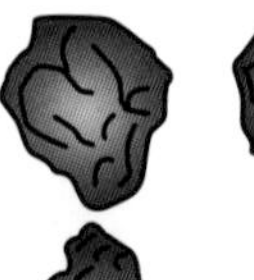

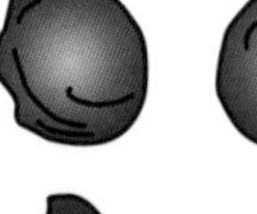

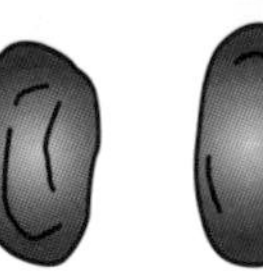

Powers' Scale of Roundness

Foundation Tier

1. State the meaning of the term drainage basin. (2)
2. What is the difference between percolation and infiltration? (2)
3. Study Resource A, which shows a sketch of a waterfall.

   Make a copy of Resource B. Complete the table by matching the correct statement to the number in Resource A. This will explain the formation of a waterfall. One has been completed for you. (4)
4. Make a copy of Resource C. Indicate with a tick how the situations stated in Resource C might affect the amount of surface run-off/overland flow. One has been completed for you. (3)
5. With reference to some flood prevention strategies (eg dams, levees and afforestation) answer the questions which follow.
   a. Choose one flood prevention strategy and explain one way in which it might help to reduce the flood risk on a river. (3)
   b. Describe one impact flooding might have on people. (2)
   c. For a named river within the British Isles, explain why it flooded giving two reasons. (1 mark for river and 6 marks for 2 reasons)
6. Some rivers are likely to flood. For a named river within the British Isles, describe **one** cause of flooding. (1 mark for river and 3 marks for one cause)

Higher Tier

1. State the meaning of the term interception. (2)
2. Explain the formation and development of a waterfall. (6)
3. A field study was carried out on a river in Northern Ireland. Study Resource D which shows information obtained.

   (i) Describe the changes in the load with distance downstream. (4)

   (ii) Explain how these changes in load occur. (5)
4. Explain how an engineering strategy might help to reduce the flood risk posed by a river. (3)
5. For a named river within the British Isles describe one physical and one human cause of flooding. (4)
6. For a named river outside the British Isles, evaluate the extent to which river management strategies used on this river can be considered sustainable. (7)

PART 4: COASTAL PROCESSES AND FEATURES

a — Constructive and Destructive waves

b — Processes at the coast
Erosion *(corrasion/abrasion, attrition, corrosion/solution and hydraulic pressure)*
Transportation *(longshore drift)*
Deposition

c — Coastal landforms
Erosional landforms (cliff, wave cut platform, cave, arch and stack)
Depositional landforms (beach and spit)

d — Skills
Using aerial photos and OS maps to identify coastal features and land uses

Key questions

By the end of this section you will be able to answer the following questions:

1. What is the difference between a destructive and constructive wave? Which type of wave is erosional?
2. What types of wave form beaches?
3. Describe the four ways that the sea erodes the coastline.
4. What type of erosion can cause compressed air to crack a rock?
5. What is the name of the process whereby beach material moves along the coastline? Why does this happen?
6. Explain the formation of a wave cut platform.
7. Using a series of diagrams show how a stack is formed.
8. What is a spit? How does it form?
9. What coastal features could you identify on an OS map?

Key words

Constructive wave
Destructive wave
Swash
Backwash
Longshore drift
Wave cut platform
Stack
Stump
Beach
Spit

Constructive and destructive waves

Our dynamic coast, where the land meets the sea, is also under constant change. The main way that our coastline is changed is through the process and action of the waves as they crash against the land.

How waves are created

Waves are created by the transfer of energy as the wind blows across the surface of the sea. The size of any wave depends on its **fetch**. The fetch is the distance that a wave travels in open water. The longer the fetch, the larger the potential wave can actually be. This means that waves that move towards Ireland across the Atlantic Ocean have the potential to be larger than waves that move across the shorter distance of the Irish Sea. However, the main influence on the size of waves is the wind speed. Strong wind speeds will bring bigger waves.

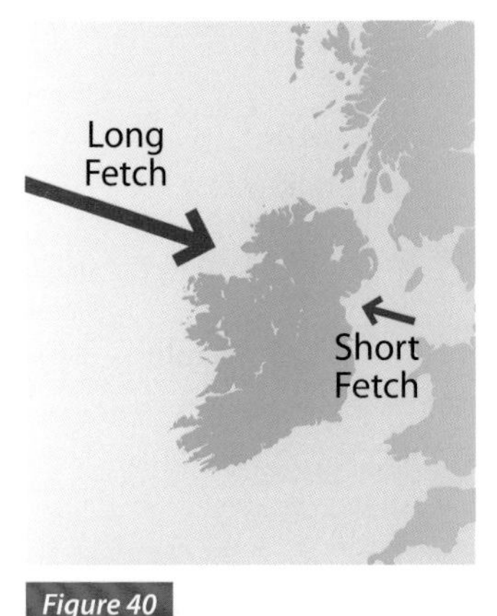

Figure 40
Fetch of waves

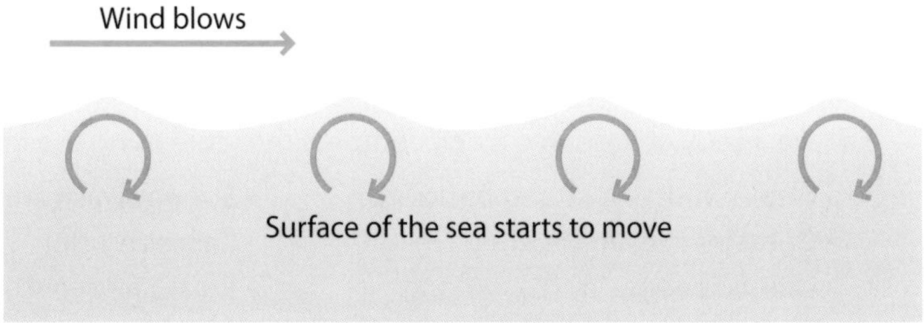

Wind blows

Surface of the sea starts to move

Figure 41
How waves are created

Out at sea the water tends to move in a circular orbit. However, any floating object would usually just move up and down (vertically), as it is only the shape of the water and the wave energy that is transferred horizontally.

Figure 42
The formation of a wave

Direction of wind
Swash
Backwash
Friction from the beach slows down the base of the wave
Beach

As the waves begin to approach the shore and the water becomes shallower, the amount of friction increases with the seabed and this slows the base of the wave down. However, the friction does not affect the top of the wave. It keeps its energy so the wave starts to be built upwards until it eventually rolls over and breaks.

The water which rushes up the beach is called the **swash**. As the wave washes up the shore, it loses its momentum and energy is transferred back to the sea. This return flow is called the **backwash**.

The force of the swash and backwash determines whether the waves are constructive or destructive waves.

How constructive waves build up the landscape

The steepness of a wave is the main influence on the shape of a beach. Constructive waves are gentle, flat and low (around one metre in height), and their energy is limited with only a few waves per minute (between six and nine). Their weak backwash does not return material to the sea so the beach builds up, generally forming gentle beaches.

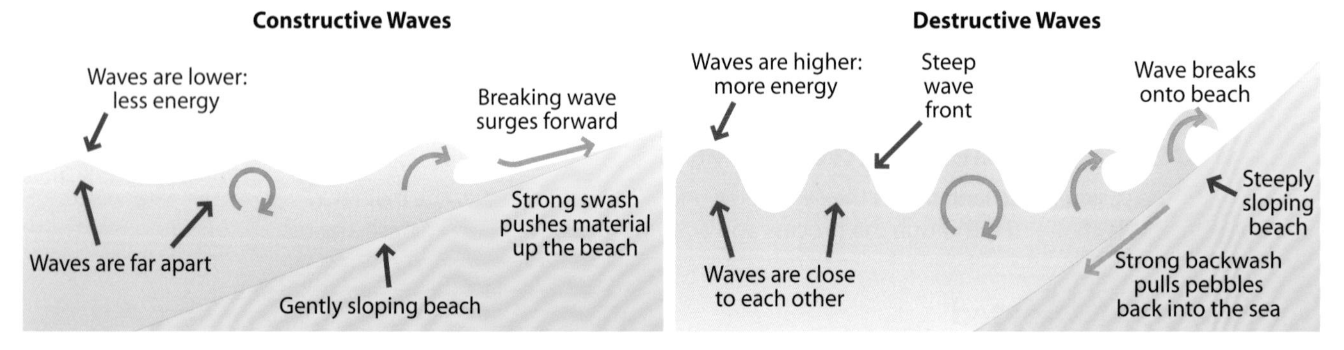

Figure 43
Constructive and destructive waves

How destructive waves erode the coastline

Destructive waves have much more energy than constructive waves. Their waves are steep (up to three or four metres high), close together and have more waves per minute (around 15). Their strong backwash pulls material back into the sea to create steep beaches.

Test yourself

1. Describe how waves break on the shore. (3)
2. What are the main differences between swash and backwash on a beach? (4)
3. Make a copy of Resource A and compete the table using the following statements:

- Weak swash and strong backwash
- Operates in calm weather
- Limited energy
- Steep beach angle
- High wave height (five or six metres)
- Build up of sediment on beach
- Removes sediment from the beach
- Gentle sloping beaches
- Strong swash and weak backwash
- Causes erosion
- Breaks frequently (around 15 per minute)
- Beach increases in size
- Low wave height (about one metre high)
- Much energy
- Operate in storm conditions
- Breaks less frequently (around seven per minute)

Resource A

CONSTRUCTIVE WAVES	DESTRUCTIVE WAVES

Processes at the coast

1. Erosion

Just like rivers, there are four main processes of erosion that operate along the coast that help to erode it. In many cases they work together to change the shape of the coastline.

Corrasion/abrasion

The force of the moving water in the sea throws stones and other eroded particles that it is carrying against the coastline and cliffs, which dislodges more material.

Attrition

Attrition takes place when stones and boulders that are being carried by the sea knock against each other and start to wear each other down. This knocks the edges of the stones and results in smaller, rounder stones/pebbles.

Solution/corrosion

This is when the salts and acids in the seawater slowly dissolve coastal cliffs (particularly limestone and sandstone).

Hydraulic action

This is when the force of the water pounds into the cliffs and the cracks in the rocks, and dislodges more material.

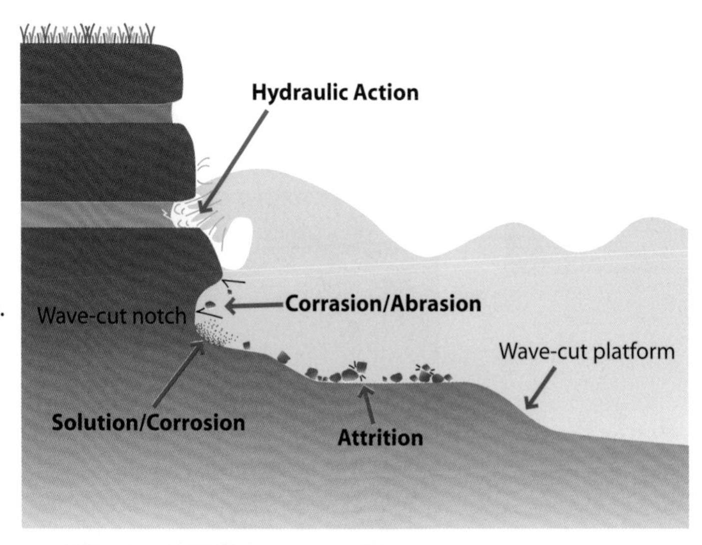

Figure 44 Processes of erosion at the coast

2. Transportation

The sea transports material using the same four methods as rivers (traction, saltation, suspension and solution). However, most of the eroded material is not carried up and down the beach. Instead it is carried in a process called longshore drift.

Waves usually approach the coast in the same direction as the wind is coming from. As the wave breaks, the swash will carry sediment diagonally up the beach and as the swash dies away, the backwash returns the material straight back down the beach at a right angle to the sea. Eroded sediment is therefore slowly moved along the coastline in a zigzag course.

In places where there is a large amount of longshore drift, wooden groynes are often built to trap the sand and prevent it from moving further along the coastline.

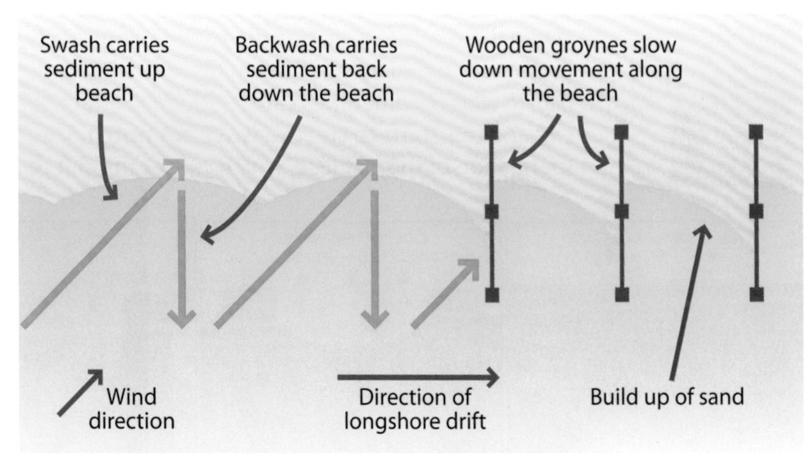

Figure 45 Longshore drift transporting material along the coast

Resource A

Source: Rev Stan

3. Deposition

Eroded material is continually transported along the coastline and the process of longshore drift will eventually result in the material being temporarily deposited to form a beach. These beaches are not permanent features as their shape is constantly being changed every time the waves wash up the shore.

Test yourself

1. Describe how you think each of the four processes of transportation might work to move beach material along the coast. (3 for each process = 12)
2. How does Resource A show longshore drift? Which direction do you think the waves usually approach the beach? (4)

Coastal landforms

1. Erosional landforms

Some of the landforms created by the processes of erosion at the coastline are quite spectacular. The sea can carve through hard and soft rock, and as it does so, it creates amazing shapes.

Figure 46
Bays and headlands at Lulworth Cove

Source: Tom Maloney

One of the most spectacular features of the coastline is the formation of bays and headlands. These are formed when outcrops of harder, more resistant rock (such as basalt) and softer, less resistant rock (such as limestone) are found in the same areas. Waves gradually erode the softer rock away, leaving the harder, more resistant rock sticking out into the sea.

When the sea comes up against a solid barrier such as a cliff face, the four different types of erosion go to work trying to break down the rocks. The greatest pressure of erosion is at the base of the cliff, in what is sometimes called the 'wave attack zone'. The waves start to undercut the foot of the cliff and this starts to create a **wave cut notch.** The notch continues to widen and undermines the foundation of the cliff face, which causes the cliff to collapse and retreat backwards.

Figure 47
The formation of bays and headlands

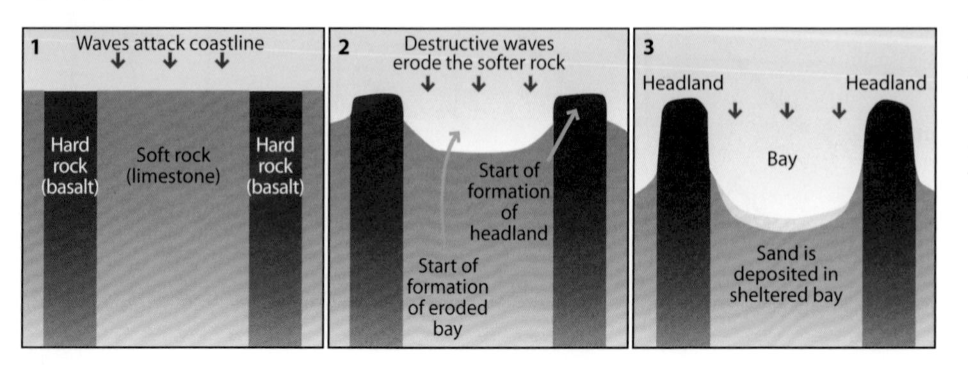

A **wave cut platform** gets left behind as the cliff retreats further away from the original position. These are generally gently sloping or flat platforms that can be seen at low tide.

Figure 48 Wave cut notches and platforms (for example, Flamborough Head, Yorkshire)

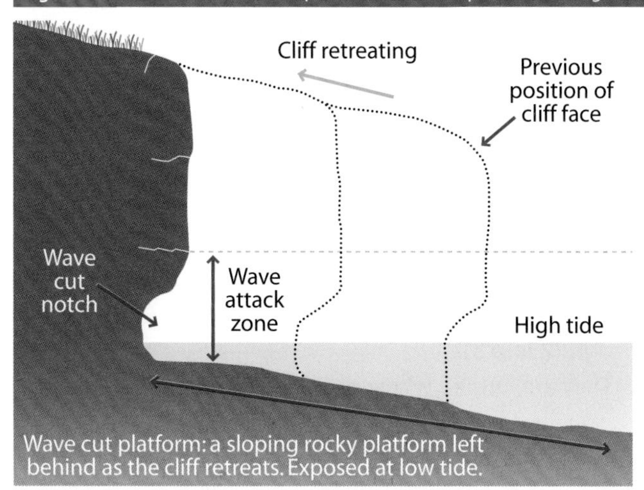

Figure 49 Flamborough Head

Any area of weakness in the rock (such as cracks or joints) will start to come under pressure due to the relentless impact of erosion. The **cracks (1)** in the rock will face erosion through abrasion. The continuous impact of the water will widen any weaknesses in the rock and split the crack into a **cave (2)**. If this cave is being formed in a headland, the cave might continue to be eroded to form an **arch (3)**. The waves continue to erode (especially at the base of the foot of the arch) until the weight of rock in the arch becomes too heavy. When the roof of the arch collapses, it will leave a stack exposed to the elements. Waves will continue to attack the **stack (4)** and eventually it will become undercut and collapse to leave a **stump (5)**.

Test yourself

1. With the use of a diagram, explain how bays are formed. (4)
2. What is the difference between a wave cut notch and a wave cut platform? (3)
3. What erosion processes do you think are responsible for each of the stages of creating cracks, caves, arches, stacks and stumps? (1 mark for each = 5)

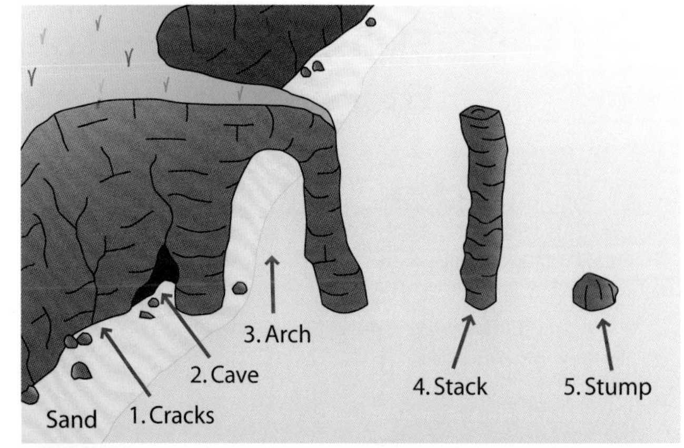

Figure 50 Cliffs, caves, arches and stacks (for example, White rocks)

2. Depositional landforms

Many people are more familiar with depositional landforms along the coastline than they are with erosional landforms. Every year, millions of people use beaches for recreational activities. However, fewer people are familiar with some of the other coastal deposition features.

Beaches

Beaches are usually described as being gently sloping areas of land that are found between the high and low water tide marks. They are built up by constructive waves moving deposited material (sand, shingle and pebbles) up the beach. They are generally fully covered by water at high tide and can be fully exposed at low tide.

Figure 51 The White Rocks near Portrush

Figure 52 Deposition at Chesil Beach

Figure 53
The formation of a spit

Spits

Spits are less common depositional features. A spit is a long, narrow ridge of land that is made up from deposited material (sand and shingle) along a coastline.

There are a few conditions in which spits can form:

- They are usually found in areas where there is an easily eroded coastline such as boulder clay **(1).**
- They are formed when prevailing winds help to transport material down the coastline due to longshore drift **(2 and 3)**.
- They are formed where the coastline changes direction (usually where a river estuary meets the sea).
- The velocity of water from a river and the power of the sea meet at a certain place, causing a loss of energy and deposition **(4)**.
- Many spits have a hooked or curved end as it is gradually shaped and re-shaped by the sea. **(5)**
- Spits bend because of the direction of waves and wind.
- Often the area behind the spit will become a salt marsh due to deposition.

Skills: Using aerial photos and OS maps

Erosion and deposition landforms at the Holderness coast (Yorkshire)

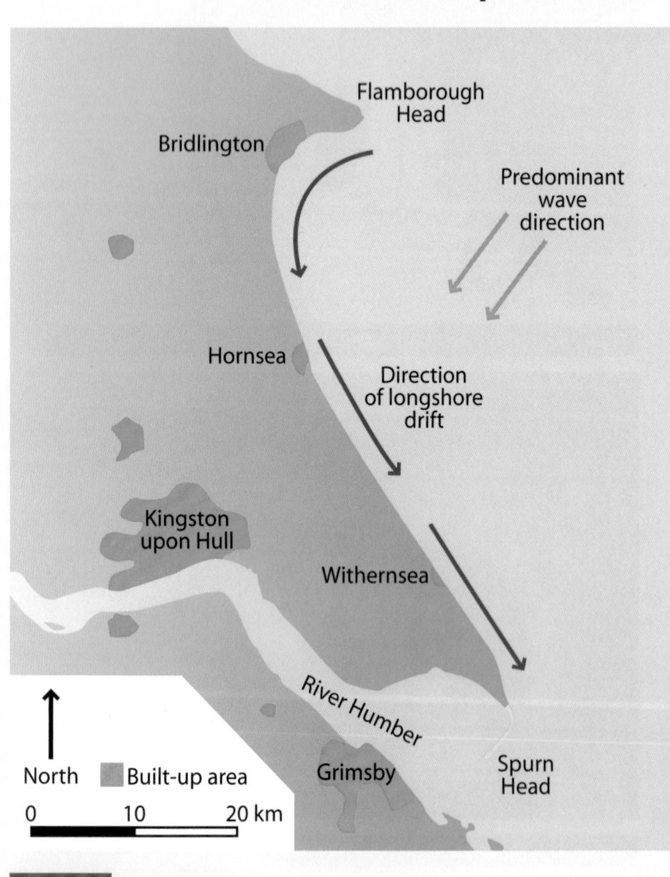

Figure 54
Erosion and deposition at the Holderness coast

The Holderness coast on the east coast of England is a 61 km long coastline that stretches from Flamborough Head in the north to Spurn Head in the South. This stretch of coastline provides us with some very good examples of how erosion and deposition can work together in an area to radically change the landscape.

The Holderness coast is one of Europe's fastest eroding coastlines. In some areas it is eroding at around two metres per year, with over two million tonnes of material being removed annually. The majority of the coastline is made up of soft boulder clay that is easily and rapidly eroded.

Using aerial photographs

1. Take a look the photograph of Flamborough Head in Resource A and make a drawing of the photograph in your notes. (2)

Resource A

Source: Fearlesspunter

a. Use the following to correctly label your diagram. (1 mark each = 4)
   - Headland
   - Limestone
   - Cracks
   - Arch

b. Do you think there is likely to be more erosion or deposition in this area? Give at least one reason to support your answer. (3)

c. Choose one of the erosional features shown in the photograph and describe how the sea has eroded this. (4)

2. Take a look at the photograph of Spurn Head in Resource B. Answer the questions which follow.

   a. Do you think there is likely to be more erosion or deposition in this area? Give at least one reason to support your answer. (4)

   b. Spurn Head is a good example of a spit. Explain detail why you think a Spit has formed in this location. (4)

   c. Towards the end of Spurn Head there is a lifeboat station. Why do you think that putting any permanent buildings in this area might be a bad idea? (3)

Resource B
Spurn Head

Using OS maps to identify coastal features

As our coastline is the buffer zone between the land and the sea, there is often much activity at these fringes of our landscape. Use your OS map symbols to help you identify coastal features on the following OS maps from within the Holderness coast.

Resource C
Map of Withernsea

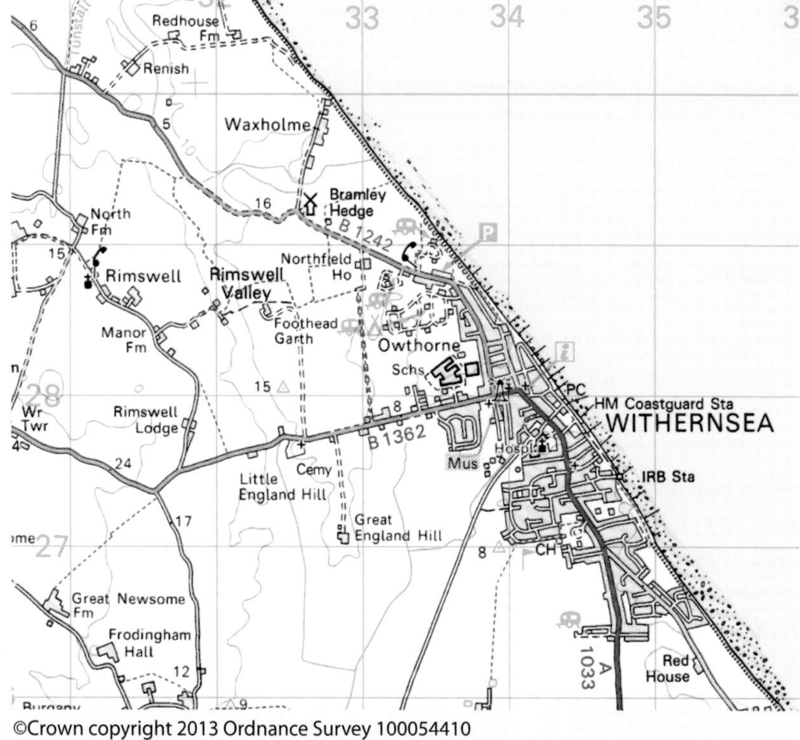

Resource D

Map of Spurn Head

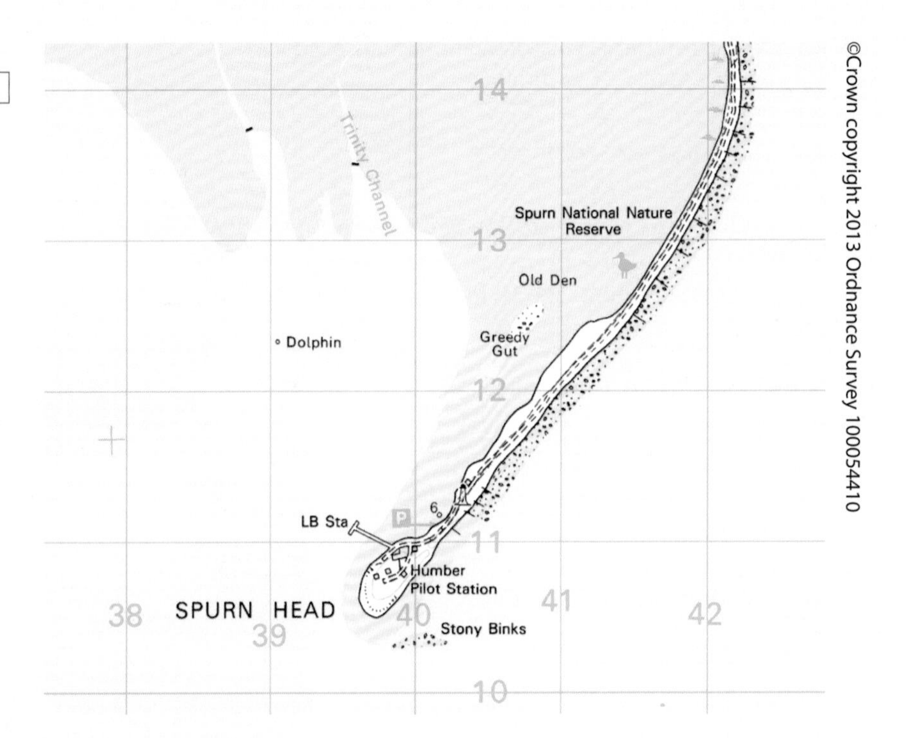

Test yourself

Basic map skills

1. In Resource C, what is found at grid reference 312289? (1)
2. In Resource C, what do you find at grid reference 327275? (1)
3. In Resource D, what is found at grid reference 415128? (1)

Understanding coasts on maps

Using Resource C, answer the following questions:

4. What evidence is there of tourist activity in Withernsea? Make a copy of Resource E and complete it by giving two pieces of evidence and noting their grid references. (6)

Resource E

TOURIST ACTIVITY	6 FIGURE GRID REFERENCE	BRIEF DESCRIPTION

5. Why do you think the feature at 328293 has been built here? (2)
6. What coastal map feature is found at 339279? Do you think this is an odd location? Why do you think it has been located here? (3)
7. What two things can be noted about the coastline to the north of Withernsea? (2)
8. In the town of Withernsea what additional coastal feature do you think is found here? (1)
9. Is there any evidence on the map of whether there is more erosion or deposition happening at this piece of coastline? (3)

Using Resource D, answer the following questions:

10. What is the six figure grid reference for the Humber Lifeboat Station? (1)
11. The green area to the west of the finger of land that is Spurn Head Spit is Kilnsea Clays. What does this indicate about what is happening at this location? (1)
12. What do you notice about the width of Spurn Head Spit from the map? Can you think of an explanation for this? (3)
13. There is evidence of some form of coastal defences being built at different positions along the coastline in this area. Why do you this might be needed? (3)

Check your learning

Now that you have studied Part 4: Coastal Processes and Features, return to page 41 and answer the Key Questions for this section.

PART 5: SUSTAINABLE MANAGEMENT OF COASTS

(a) Human activity in the coastal zone and conflict
Residential
Tourism
Transport
Industry

(b) The need for coastal defences

(c) Coastal Management Strategies
Keep the sea out *(sea walls)*
Retain cliffs and beaches *(groynes, gabions and beach nourishment)*

(d) Case Study:
Newcastle, Co Down *(a coastal management strategy from the British Isles)*
Evaluate the coastal management strategy used with reference to the principles of sustainable development

Key questions

By the end of this section you will be able to answer the following questions:

1. State four different uses of the land at the coast. For two of these explain how there is conflict between them.
2. Suggest two problems caused by a large number of visitors to a beach.
3. Name and explain four methods of protecting the coast.
4. Describe two coastal management strategies used in a place you have studied.
5. Explain why the coastal management strategies were needed in Newcastle, Co Down.
6. Evaluate the extent to which the coastal strategies used in Newcastle are sustainable.

Key words

Residential
Tourism
Transportation
Industrial activity
Conflict
Erosion
Coastal defences
Sea walls
Groynes
Gabions
Beach nourishment
Rock armour
Newcastle
Sustainable development

Human activity in the coastal zone and conflict

The coastal zone has always been attractive economically as a place where ports can be built, castles can provide defence and beaches can provide recreational opportunities for visitors. The problem with the coastal zone is that over the last 150 years it has rapidly become a highly desirable location for people to live and work.

Test yourself

Take a look at Resource A below and answer the questions that follow.

Resource A

HUMAN ACTIVITY	QUESTIONS
Residential Within Northern Ireland, house prices are highest at the coast. A house with a nice view, in an attractive location can command big revenues. However, in other places in the world people are reluctant to live close to the coast, as coastal flooding can bring problems (for example, Bangladesh).	1. Why do you think people in Northern Ireland like to live close to the sea? (2) 2. What are the main advantages and disadvantages of living at the seaside? (4) 3. Why might a subsistence farmer in Bangladesh NOT wish to live close to the sea? (3)

HUMAN ACTIVITY	QUESTIONS
Tourism The tourism industry in Northern Ireland is the fastest growing sector of our economy. Many people will visit the coast for day trips and longer holidays in caravans, bed and breakfast accommodation and hotels. However, this influx of people also brings pollution, crime and vandalism, congestion and overcrowding.	1. What are some of the things that might draw a tourist to the coast? (4) 2. Many local residents who live in a tourist resort might say that the negatives outweigh the positives for tourism. Explain in detail why they might argue that. (6)

HUMAN ACTIVITY	QUESTIONS
Transport In the past, the easiest way to transport goods around the country was by sea and many ports were set up around the coast. In later years, roads were built around the coast and railway lines carried people (mostly tourists) to their destinations. Coastal erosion often caused problems, cutting lines and eroding roads, which cost a lot of money to repair.	1. What are the main arguments against the building of transport facilities in a coastal location? (4) 2. The demands for more efficient transport at the coast mean that ships need deeper harbours, trains need straighter, more modern lines and roads have to be converted to dual carriageways. What do you think the arguments for and against these measures might be? (4)

Source: Library of Congress Prints and Photographs Division Washington, D.C. Reproduction number: LC-DIG-ppmsc-09840

HUMAN ACTIVITY	QUESTIONS
Industry The Irish fishing industry barely exists any more in Northern Ireland. In the past, the coast of Northern Ireland was littered with lime kilns, quarrying, mining and farming. Many of these required materials to be shipped away. Shipbuilding was an important source of employment for men in Belfast from 1850 to 1995. The Harland and Wolff shipbuilding yard continues to manufacture maritime vessels today, this time large wind turbines for offshore energy production.	1. Make a list of some of the industries that you might expect to see close to the coast in Northern Ireland. (3) 2. In the nineteenth century Belfast was famous for linen, rope making and shipbuilding. It was said that 90% of the rope on all ships across the world was made in Belfast. Why do you think that was? (2) 3. The building of wind turbines at Harland and Wolff has created a new focus on renewable energy in Northern Ireland. Can you think of any other revenue streams that they should consider? (3)

Conflict at the coast

We can identify four main human activities that take place at the coast and the action of each of these can create problems with others in the coastal zone. There is not always a smooth transition between these activities and this can lead to conflict in relation to how land is used.

Residential activities

People like living beside the sea and there is a huge demand for houses with access to the sea and sea views. People who already live there want the area to remain attractive, they want employment opportunities and not just investment in tourist facilities. They do not want many more houses making the area overcrowded and creating empty holiday homes.

Tourism activities

Tourism is a fast growing source of employment and there is an ever-increasing demand for hotels, restaurants, accommodation, entertainment and things for tourists to do whilst on holiday. However, many residents do not want more tourists, as this can also bring increased crime, vandalism, pollution and congestion in small places.

Transport activities

The more people who live in a place, then the more transportation links are needed for that area. Ships are the most obvious way to transport people and material at the coast but this may require new facilities to be built at a huge cost. Boats can cause pollution which might affect the leisure users of the beaches.

Industry

Some industries use the coast. Sand can be extracted from beaches and used in building, oil refineries and sewage treatment works can be built, and sea water can be used to cool power stations. However, most industrial land uses at the coast are unattractive, and tourists and residents are often unhappy about their development.

Structured notes

Make a copy of the table below. Identify at least four different examples that explain the land use connected with each human activity. The first has been done for you.

	RESIDENTIAL	TOURISM	TRANSPORT	INDUSTRY
LAND USE	1. Homes 2. Shops 3. Local services: schools, doctors, community centres 4. Recreational areas	1. 2. 3. 4.	1. 2. 3. 4.	1. 2. 3. 4.

Test yourself

Resource A is a photograph of the Spanish tourist resort of Benidorm.

Resource A

©iStockphoto.com

1. What do you think are the competing demands for land in places such as Benidorm? (4)
2. List two advantages of development along a coastal area. (4)
3. List two disadvantages of building in the coastal zone. (4)
4. Can you think of any strategies that could have been taken to reduce conflict between interest groups in a place like this? (4)

Resource B

***Shifting Shores:* Living with a changing coastline**

The National Trust in Northern Ireland recognises the importance of the coastal zone. They have produced a booklet called Shifting Shores: Living with a changing coastline.

They write: "we have always recognised that our coastline is a dynamic and constantly changing environment and we need to take a long term approach to its management." They also outline some of the key changes to the Northern Ireland climate that they expect to take place over the coming century. These include:

- Warmer annual temperatures
- Wetter winters and drier summers
- Sea levels rise (between 85 cm and 1 m likely by 2100)
- Increased frequency of extreme storm surge events
- Greater frequency of extreme wave events

Whilst climate scientists might challenge many of these predictions, they also go on to identify some of the key impacts that they anticipate at their coastal sites over the next 25–100 years. These include:

- Increased coastal erosion and increased flooding
- Changes to, degradation of and even loss of important habitats and wildlife
- Increased land instability
- Roads and paths becoming unsustainable
- Risk to settlement on or close to coastal areas

For the National Trust, a conservation organisation, the proposed changes to the coast over the next few years creates a dilemma; should they 'hold the line' or adapt? They write:

"In summary we can either try to 'hold the line' or to adapt to change, either immediately or through buying time with interim measures.

The traditional response has been to resist change through engineering works and hard defences of rock or concrete. Through growing evidence and experience, we are gaining a better understanding of the forces of nature and the consequences of working against them. As sea levels rise and severe storms increase, it will become increasingly difficult and expensive to build and maintain strong defences and ultimately their long term effectiveness will be diminished. They can also disfigure the coast and cause environmental harm by moving the problem to another location. We therefore believe that hard defences should only be used as a last resort.

The National Trust's preferred approach is to take the long term view and seek to work with natural coastal change wherever possible. However, we recognise how important it is to take account of the social and economic impacts, as well as the environmental impacts, of our decisions. This adaption approach will give the time and space for us to adjust with the coast's forecast behaviour in the future."

Source: Shifting Shores: Living with a changing coastline, National Trust publication, 2007

5. In what ways does the National Trust expect the coast to change in the future? (5)
6. What are the main factors that the National Trust expects to cause these changes to the coast? (5)
7. Explain the difference between the two approaches that the Trust might take to respond to coastal erosion. (4)
8. The National Trust's preferred approach might be described as 'sustainable', as it takes note of social, economic and environmental factors. How do you think this will influence their actions? Will it protect the coast or not? (6)

The need for coastal defences

Many scientists and conservationists believe that the coastline is going to continue to change over the next 100 years. Fragile environments are going to be under serious threats from flooding, erosion and environmental change. With this in mind, environmental, social, moral and political decisions will need to be made about whether to protect the coast or allow nature to take its course.

In the UK there are two main issues in relation to coastal management:

1. How to prevent coastal erosion?
2. How to manage the coast for tourist use?

It can be difficult to manage the coastline in such a way as to have a balanced approach to each of these issues. For example, in some places in the UK, coastal erosion is removing between 3 and 5 m of land every year. Erosion processes take place on areas with less resistant rock, such as the Holderness coast in Yorkshire, or where there is a build up of sand and shingle at the base of cliffs. As a result huge concrete barriers have been built. These sea walls reflect the energy of the sea and reduce the erosion of the soil and rock, but they make it difficult for tourists to access the beach.

Sometimes coastal areas come under threat of coastal flooding. Low lying areas can be flooded, causing huge problems to the people who live there. Sand dunes are also susceptible to erosion by the wind and the sea. The more people that trample over the sand dunes, the faster the vegetation will be reduced, which can make the sand unstable and easily eroded.

Local government authorities and councils have to make some difficult decisions about what is the most appropriate response to coastal erosion. Should it be protected, prevented or should nature be allowed to take its course?

Coastal management strategies

There are two main priorities for the implementation of coastal management strategies:

1. To keep the sea out (sea walls)
2. To retain cliffs and beaches (groynes, gabions and beach nourishment)

1. Keeping the sea out

Sea walls

Figure 55 *Sea walls*

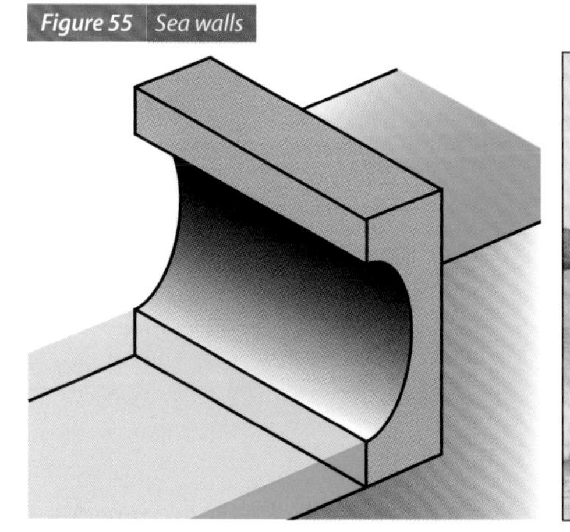

A sea wall in Portrush

Description

- A solid wall that separates the land from the sea.
- This is a traditional 'hard engineering' method usually constructed of concrete.
- It supports the land while holding back the sea.
- The wave action can beat against the sea wall without eroding the coast.
- It can be designed to absorb and deflect wave energy.

Evaluation

- Sea walls can be expensive (over £10,000 per metre).
- They need very deep foundations so that they do not erode away.
- They can sometimes be an ugly addition to a beautiful area.
- They can reflect wave energy back into the sea, which can cause erosion either further out to sea or somewhere else along the coastline.
- Sometimes sea walls can actually cause the removal of a beach.
- They might need to be replaced every 25–30 years.

2. Retaining cliffs and beaches

Groynes

Figure 56 Groynes

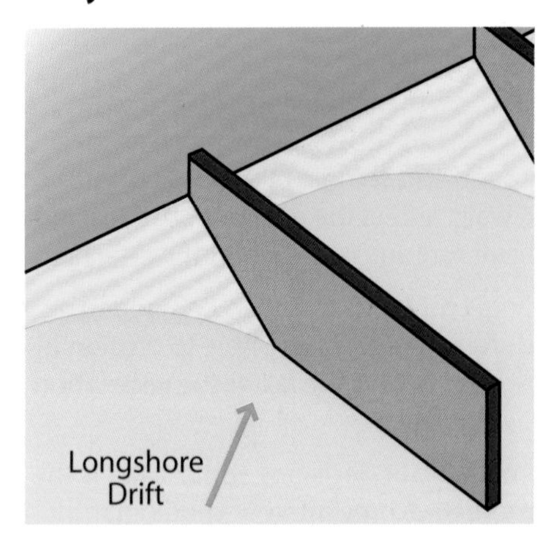

Two metre high groynes at Lyme Regis in England

Description

- Groynes are wooden, concrete or rock barriers than can be built out into the sea.
- They can trap the sand carried by longshore drift and help to increase the build up of a deposition beach.
- Usually these are seen as 'hard engineering' but it depends on the material being used. Some consider them to be 'soft' measures as they enhance the existing beach.

Evaluation

- Groynes can trap sediment that is supposed to be going to another place along the coastline. This means that another area of the coastline is going to be more vulnerable to erosion.
- They are very cost effective.
- They need continual maintenance and replaced regularly.
- Some people see groynes as unattractive and an obstacle for people who like to walk along the beach.

Gabions

Figure 57 Gabions

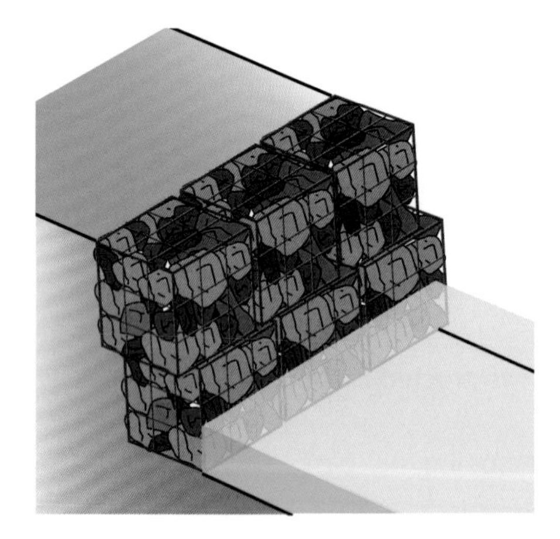

Gabions used to protect the coastline at Newcastle, Co Down

Description

- Gabions are metal cages that are filled with rocks.
- They are usually stacked together to create a wall of rock.
- As the waves crash against the gabion cages, the energy of the water is forced between the spaces in the rocks and the energy is absorbed inside the cage. This stops the cliffs from taking the brunt of the erosion.
- Water then percolates through the rocks and back into the sea.

Evaluation

- Gabions can be a good short term solution. However, they rarely make an effective long term solution, as the cages can split and release the rocks inside, which can be used by the waves to erode the coast.
- They can be relatively cheap to build (as little as £11 per metre).
- The cages are often described as unattractive.
- The cages and rocks can trap debris and pollution, which can rot leaving bad smells and health concerns.
- The cages can provide an ideal habitat for rats.

Beach nourishment

Description

- This is when sand or pebbles are added to a beach, replenishing it or building it up.
- It is technically a 'soft engineering' solution but this depends on where the material is coming from. Soft engineering should not change the structure of the beach through building works. However, the size of the beach is often increased as the beach is made much wider.
- A large, wide beach naturally protects the coastline because the wave energy is absorbed through the sand.

Source: U.S. Army Corps of Engineers photo by Ann Cameron Siegal

Figure 58
Beach nourishment in Delaware, USA

Evaluation

- The sand used must be of a similar quality to the existing beach material so that it can integrate with the natural processes occurring there, without causing any adverse effects.
- Sometimes beach nourishment is used alongside groyne schemes.
- Nourishment requires constant maintenance and will need constant recharging of the beach material.
- It can be expensive at between £5,000 and £20,000 per 100 m.

Test yourself

1. Which of the coastal management strategies listed is seen as 'soft engineering'? Explain why you think this is the case. (4)
2. Which of the coastal management strategies do you think is the most sustainable? Use the following questions to help work out your answer.
   a. Economically, which of the measures listed do you think is the most practical? (3)
   b. Socially, which of the measures listed do you think that people will find the most acceptable? (3)
   c. Environmentally, which of the measures listed do you think causes the least amount of environmental damage and helps to protect the environment from further disaster? (3)
   d. Overall, which of the measures do you think is the most sustainable? (3)
   e. Is this approach a short or a long term measure? What are the main drawbacks of using an approach such as this? (3)
3. Are there any other coastal management strategies that you can find information on which might be appropriate for consideration in protecting the coastline around Northern Ireland? (3)

Case study: Newcastle, Co Down

(a coastal management strategy from the British Isles)

Sometimes places along our coastline are under attack from the sea and decisions need to be taken to make sure that economically, socially or environmentally important areas are protected and conserved for future generations.

In Northern Ireland, one such place is the coastal resort of Newcastle, Co Down. The erosion of the coast there was occurring at such a rate that it caused concern.

Background

Newcastle is a small town with a population of 7,500. It lies at the base of Slieve Donard, the highest peak in the Mourne Mountains and in Northern Ireland. Tourism is the main industry in the town and it is a popular seaside resort. In the past a railway line brought holiday-makers from Belfast to holiday on the coast. Many visitors use Newcastle as an access point to the Mourne Mountains, to play golf at Royal County Down Golf Course, to visit the Slieve Donard Hotel or to stroll along the promenade.

The beach itself is about 8 km long. Most of the waves are gentle, constructive waves due to the limited fetch. The sediment that makes up the beach has been washed down from the mountains using the Shimna river. Over the years there have been numerous attempts to manage the coast at Newcastle.

Figure 59

Newcastle in the early Twentieth Century

Source: Library of Congress Prints and Photographs Division Washington, D.C. Reproduction number: LC-DIG-ppmsc-09872

Phase 1: The nineteenth and early twentieth centuries

In the early years, as Newcastle was becoming established as a tourist resort, there was very little management of the coastline. Newcastle had a small harbour with some fishing boats. As visitors started to come from Belfast, gardens were established along the sea front and a very basic footpath. Winter storms frequently washed material away and the shape of the beach was constantly changing.

Phase 2: The late twentieth century

From the 1950s to the late twentieth century, Newcastle continued to develop as a popular place for tourists. The local urban council developed some plans to try to retain certain aspects of the beach and built some small **groynes** to try to collect the beach material as it was moved along the coast.

In the 1980s some further groynes were built in an effort to build up the beach, as it was gradually becoming too thin for visitors to enjoy. These did help to stop erosion of the beach and were easier to build than a sea wall.

Figure 60

Groynes in Dundrum Bay, Newcastle

The Mournes area has been an Area of Outstanding Natural Beauty since 1966, which means that protective measures are in place to protect and manage the countryside and coastline. One area of concern in the 1990s was the area between where the Shimna river enters Dundrum Bay and the Slieve Donard Hotel. Some measures needed to be taken and **gabions** were built up to protect the coast at this point. However, these were not very effective and had to be replaced fully about 10 years later.

Additional coastal protection was added further along the beach at the Royal County Down Golf Course. **Rock armour** was used to protect the coastline from eroding. This consisted of big rock boulders placed at the foot of cliffs and sand dunes to try and reduce the amount of wave energy hitting the coastline.

Figure 61 *Rock armour protecting the new coastal path in Newcastle*

Phase 3: The early twenty-first century

Following a couple of major storms at the tail end of the twentieth century and the start of the twenty-first century, it was recognised that Newcastle needed further coastal management measures. The beach has been badly eroded over the last few years but still remains a popular attraction for the area.

In 2007 the new Newcastle Promenade development won a top UK design award. A new wave return sea wall was built 1 m above the former level of the promenade and the gardens were raised to reunite the shore with the street.

Newcastle's improvements cost £4 million. One of the biggest expenses in the project was the innovative curved **sea wall** defence, which is designed to protect the town from the winter gales that formerly caused flooding in the Main Street. However, some people have noted that the sea wall has damaged the natural beauty of the area and created a more urban landscape with no sand dunes and no habitat for plants and animals to survive.

Figure 62 *The new seafront promenade at Newcastle, built in 2007*

The Victorian promenade and seafront was completely redesigned to integrate the town with the beach and make it a welcoming space for tourists. The promenade was raised by 1 m to allow access from the street and splash has been considerably reduced by the new wave return wall. The footpath has been widened and incorporates new seating, modern stainless steel railings, a 'watery' wall in keeping with the sea theme and a bold new lighting scheme. Overall the new promenade has greatly improved the quality of the environment for tourists and local residents.

Further gabions were built up again at the mouth of the Shimna river and Down District Council is still carrying out tests to see if it is worthwhile to build new groynes along the beach to continue to trap deposited material.

The Mourne Heritage Trust has composed specific policies for Sustainable Tourism in this area. On their web site (www.mournelive.com), they write:

"Sustainable Tourism is a way of positively managing tourism for the benefit of the visitors, residents and the environment. This ensures the long term survival of the environment, to be enjoyed by future generations. Sustainable Tourism helps local people create and maintain quality long term jobs. It also aims to buy goods and services locally and generally fit in with the character of the area."

Evaluation of coastal management techniques used in Newcastle

The new promenade development in Newcastle has won many awards since it was started. It certainly has made a difference to the seafront in Newcastle and has helped to revive what was a dilapidated part of the town. Many local residents hope that it will attract more visitors and rejuvenate tourism. The new development aims to combine a focus on sustainable tourism with a sustainable approach to the management of coastal erosion.

However, many environmentalists are concerned that using hard engineering methods and changing the landscape in such a radical way has altered the environment of the Newcastle seafront and made it more difficult for birds and wildlife to survive there.

Test yourself

1. So far, groynes, gabions and a sea wall have all been used to attempt to manage the coastal erosion at Newcastle. Which of these measures do you think has been the most effective? (5)
2. The council are still thinking about using groynes along the beachfront in Newcastle. What are the main advantages and disadvantages of using groynes here? (3 marks for advantages, 3 marks for disadvantages = 6)
3. Should the council consider implementing a beach nourishment programme? How could you argue against them starting this? (3)
4. How do you think the following interest groups might respond to the building of the new coastal management promenade in Newcastle?
   a. The National Trust at Murlough Bay (an ASSI – Area of Scientific Special Interest – that has noticed that there is less sediment deposited on their beach since the development of the Promenade). (3)
   b. The residents and ratepayers in Newcastle. (3)
   c. The tourists. (3)
5. With reference to the new Newcastle Promenade development in 2006, do you think this has been a sustainable solution? Use the following questions to help you with your answer:
   a. Economically, which of the measures listed do you think is the most practical? (3)
   b. Socially, which of the measures listed do you think people will find the most acceptable/has brought the biggest impact and most satisfaction to the residents? (3)
   c. Environmentally, which of the measures listed do you think causes the least amount of environmental damage and helps to protect the environment from further disaster? (3)
   d. Overall, which of the measures do you think is the most sustainable? (3)
   e. Is this approach a short or a long term measure? What are the main drawbacks of using an approach such as this? (3)

Test yourself

Read the following article, which was printed in the *Down News* on 4 October 2010 and answer the questions which follow.

Resource A

Clarke Calls for Action as Coastal Erosion Hits Newcastle

With global warming, the rise in sea levels and the forecast of much more severe storms in the years ahead, our coastal areas are under pressure from the inexorable march of the sea. With a rise of just a few feet in its level, the sea can become a serious threat to our way of life. Already there are incursions by the sea into the land in various points around the County Down Coast.

On the back of this serious environmental threat, South Down Assembly member councillor Willie Clarke believes there is a serious local risk to the A2 Newcastle to Kilkeel Road being undermined because of damage to the sea defence along the beach at the Black Rock.

Councillor Clarke said, "My initial assessment would indicate a risk of further erosion and undermining in the winter months ahead if the problem is not addressed immediately. My evaluation of the structural condition of the sea defence indicates a significant erosion and the loss of armour/stone pitching in at least two areas.

"Apart from the implications of future damage to the sea defence, there is a great risk associated to the undermining of the Road itself. Moreover, because of the movement of the rock armour there is also a risk to the Down District Council car park itself, which was previously severely damaged by heavy seas a number of years ago.

"I will be arranging a meeting between the relevant statutory bodies to see what can be done to address this urgent problem. I believe a small amount of stone pitching and replacement of rock armour will save much needed resources in the long term.

"Also, I will also be contacting Roads Service maintenance personnel to repair the stone wall at the Rock area which has a number of coping stones damaged and will have weakened the structure," he added.

Other areas around the Down coast such as the sea bank on the Ballyhornan road to Killard are also being eroded.

Source: www.downnews.co.uk

1. What are some of the causes of increased concern regarding coastal erosion in Newcastle in recent years? (3)
2. What is the problem that Councillor Clarke is trying to raise? (3)
3. What solutions other than rock armour might you suggest for this place? (4)
4. In your opinion, what is the most sustainable solution for this location? (3)

Check your learning

Now that you have studied Part 5: Sustainable Management of Coasts, return to page 49 and answer the Key Questions for this section.

ORGANISE YOUR REVISION

Draw a mind map to summarise Unit 1A The Dynamic Landscape.
You might want to read the 'Organise your revision' section on page 14 again to help you.

EXAM PRACTICE QUESTIONS

Some of these questions are from previous CCEA GCSE examination papers and others have been written in the same format to give you practice at answering 'exam style' questions.

Try to answer the questions with as much detail as possible. Also consider the number of marks that each question receives, as this will give you a good indication of the amount of depth that your answer needs.

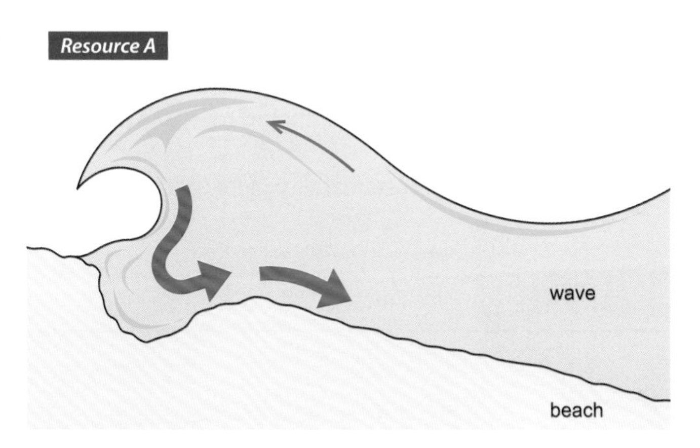

© Dawlish Warren Tourism

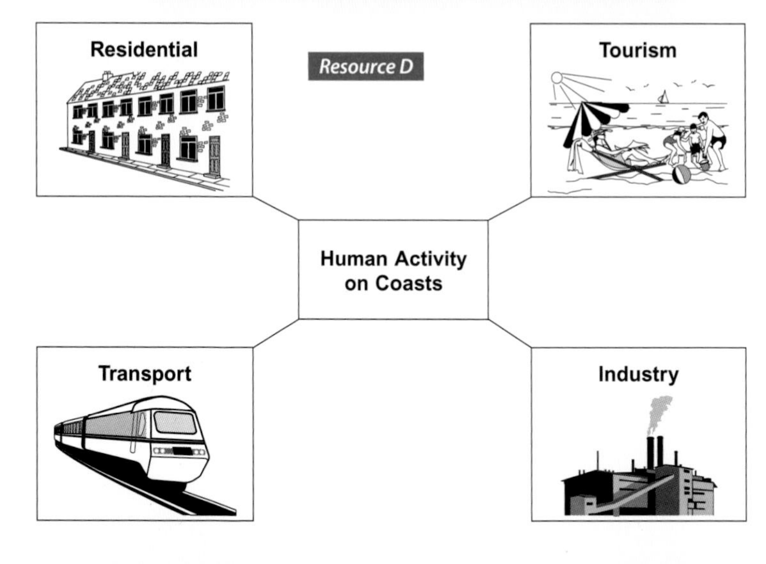

Foundation Tier

1. Study Resource A which shows a destructive wave.
   (i) Describe how destructive waves erode coasts. (3)
   (ii) Describe **one** of the coastal processes given in the list below. (2)
   Attrition, Abrasion, Hydraulic pressure
2. Study Resource B which is a photograph of a spit in Devon.
   Explain how a spit such as the one shown in Resource B was formed. (6)
3. Study Resource C which is a photograph of an arch in Dorset.
   Explain how an arch is formed. (6)
4. Sea cliffs are formed by erosion. Name **two** processes by which the coast is eroded. (2)
5. Describe how a groyne works to protect the coast. (3)
6. Name an area where the coast has been defended.
   a. Area:______________ (1)
   b. Describe **two** coastal management strategies used. (4)
   c. Explain why the coastal management strategies were needed. (4)

Higher Tier

1. Study Resource A which shows a destructive wave.
   Explain why destructive waves erode coasts. (3)
2. Study Resource C which is a photograph of an arch in Dorset.
   Explain how an arch such as the one was formed. (6)
3. Suggest a possible conflict which might arise between tourism and industrial land-use in the coastal zone. (2)
4. Explain how sea walls and groynes work as coastal management strategies. (6)
5. Study Resource D which shows some human activities in coastal areas. Select **two** of these activities and suggest how these activities might be in conflict. You should refer to a place to illustrate your answer. (3)
6. For one area in the British Isles, describe and evaluate the sustainability of its coastal management strategy. (9)

1B Our Changing Weather and Climate

"Climate is what we expect, weather is what we get"
Mark Twain

Weather is one of the most fascinating aspects of our study of Geography and many people have tried to get to grips with it and understand its patterns. Much time has been spent trying to work out how it is going to change on a daily basis. However, in recent years, a new focus has come to the study of weather: climate change and the question of whether human activity has caused or has sped up this process of change.

This theme takes a look at the ways that weather is measured and then at the different weather systems we should expect to see over the British Isles. Finally, it looks at climate change and asks what consequences this might bring.

The theme is divided up into three parts:

1. Measuring the Elements of the Weather
2. Weather Systems Affecting the British Isles
3. The Causes and Consequences of Climate Change

PART 1: MEASURING THE ELEMENTS OF THE WEATHER

(a) **The difference between weather and climate**

(b) **Measuring the elements of the weather:**
Temperature (°C): *minimum and maximum thermometers*
Precipitation (mm): *rain gauge*
Wind direction (8 compass points): *wind vane*
Wind speed (knots): *anemometer*
Air pressure (mb): *barometer*
Cloud types: *stratus, cumulus, nimbus, cumulonimbus and cirrus*
Cloud cover: *oktas*

(c) **Factors to be considered when locating weather instruments:**
Thermometers
Rain gauge
Wind vane
Anemometer

(d) **Sources of weather data used to create a weather forecast:**
Land-based stations
Balloons
Buoys
Weather ships
Geostationary and polar satellites

Key questions

By the end of this section you will be able to answer the following questions:

1. State the meaning of the term weather.
2. What is the unit of measurement for pressure?
3. Explain why thermometers must be placed in the shade.
4. Explain how satellites can help create weather forecasts.
5. What is the difference between weather and climate?
6. Outline one factor that would need to be taken into account when locating a wind vane instrument.
7. Name the instrument used to measure rainfall and explain two factors which must be taken into account when placing this instrument within a land-based weather station.

Key words

Climate
Weather
Temperature
Range of temperature
Maximum thermometer
Minimum thermometer
Precipitation
Rain gauge
Wind speed
Wind vane
Wind direction
Anemometer
Atmospheric pressure
Barometer
Millibar
Cloud cover
Okta
Cloud types
Stratus
Cumulus
Nimbus
Cumulonimbus
Cirrus

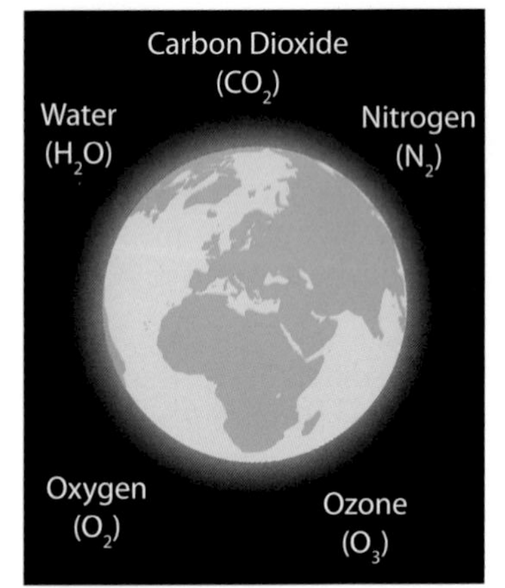

Figure 63

The main gases that make up the atmosphere

The difference between weather and climate

Weather is the day-to-day state of the atmosphere (the layer of gases that surrounds our planet). It is a dynamic process which is constantly changing. It is difficult to describe the weather of Northern Ireland in one sentence, as the different weather elements are always changing and show no daily pattern.

Climate is more long term. It is the average weather taken over a long period of time (usually over 35 years). Climate is a less dynamic process and does not change as quickly as the weather … in theory anyway! The climate of Northern Ireland is often described as being mild and damp. This is because Northern Ireland is within a 'temperate' zone, where the temperature is neither extremely hot nor extremely cold, and neither extremely wet nor extremely dry.

Measuring the elements of the weather

The amazing thing about the weather is that is always changing. It can alter from one moment to the next and from one place to another. It can also be different on a very small scale, for example, the weather in Belfast can be very different from the weather in Bangor. Beyond this, the weather can actually be different in one part of Bangor compared to another. In fact, it can even be different at the front of your house compared to the back of your house. This is because the different elements of the weather act in a very localised way. It is their interaction with the local environmental conditions and features that determines how we experience the different weather elements.

Measuring the weather used to be something that many people enjoyed doing, with many amateur meteorologists taking their own observations perhaps 2–4 times a day. They would have kept a notebook log each day and used thermometers, rain gauges and a range of weather equipment to keep track of the changes in the atmosphere. However, in this digital age we have many more electronic gadgets that help us to take readings and estimates of the weather without having to take out the recording equipment.

The following table lists some basic information about each of the six main elements that together make up the weather.

Element	Description of element	Method of measurement	Unit of measurement
Temperature	Temperature is a measure of the amount of heat in the atmosphere around us.	Max and min thermometer (instrument)	Degrees centigrade (°C)
Description of measurement equipment	A max and min thermometer helps us to record the highest daily (maximum) temperature and the lowest daily (minimum) temperature. The thermometer is usually made in a U-shape, with a separate temperature scale up each arm of the U. One (usually on the right) is used for recording the maximum temperature. The glass tube on this side is usually filled with mercury. The other side of the tube (usually on the left) is used to record the minimum temperature. The glass tube on this side is filled with alcohol. The maximum and minimum are recorded using two small steel markers (called indexes), which are sprung into the tube so that they cannot easily slide unless a small force (the moving liquid) is applied to them. It is essential that the markers are returned to the top of the mercury using a magnet after each reading is taken.		

Figure 64

Three different types of thermometer

Element	Description of element	Method of measurement	Unit of measurement
Precipitation	This is the amount of moisture in the atmosphere and usually involves water in all of its states: liquid, solid and gas (vapour). It includes water, dew, hail, rain, sleet and snow.	Rain gauge (instrument)	Millimetres (mm)

Description of measurement equipment

Figure 65 Rain gauge

Precipitation (mostly in the form of rain) is usually collected in a rain gauge. Gauges can come in all sorts of different shapes and sizes. Some sit on top of the surface and others are graduated cones with a spike at the bottom, allowing them to be easily secured in grass/soil areas.

The most common rain gauge for measuring the weather is the 'standard rain gauge'. This is made up of a copper cylinder, which is dug into the ground with about 30 cm sitting above the ground surface. Any precipitation that is collected runs down into a funnel and into a collection bottle. Once a day (usually at 9 am), the weather observer will come and pour any collected water into a graduated measuring tube, so that the amount of daily precipitation can be recorded to the closest 0.1 mm.

Element	Description of element	Method of measurement	Unit of measurement
Wind direction	Wind is the air in motion in a horizontal direction.	Wind vane (instrument)	8 compass points

Description of measurement equipment

Figure 66 Wind vane

Source: Amanda Slater

The air in the atmosphere moves from one place to another. The direction from which the wind is coming from can help to tell us important details about what type of weather we can expect.

A wind vane is usually a specially made piece of engineering. It has four main compass points secured in their respective positions, with an indicator dial free to move. The arrow moves to point into the wind, which shows the direction that the wind is coming from.

Element	Description of element	Method of measurement	Unit of measurement
Wind speed	The speed of wind can range from calm to hurricane force.	Anemometer (instrument)	km per hour (kph)

Description of measurement equipment

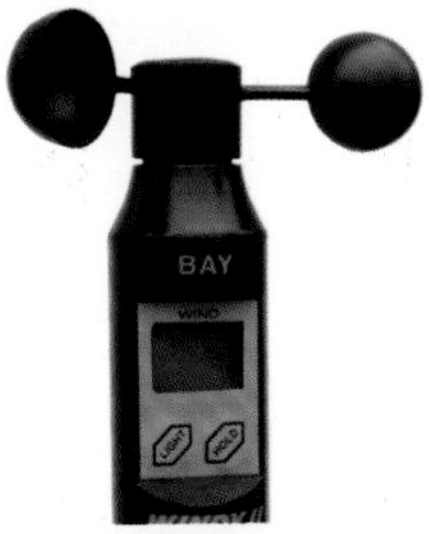

Figure 67 Anemometer

The speed of air movement in the atmosphere helps us to understand how quickly a weather system is moving into an area.

An anemometer is a device which helps to measure the speed of the air. This is usually either handheld or attached to the top of a securing frame. A cup anemometer will usually have three hemispherical cups at its top. The wind will cause the cups to turn round and this allows the observer to record the speed of the wind.

Element	Description of element	Method of measurement	Unit of measurement
Air pressure	Air pressure (or atmospheric pressure) is the pressure exerted by the weight of the atmosphere on the earth's surface. Normal pressure is set at 1000 mb. Low pressure is defined as anything below this and high pressure is anything above.	Barometer (instrument)	Millibars (mb)

Description of measurement equipment

Figure 68 Aneroid barometer

Source: Andres Rueda, http://www.flickr.com/photos/andresrueda/

In many cases, if you can measure the pressure at a place and know whether the pressure is rising or falling, this is a very good indicator of how the weather is likely to change in the near future. Falling pressure usually indicates a depression moving in (with bad weather) and rising pressure indicates anticyclonic conditions (and better weather).

The main piece of equipment for measuring air pressure is the aneroid barometer. Aneroid means without air. The pressure is measured when a small metal capsule is squashed when there is lower pressure and is allowed to expand when there is higher pressure. This is linked to a needle which then indicates the amount of pressure in millibars on the dial.

Element	Description of element	Method of measurement	Unit of measurement
Cloud types	A cloud is a visible mass of tiny particles floating in the atmosphere, consisting of ice crystals or water formed from the condensation of water vapour.	Observation	Cirrus, cirrocumulus, altocumulus, altostratus, nimbostratus, stratocumulus, stratus, cumulus, cumulonimbus

Description of measurement equipment

Figure 69 Cumulonimbus cloud

Source: Osa Mu

Usually clouds are described as being found at one of three layers in the sky: low, middle or high. Which layer the cloud is found in gives further information about the characteristics of the cloud.

High clouds: cirrus, cirrocumulus, altocumulus, altostratus
Middle clouds: nimbostratus, stratocumulus
Low clouds: status, cumulus and cumulonimbus

Main cloud types

Cirrus: these are very high wispy clouds that are usually made up of ice particles.

Nimbostratus: Nimbus comes from the Latin word for shower and stratus reminds us that these clouds are stratified into layers. These can be quite thick layers of rain-bearing cloud.

Stratus: The name comes from the Latin word for layer. These tend to be lower and less grey than nimbostratus but they still bring a lot of rain, especially when close to the surface (and are responsible for drizzle).

Cumulus: These clouds always seem to be moving faster than the others. They are mostly white and have a fluffy appearance. They can bring rainy intervals, especially when they turn grey at the base.

Cumulonimbus: These are cumulus clouds that keep gathering more and more moisture. They start to tower as air rises and forces the cloud to develop vertically. Sometimes these clouds can stretch up as high as 10 km and can trigger hail, thunder and lightning, and can bring intensive thunderstorms.

Figure 70 Cloud map

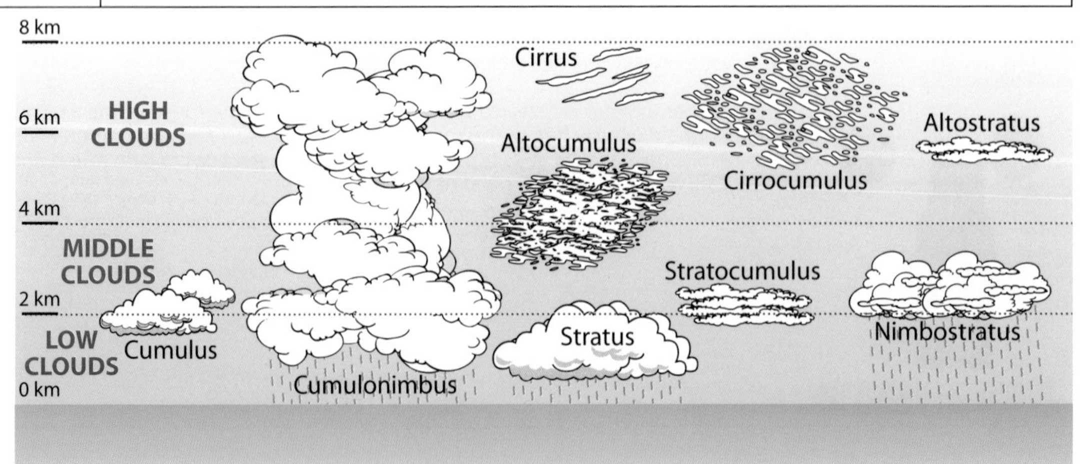

Element	Description of element	Method of measurement	Unit of measurement
Cloud cover	Due to an international agreement, the amount of cloud cover (or amount) is reported as eighths (or oktas) of the sky covered.	Observation	Oktas (eighths)
Description of measurement equipment *Figure 71* Cloud cover Source: the aucitron	An observer should estimate the amount of the blue sky that is visible and identify how much of the sky is covered in cloud. The results of individual observations will be shown on weather maps using the key in Figure 86 on page 74.		

Test yourself

1. Take a look at the reading on the max/min thermometer in Resource A.

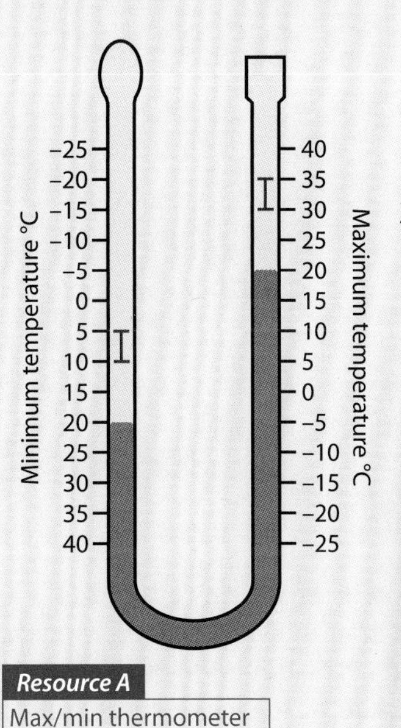

Resource A Max/min thermometer

   a. What is the current temperature? (1)
   b. What was the maximum temperature recorded? (1)
   c. What was the minimum temperature recorded? (1)

2. Name two advantages and two disadvantages that there might be when comparing the use of a copper standard rain gauge with a more modern plastic spike gauge. (4)

Resource B Copper rain gauge

Source: Alan Sim

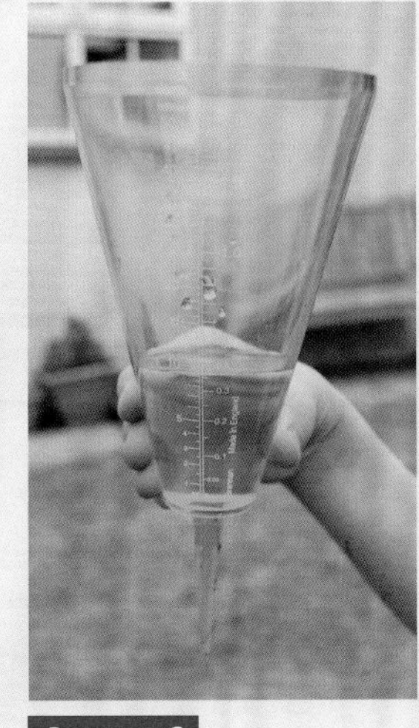

Resource C Plastic rain gauge

3. Wind vanes and anemometers are not as precise when conditions are calm, with little wind. Why do you think this is the case? (2)

4. Describe some of the problems associated with collecting weather records by observing cloud type and cloud cover. Evaluate the accuracy of this recording method. (Description = 3, Evaluation = 3)

Factors to be considered when locating weather instruments

Taking weather measurements requires control and consistency. Measurements need to be taken at the same time every day. There are some other factors that weather observers must take careful note of when collecting weather data:

Source: Alan Sim

Figure 72
A Stevenson screen

Thermometers and the use of a Stevenson screen

The Met Office advises that the standard method of measuring air temperature is to house the thermometer within a Stevenson screen. The screen is designed to make sure that only the air flowing through the box is recorded. The surface of the screen is painted white to reflect any heat. The shade within the box ensures that direct sunlight cannot give a false reading. This means that any recording taken inside a screen will be more accurate.

Figure 73
Rain gauge

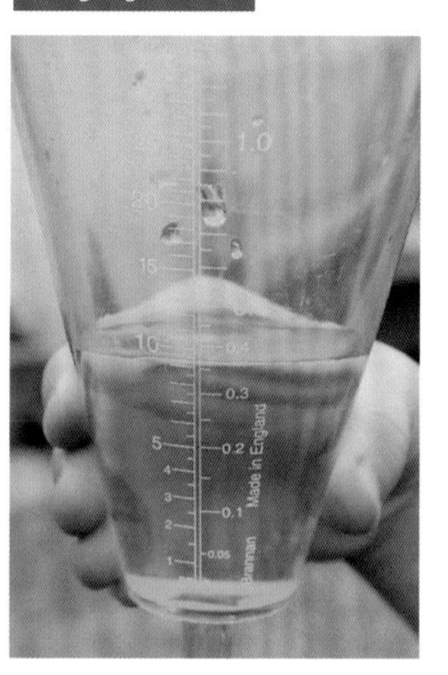

Rain gauges

Rain gauges are usually made of copper or plastic so that they do not rust when they are secured in the ground. In order to get an accurate result from a rain gauge, it is important to make sure that it is sited in the best possible place:

- They need to be anchored securely so that they do not get blown over by the wind.
- The top section of the gauge (where the funnel for collecting the water is) needs to be more than 30 cm above the ground to avoid rain splash when the ground is saturated.
- The rain gauge location needs to be clear of any building, guttering or any other source of potential water that might affect the results.

Figure 74
Wind Vane in Kobe, Japan

Source: Arjan Richter

Wind vane

Many of the wind vanes that adorn buildings today are decorative features rather than functional instruments. Care needs to be taken to make sure that the vane is actually capable of providing an accurate direction. For example, the position of the direction arrows can be verified using a compass and the wind direction can be checked by throwing some grass into the air to see if its movement correlates with that shown by the gauge.

To work efficiently, wind vanes need to be placed on the top of a roof and must be free from any shelter from chimneys or nearby buildings. In many cities, church wind vanes have become less accurate, as new high rise buildings have created wind tunnels that do not represent the true wind direction.

Figure 75 Hand held anemometer

Anemometer

An increasing number of anemometers are being permanently mounted on stands or to the top of buildings rather than being held by an observer. This should help to eliminate problems with data collection.

The best place to take a wind speed reading is on top of a building, again avoiding shelter (as per wind vanes). If being held by the observer, the anemometer needs to be held at full stretch above the head of the observer, making sure that the flow to the cups is not disrupted in any way. The big problem with the anemometer is that it usually needs more than 10 knots of air flow to make sure that it is accurate.

Test yourself

Make a copy of the table below and complete it in as much detail as you can.
(1 mark for each statement = 12)

ELEMENT	LABELLED DIAGRAM OF EQUIPMENT (DRAW)	EXPLANATION (HOW DOES IT WORK?)	ANY LOCATION ISSUES? (DOS AND DON'TS)
Temperature			
Precipitation			
Air Pressure			
Wind direction and speed			

Sources of weather data used to create a weather forecast

Surface observations have been a very important source of data to model how the weather changes over time across the UK. They have helped meteorologists to develop a better understanding of the different ways that weather systems will affect the landscape that they come into contact with. In the past, major weather stations phoned their data in to the Met Office headquarters but now the 'Weather Observations Website' (WOW) allows weather observers to share their recordings digitally.

The production of a 'weather forecast' allows meteorologists to produce a prediction of how the weather is likely to change and affect people over the next few hours. Observations taken today are used to analyse the weather patterns and a forecast of the weather pattern can be made (usually with the aid of a computer, which is used to model the data). The weather forecaster then uses the output from the computer to make decisions about how the weather will change.

In addition to surface observations, there are a number of other sources which can be used to help generate an accurate picture of the weather across the UK.

1. Land-based weather stations

Across the UK there are 30 major weather observation stations which help to develop the detail needed for a forecast. These are staffed by professional meteorologists who take observations every hour. There are an additional 100 auxiliary stations manned by coastguards, along with another 100 or so fully automated stations.

Figure 76 Sources of weather data

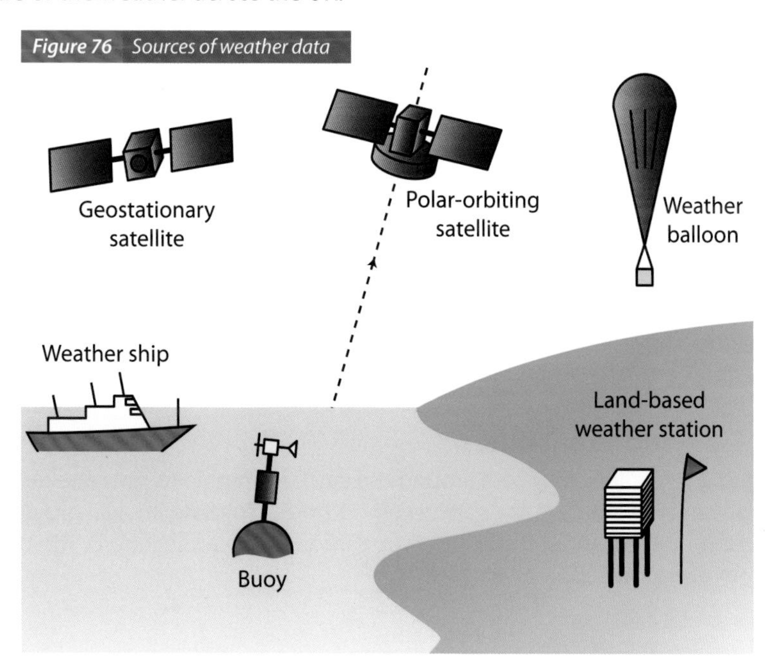

Source: National Oceanic and Atmospheric Administration (NOAA)

Figure 77 Balloon which holds the radiosonde package

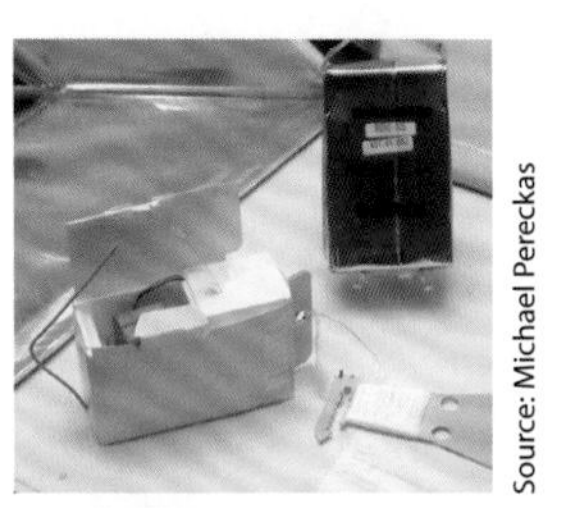

Source: Michael Pereckas

Figure 78 Radiosonde package

2. Balloons

Sometimes it is important to find out what is happening in the upper parts of the atmosphere. To collect information at this level a weather balloon is used (radiosonde). A rubber balloon is filled with helium and this lifts the recording package up into the atmosphere. The data is transmitted back to computers on the surface. Eventually, the balloon will burst (about 20 km up) and the radiosonde package will fall back to earth. In the UK there are eight radiosonde sites and measurements are taken twice a day.

Figure 79
Weather buoy in Belfast harbour

Buoys

Weather buoys are sometimes moored and sometimes free floating (drifting) in the ocean. Information is sent back to weather centres using satellite communications. These are very useful for helping airline pilots know what weather systems they are likely to face during long flights over the ocean. They also help us understand more about the impact that climate change might have on sea temperatures and sea currents. You might have seen these in action in the film 'The day after tomorrow'.

Weather ships

Most ships today carry a host of weather recording devices to allow the crew to avoid big storms at sea. These can also transmit the data to the Met Office for collation. In previous years, there used to be a network of weather ships that specialised in collecting meteorological information. The last remaining weather ship in the world was the Norwegian MS *Polarfront*, which was removed from service on 1 January 2010, as its usefulness has been replaced by buoys and satellites. However, some organisations such as the US National Oceanic and Atmospheric Administration (NOAA) still have some ships which continue to carry out atmospheric investigations.

Source: NOAA's National Ocean service

Figure 80 The NOAA ship Thomas Jefferson

Geostationary and polar satellites

In recent years satellites have become increasingly important in the formation of weather forecasts, as they help to give a 'big picture' of what is happening in the atmosphere. A satellite is a small spacecraft that is launched to orbit the earth and record the weather. The first satellite was launched in 1957. Satellites use radiometers to view the earth and measure radiation in different wavelengths. Some images are visible light images, which help locate cloud formations (only in the daylight), and other images are infrared images, which can be taken at night as well as in the day.

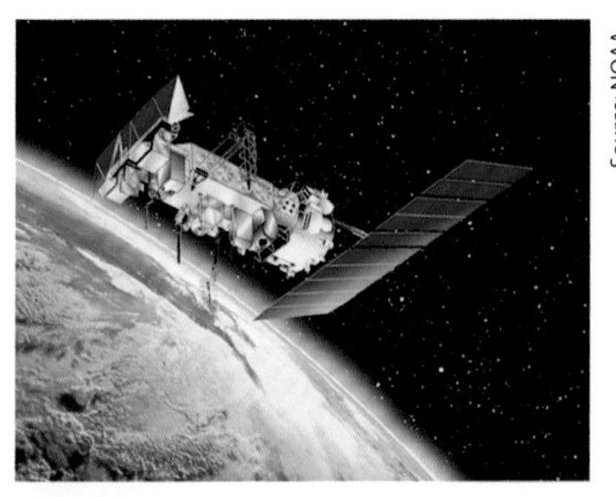

Source: NOAA

Figure 81
NOAA polar-orbiting weather satellite

Geostationary satellites

These are satellites that hover over the same spot on the earth, moving at the speed of the earth's rotation. They remain over the equator at an altitude of 36,000 km. The European Meteosat satellite is a good example of this.

Polar-orbiting satellites

These are satellites that pass around the earth from pole to pole at a height of 850 km. They will pass the same point on the earth every 12 hours. The satellite will provide pictures of the clouds and information about the temperature through the atmosphere. The US NOAA satellites are a good example of this.

Check your learning

Now that you have studied Part 1: Measuring the Elements of the Weather, return to page 61 and answer the Key Questions for this section.

Test yourself

1. Why is it important to use weather balloons alongside land-based weather observations? (3)
2. Describe and explain why buoys have been replacing weather ships in recent years as sources of ocean weather observation? (4)
3. What are the main differences between geostationary and polar satellites? (3)
4. Describe the main features of the images that satellites can generate:
   Visible images (2) Infrared images (2)

PART 2: WEATHER SYSTEMS AFFECTING THE BRITISH ISLES

(a) The temperature and moisture characteristics of the air masses affecting the British Isles and their seasonal variation:
Tropical maritime
Tropical continental
Polar maritime
Polar continental

(b) Show knowledge (with reference to places) of:
the weather patterns and sequence of change associated with a frontal depression as it moves across the British Isles (weather at the warm front, in the warm sector and at the cold front)
the weather patterns associated with anticyclones in the British Isles during the summer and winter

(c) Interpret synoptic charts and satellite images and understand the limitations of forecasting (range and accuracy)

(d) Evaluate the effects (positive and negative) of depressions and anticyclones on the economy and people

Key questions

By the end of this section you will be able to answer the following questions:

1. State the meaning of the term air mass.
2. Name the air mass that brings warm and moist air to the British Isles.
3. State two differences between polar continental and tropical maritime air masses.
4. Describe the sequence of weather experienced with the passage of a depression.
5. Explain why two elements of the weather change as a depression passes over Belfast.
6. Explain why anticyclones might bring sunny weather during the daytime but cold weather at night in the winter.

Key words

Air mass
Tropical maritime
Tropical continental
Polar maritime
Polar continental
Depression
Front
Cold front
Warm front
Warm sector
Anticyclone
Synoptic charts
Satellite image

The temperature and moisture characteristics of the air masses affecting the British Isles and their seasonal variation

Unlike many other regions of the world, the weather across the British Isles is constantly changing. The weather is rarely the same from one day to the next. The reasons behind the dynamic nature of our weather are to do with the four different air masses that affect us. These air masses control our weather and dictate which weather system is going to affect us at a particular time.

What is an air mass?

An air mass is a large parcel of air (often thousands of kilometres wide) which stays still over a place for a long period of time, picking up the area's temperature and moisture characteristics. It can blow steadily across the country from one direction, bringing with it fairly constant temperature and humidity conditions.

The air masses affecting the British Isles can be broadly categorised in terms of their source and their path:

1. **Tropical maritime** – warm and moist
2. **Tropical continental** – warm and dry
3. **Polar maritime** – cold and (fairly) moist
4. **Polar continental** – cold and dry

The air masses are classed according to:

- the **latitude** where they start, as this affects the temperature of the air.

  Air Masses from the cold north are classified as polar (P).
  Air Masses from the warm south are classified as tropical (T).

- the **surface** over which they develop and move, as this affects the amount of moisture in the air.

  Air Masses that move over the sea will have a higher humidity and will produce more precipitation. They are described as maritime (m).

  Air Masses that develop over a continental land mass will have less opportunity to pick up moisture. They are described as continental (c).

Polar maritime (Pm)

- This is the most common air mass affecting the British Isles.
- It originates over the North Atlantic Ocean.
- It reaches the UK from the west or north west.
- It produces unstable air that creates cumulus and cumulonimbus clouds but has good visibility between showers.
- It can also cause convectional rainfall in the summer months.

Polar continental (Pc)

- This is more prevalent in the winter than the summer.
- Its air originates over Northern Europe (Norway/Sweden) and moves from the east/north east.
- It produces very cold conditions in the winter but milder conditions in the summer.
- Its air can be unstable in the winter and can bring snow showers along the east coast of England.
- It usually brings dry but cool conditions (with Stratus cloud).

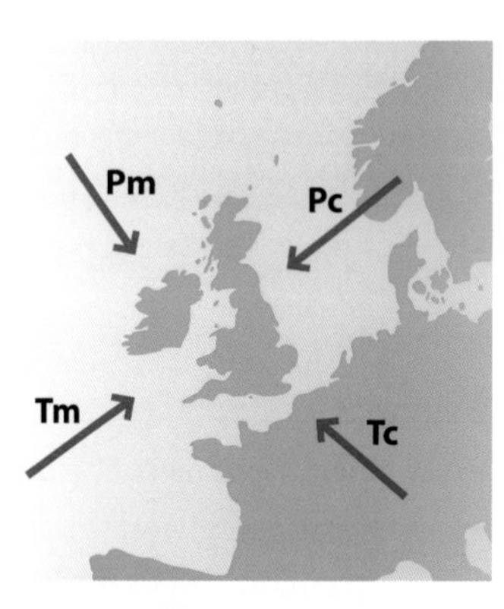

Figure 82
Air Masses affecting the British Isles

Tropical maritime (Tm)

- This is a very common air mass over the British Isles.
- Its air travels from the warm southern Atlantic Ocean and moves over the south west of the British Isles.
- It brings mild conditions in the winter and warm but wet weather in the summer.
- It is responsible for bringing dull skies (nimbostratus clouds), drizzle and fog (poor visibility).

Tropical continental (Tc)

- This is the least common air mass affecting the British Isles and usually only in the summer.
- Its air travels from North Africa and the Mediterranean.
- It brings very warm and dry air from the south and south east.
- It brings mild conditions in the winter but hot weather (heat wave) in the summer.
- It can cause thunderstorms to develop if the temperatures rise.

Test yourself

1. Which air mass do you think might be dominant in the following weather conditions? Note down the air mass, what time of year you think it is and a reason for your answer. (1 mark for columns 2 and 3, and 2 marks for column 4 = 16 marks in total)

STATEMENT	WHICH AIR MASS?	SEASON?	REASON?
a. Snowploughs are on alert in the east of England			
b. Temperatures are freezing but there is not a cloud in the sky			
c. Sahara dust is found in the south of England			
d. Sleet and snow showers are forecast over the whole of the UK			

2. Describe and explain how the Polar Maritime air mass affects the weather of the British Isles. (4)
3. Describe the main differences between the Tropical maritime and Tropical continental air masses, and note the different types of weather that they bring over the UK. (4)

The weather patterns and sequence of change associated with a frontal depression as it moves across the British Isles (weather at the warm front, in the warm sector and at the cold front)

Across the British Isles there are two separate weather systems that control our weather:

1. A low pressure system is called a **depression**. This is usually associated with unsettled weather and is characterised by bands of wind and rain, and even snow in the winter.
2. A high pressure system is called an **anticyclone**. This usually happens when the weather is settled and produces cloudless skies.

Before trying to further understand the weather patterns associated with depressions and anticyclones, it is important to get to grips with how the air masses move to create and influence these weather systems.

The influence of pressure and wind

Atmospheric pressure is one of the most important driving factors on the weather. Pressure is recorded frequently onto what is called a surface pressure chart, as shown in Figure 83.

Figure 83
A Met Office surface pressure chart for 12 February 2013

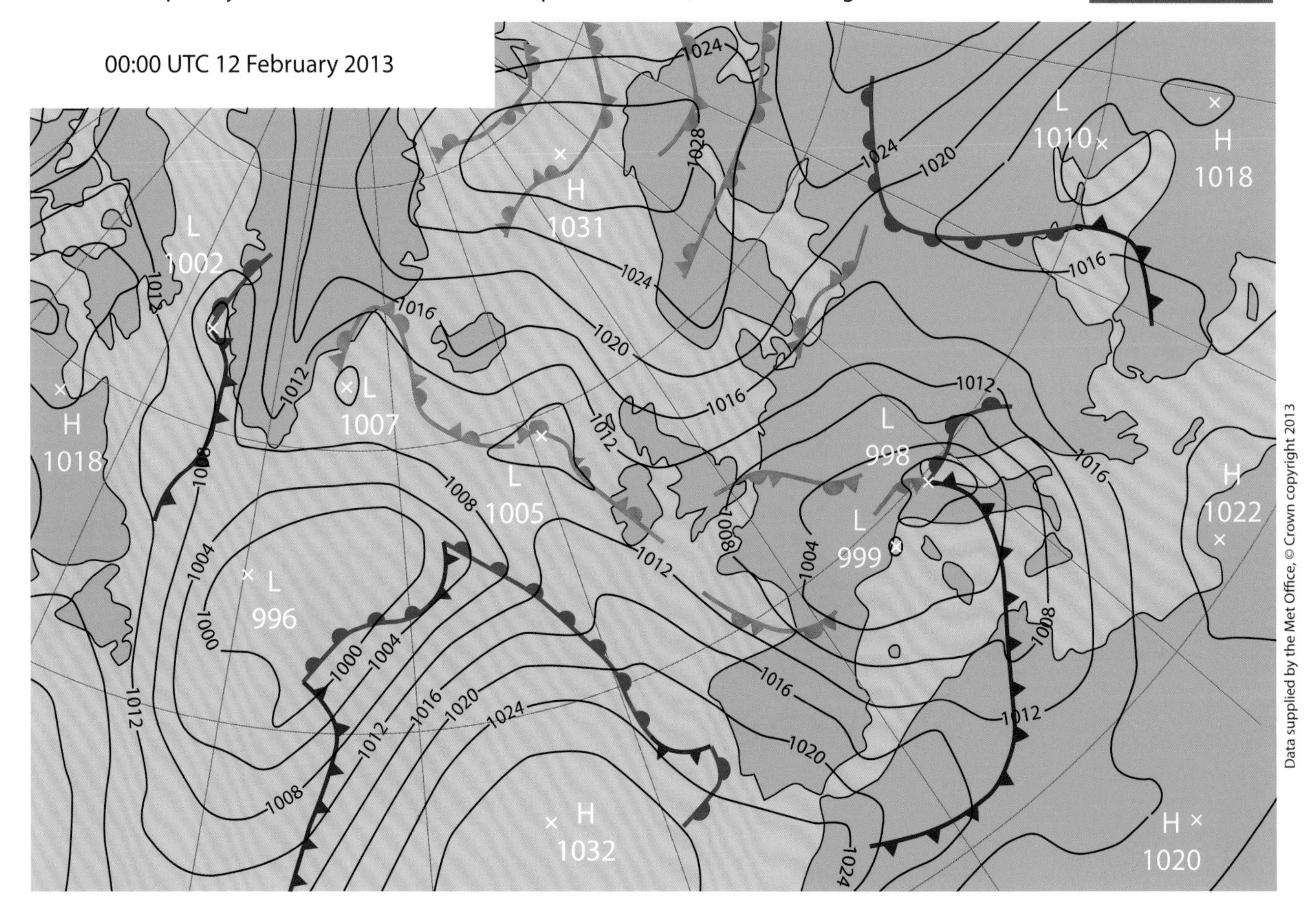

A description of this surface pressure chart could read: "A large area of pressure dominates the eastern Atlantic, which will bring increasingly unstable south to south westerly flow to the UK. A depression (with leading warm front) is moving eastwards and northwards across the country. This will bring changeable weather, high winds and heavy rain over the next few days."

The black lines on this chart are called **isobars.** These are lines that join places which have the same atmospheric pressure. They are usually drawn every four millibars (with the air pressure always written along the isobar). There is a **pressure gradient** from areas of high pressure to areas of low pressure. The wind tends to blow along this gradient from places with high pressure to those with low pressure (however, they are deflected slightly due the rotation of the earth, which produces a Coriolis force and curves the winds a little to the right) as shown in Figure 84.

When isobars are close together the pressure is increased and this indicates strong winds. When the isobars are spaced apart the winds tend to be gentle.

Figure 84
Pressure and wind direction

Test yourself

On a copy of the map in Resource A, follow the instructions below to create your own surface pressure chart.

a. Complete the map by drawing in pencil the isobars at intervals of 4 mb. Make sure that you try to keep the isobars in a circular motion and label each isobar carefully. (7)
b. Add the label 'LOW' to the centre of the area where you think pressure is lowest and 'HIGH' where the pressure is highest (2).
c. Add about 10 arrows to indicate the direction that you think the wind might blow across this map. (4)

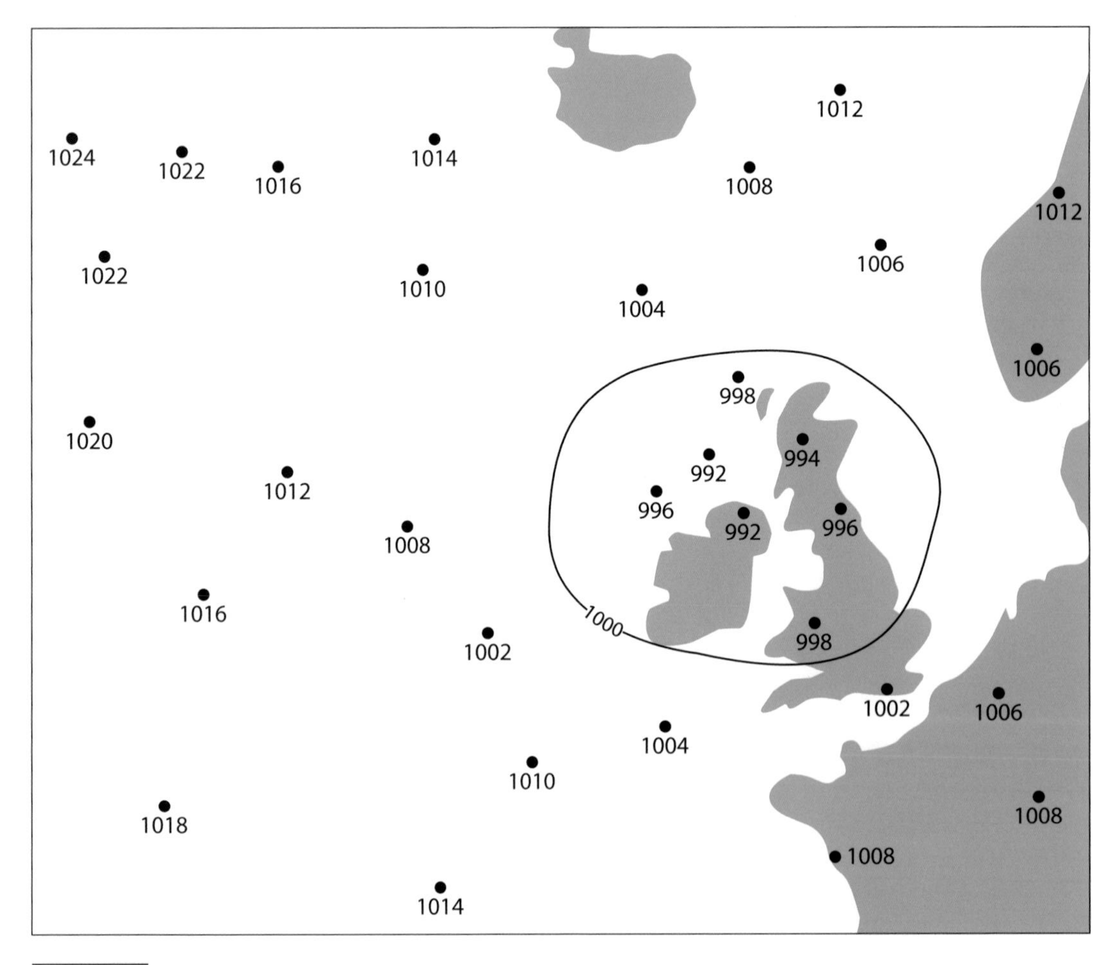

Resource A
Draw your own pressure chart

How weather fronts are formed

One of the most important features of any changing weather system is the movement of a new weather front over the land. A weather front is simply the boundary between two air masses. There are three different types of weather front:

1. A cold front (blue)
2. A warm front (red)
3. An occluded front (pink)

1. A cold front

A cold front on a synoptic chart indicates that a block of cold air is moving, which means that the weather will change with a narrow belt of rain and clouds. After the front has passed, the weather usually settles further, becoming bright and clear, though sometimes accompanied by a few showers. The atmospheric pressure will fall before the front passes and then will start to rise after it passes.

2. A warm front

As a warm front moves it brings a belt of cloud with some rain as a block of warm air is moving overhead. The rain will gradually increase as the front gets closer. Behind the front the rain will be lighter but it will still be cloudy. The warm front does not always bring a warmer temperature, especially in the summer as cloud will reduce the temperature. The atmospheric pressure will fall quickly before the front continues overhead but it will stabilise after it passes.

3. An occluded front

This is a situation where warm air has already started to rise up over cold air. Usually the weather at an occluded front will be the same as at a cold front but without much cloud or rain. However, on other occasions it can bring a band of thunderstorms.

Weather charts

A **synoptic chart** is a weather map that gives a snapshot of the weather across a region. It is something which summarises a large amount of very complicated, detailed information.

A **plotted surface chart** is a chart containing all the information that is generated from surface weather observations or from automated sites (usually shown as triangles rather than circles).

Up until very recently, most of these charts were hand drawn but now the observations are put into a computer which will then 'model' the weather forecast.

Weather observation

Weather forecasters use the measurements and observations taken across the country to help them develop the weather forecast for an area. These observations can be extremely complicated and the Met Office uses a very detailed system to help record the measurements taken at a station. However, records can be simplified in such a way that the following symbols will provide any meteorologist with the details needed to help make an accurate forecast. The weather taken at a particular weather centre/station might end up looking like this:

This can be explained in the following way:

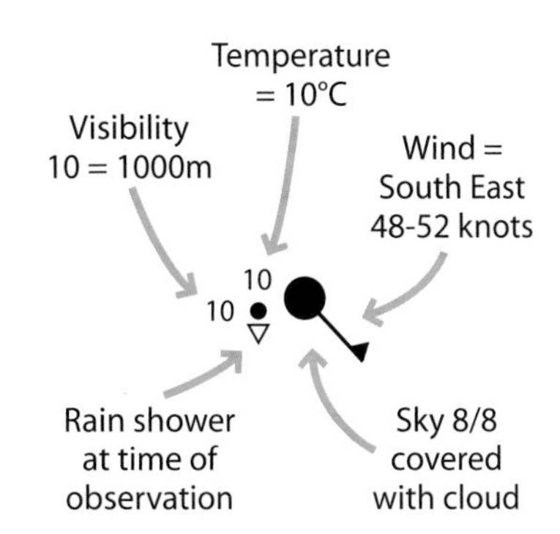

Figure 85
Weather station observations

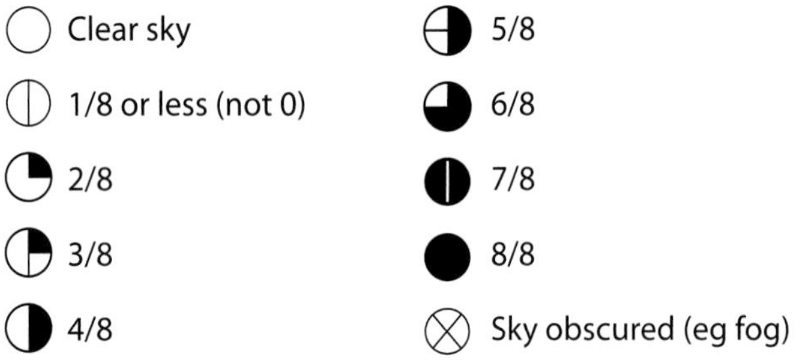

Clear sky
1/8 or less (not 0)
2/8
3/8
4/8
5/8
6/8
7/8
8/8
Sky obscured (eg fog)

Figure 86 Cloud cover

The circle that indicates the weather centre also helps to show the observed cloud cover.

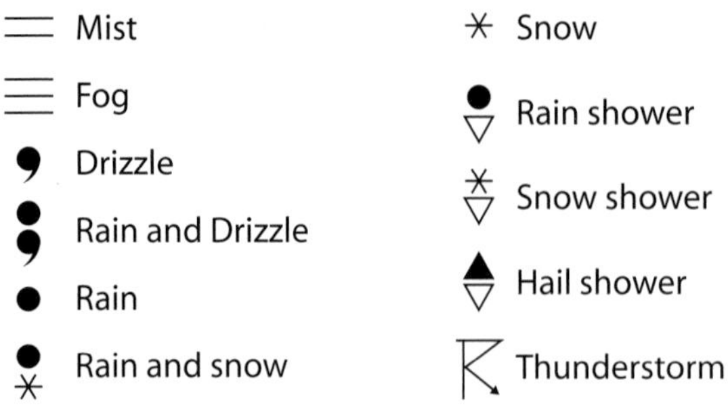

Mist
Fog
Drizzle
Rain and Drizzle
Rain
Rain and snow
Snow
Rain shower
Snow shower
Hail shower
Thunderstorm

Figure 87 Precipitation

The type of precipitation that is currently falling is usually shown underneath the temperature measurement.

Calm
1–2
3–7
8–12
13–17
18–22
23–27
28–32
33–37
38–42
43–47
48–52
53–57
57–62 etc

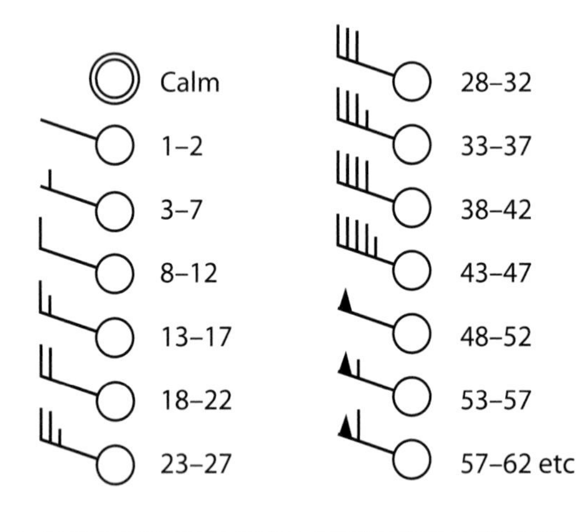

Figure 88 Wind speed

The wind speed is usually recorded as a 'shaft', which also points to the direction where the wind is coming from. The 'feathers' on the end of the shaft help to indicate the speed of the wind.

Test yourself

1. Answer the questions which follow to test your understanding of the link between pressure and winds. Each diagram in Resource A refers to a question.

A
988
992
996
1000
D
980
976
972
968
964
984
988
992
996
B
1004
1000
996
992
E
1000
1004
1008
C
980
996
F
988
992
996
1000
1004
1008

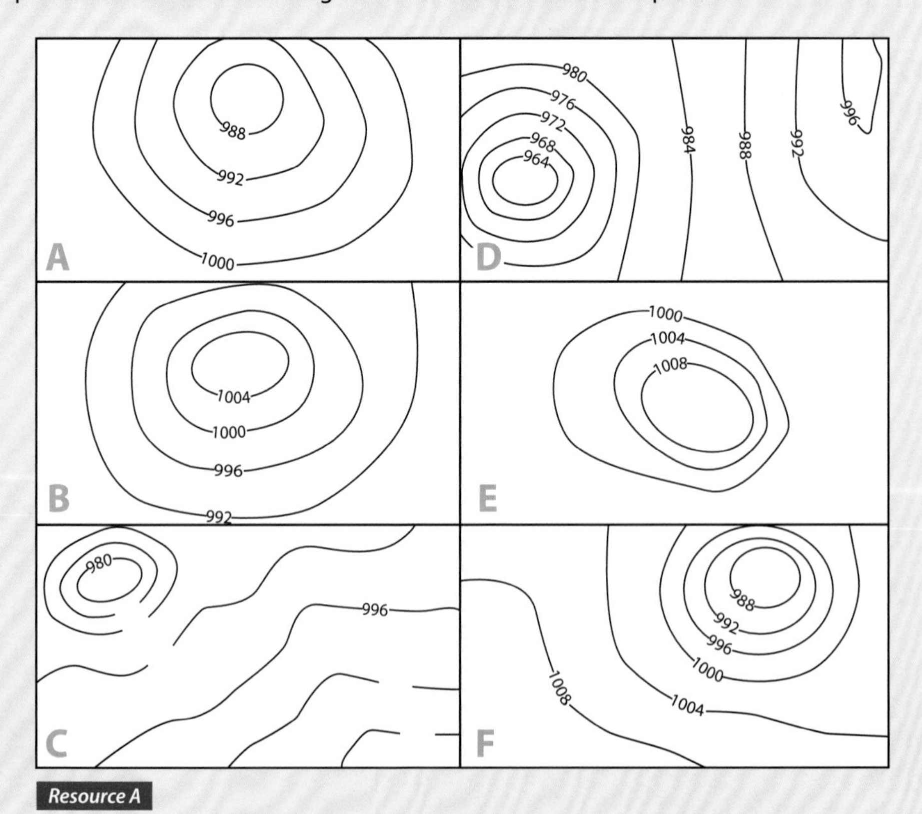

Resource A

a. Is diagram A a High or Low pressure system? (1)
b. Is diagram B a High or Low pressure system? (1)
c. Make a copy of diagram C and complete the diagram by numbering the isobars. (2)
d. Which of the two weather systems in diagram D will have stronger winds? Describe the area of low and high pressure on the map. (3)
e. Describe the wind directions in diagram E clearly. (2)
f. Describe and explain the wind direction in diagram F. (2)

2. Study the weather map in Resource B and identify the different features labelled A, B, C and D. (4)

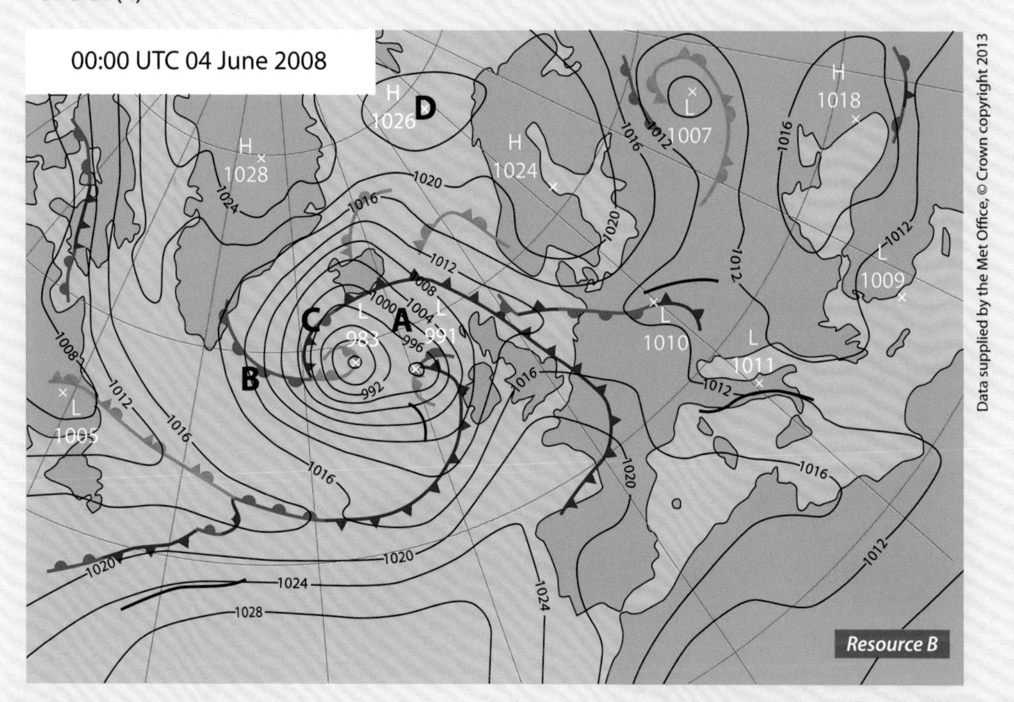

3. Using the weather observation symbols, study the map in Resource C and answer the questions which follow.

a. Describe the weather conditions in London at the time the weather chart represents. (3)
b. In what ways will the weather in Dublin be different from that in London? (3)
c. How is the weather on the west coast of Ireland different from the weather in central France? (4)
d. If you had to draw out a weather observation for Cardiff in South Wales, use the information in Resource D to show what you would draw. (3)

Resource C

4. Do you think that the map in Resource C was drawn in the summer or the winter? Use evidence from the chart to back up your answer. (3)

Temperature	26 °C
Wind direction	South
Wind speed	7–10 knots
Cloud cover	2/8 oktas
Precipitation	None

Resource D

Depressions across the British Isles

Depressions are areas of low atmospheric pressure which produce cloudy, rainy and windy weather. They are low pressure systems, which often begin far out in the Atlantic Ocean as cold polar maritime air from the north and warm tropical maritime air from the south join together and cause very changeable weather across the British Isles. The weather in one part of the British Isles can be very different from another part, depending on which front is affecting a particular place.

The formation of a depression

In the atmosphere air will always move from areas where the pressure is high to areas where the pressure is low. This horizontal air movement is called wind. Generally in the UK the depressions that we get are formed out in the Atlantic Ocean. Cold, polar maritime air from the north moves south and meets some warm, tropical maritime air which is moving up from the south.

Figure 89
The formation of a depression

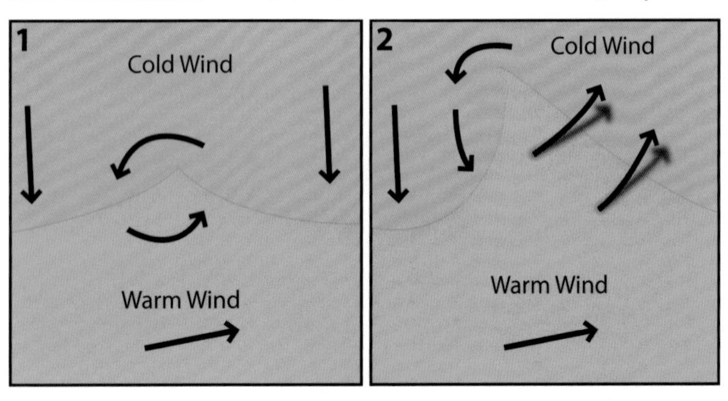

The lighter, warm air will start to rise up over the denser, cold air. This creates a disturbance called a Baroclinic instability, which, if strong enough, will develop a front.

The passage of a depression

Figure 90 shows the different weather changes that you would expect to experience as a depression passes overhead. The depression starts with a warm front and is trailed by a cold front which moves from west to east across the UK.

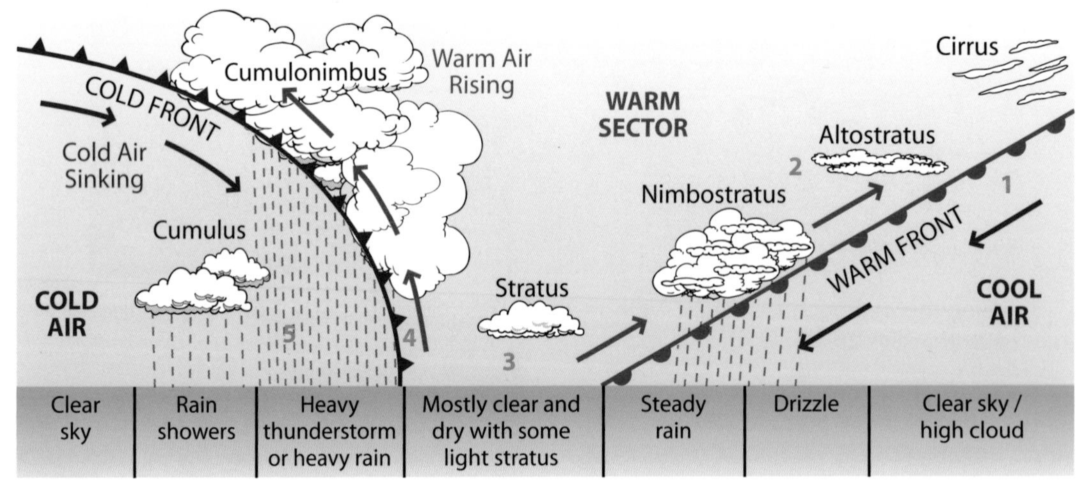

Figure 90
The passage of a depression

1. At the warm front the lighter, warmer air from the south (tropical maritime) meets colder air from the north (polar maritime) and it starts to rise over it.
2. The warm air rises over the top of the colder air and as it rises, the air cools, the water condenses and clouds start to form (mostly altostratus and nimbostratus). The result is steady rain which later turns into drizzle. An early indication of the warm front is some clear skies and high wispy cirrus clouds.
3. As the warm sector passes, the warm air slowly rises, cools and condenses over the centre of the low pressure system. The weather here can be a little more settled but this usually does not last for a very long time.
4. As the cold front starts to move overhead, the colder, heavier air from the cold front meets the warmer air. The cold front starts to undercut the warm air, forcing the warm air higher and higher into the atmosphere. The fast moving air can produce high winds and cool temperatures.
5. As warm air is forced to rise quite quickly, it cools, condenses and clouds are formed (cumulonimbus and then cumulus clouds). The result can be heavy thunderstorms and heavy rain showers that will gradually clear as the front continues to move.

Structured notes

Copy and complete the table in Resource A to indicate the types of weather you would expect to experience as a depression passes overhead.

Resource A	1. AHEAD OF THE WARM FRONT	2. PASSAGE OF THE WARM FRONT	3. IN THE WARM SECTOR	4. PASSAGE OF THE COLD FRONT	5. COLD SECTOR
Pressure	Starts to fall steadily		Steadies	Starts to rise	
Temperature	Quite cold	Temperatures rise		Temperature drops suddenly	
Cloud Cover	Cloud base drops and cloud types get thicker (cirrus and altostratus)				
Wind speed and direction	Speeds increase and direction changes			Speeds increase, sometimes to gale force	
Precipitation	Very little at first but some as warm front starts to pass		Rain turns to drizzle and often stops		Showers

Test yourself

Look at the weather map of the British Isles and use the information to give precise details of what weather you would expect to experience in different parts of the country.

1. What is the lowest pressure on the map? (1)
2. What is the highest pressure on the map? (1)
3. Describe the location of the warm front across the British Isles? (2)
4. This weather map was for June. Describe the weather that you would expect in Edinburgh. (3)
5. How would you expect the weather in London to be different to Edinburgh? (3)
6. What would the difference be between the weather in Belfast and Bristol? Explain carefully why you think these differences would exist. (4)
7. If this weather map was for December rather than June, how might the weather forecast be different? (3)

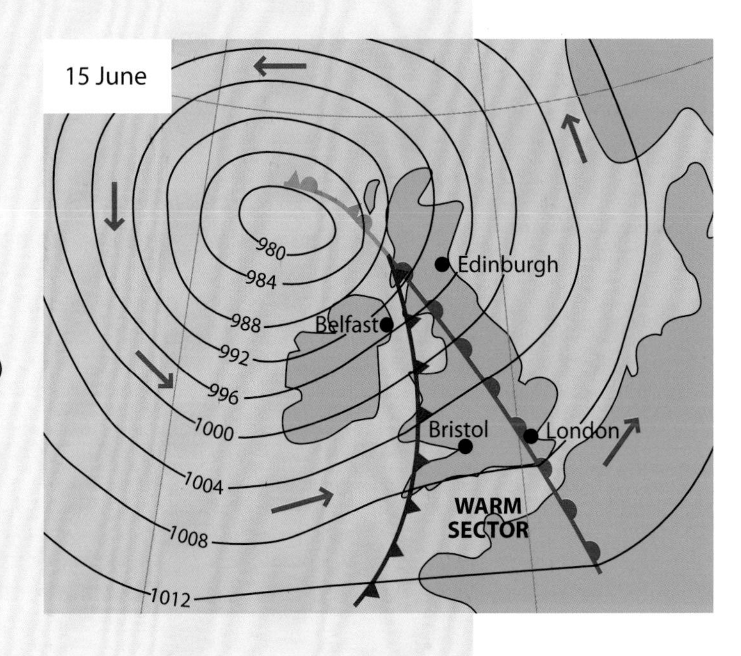

Key features of a depression

Pressure	Low pressure and falling (below 1000 mb).
Temperature	Temperatures in a depression will vary depending on the type of air that is passing overhead.
Cloud cover	A wide and varied selection of clouds can be found as the depression passes. Generally, the same sequence of clouds will be evident as the depression passes overhead.
Wind speed and direction	Isobars can be quite close together, which indicates strong winds. Winds blow in an anticlockwise direction. A typical depression will pass over the UK, moving in a NE direction (from the SW) and will usually take three to five days to pass completely over the UK.
Precipitation	Depressions bring a variable selection of precipitation. They often bring snow in winter conditions and can bring a lot of rain over a three to five day period. When the cold front passes, depressions can bring thunderstorms and lightning (especially in summer months), and can also cause hail.

The weather patterns associated with anticyclones in the British Isles during the summer and winter

Anticyclones in the summer

Anticyclones are areas of high atmospheric pressure which are usually much bigger than depressions and produce calm, settled weather where there is little cloud cover or precipitation. Temperatures in the summer are usually quite high and anticyclones are associated with good weather.

The weather associated with an anticyclone will affect a much larger area than the different parts of a depression. An anticyclone will often bring the same weather conditions to the whole country and different places across the British Isles will experience the same weather. There are no fronts in an anticyclone.

Figure 91
The formation of an anticyclone

Summer
Polar Continental High Pressure
Tropical Continental High Pressure

The formation of an anticyclone

The high pressure system which makes up an anticyclone usually comes from the south of the UK. The tropical continental air mass brings warm and dry weather north into the British Isles. The air sinks from high altitude and as it descends, it absorbs any moisture and stops condensation from taking place (remember that clouds and rain can only be formed through rising air and not sinking). In addition, in summer dry air from the polar continental area often brings high pressure and sinking air.

The features of an anticyclone

Figure 92 shows some of the different weather features that you would expect to experience if an anticyclone was overhead. Anticyclones can move into the airspace above the British Isles and stay for as little as one or two days, or as long as three weeks.

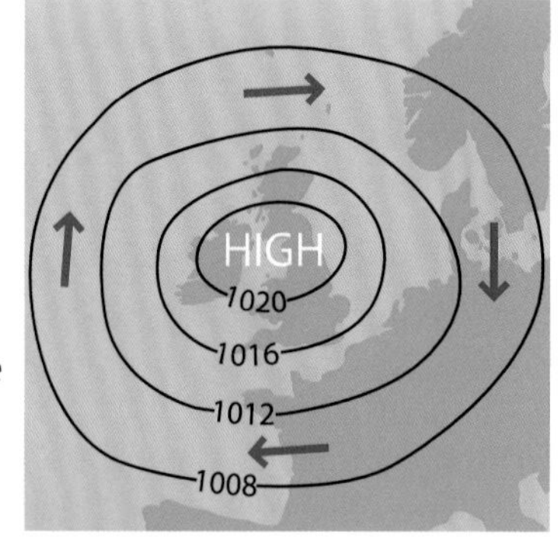

Figure 92
Typical synoptic chart for an anticyclone

Key features of a summer anticyclone

Pressure	High and increasing (over 1000 mb)
Temperature	In the summer the temperatures can be warm, as air is coming from the warm south. The hot, sunny days (when there is a lack of clouds) allows the air temperature to be high (around 24 °C on average).
Cloud cover	Sinking air means that there are very settled conditions with few clouds and clear skies. This lack of cloud cover allows daily temperatures to be increased. However, at night, clear skies mean that heat cannot be stored in the atmosphere and it will escape back into space, and overnight temperatures will drop rapidly.
Wind speed and direction	Isobars are far apart which indicates light winds and often leads to very calm conditions. Wind direction is clockwise around the high pressure.
Precipitation	There is little or no precipitation during the day but clear skies can sometimes result in dew and fog occurring, especially in the mornings. Thunderstorms can be triggered on very hot days, when the hot air starts to rise and produces convectional rainfall in some parts of the British Isles.

Anticyclones are very powerful, large weather systems. If a strong anticyclone moves into position across the British Isles, it can actually 'park' itself above the isles for a long period of time and will proceed to 'block' any other weather system (like depressions) from moving the anticyclone on further. In some cases the depressions might be forced north or south to go around the 'blocking anticyclone'.

Anticyclones in the winter

The formation and features of an anticyclone are largely the same in the summer and the winter. However, the impact can be a little different.

Key features of a winter anticyclone

Pressure	High and increasing (over 1000 mb).
Temperature	In the winter temperatures are much lower than during the summer, as the sun is low in the sky, reducing the heat received from it.
Cloud cover	Sinking air means that there are very settled conditions with few clouds and clear skies. This lack of cloud cover allows daily temperatures to be increased. However, at night, clear skies mean that heat cannot be stored in the atmosphere and it will escape back into space, causing overnight temperatures to drop rapidly.
Wind speed and direction	Isobars are far apart which indicates light winds and often leads to very calm conditions. Wind direction is in a clockwise direction around the high pressure.
Precipitation	There will still be very little direct atmospheric precipitation in a winter anticyclone, however, the rapid loss of any heat means that nights can be very cold and the temperature will dip below freezing. Condensation caused by a temperature inversion near the ground level can cause serious fog and frost is common first thing in the morning.

Test yourself

1. Describe and explain two of the main differences between winter and summer anticyclones. (4)
2. Look at Resource A, a weather map for the UK and use the information to answer the questions which follow.

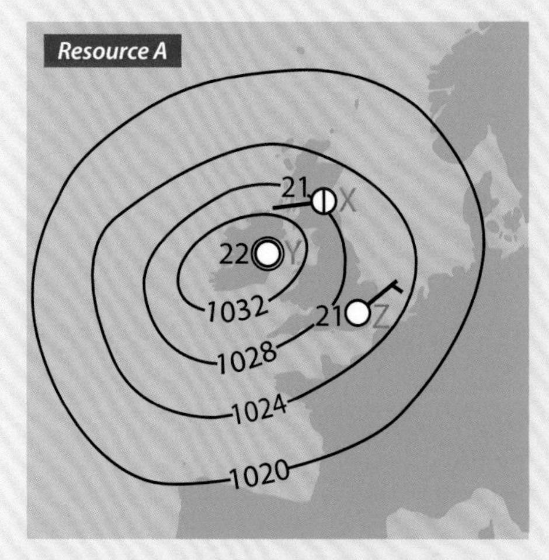

a. Is an anticyclone an area of high or low pressure? (1)
b. Where do you think the wind speeds will be least, at the centre of an anticyclone or towards the outside of the anticyclone? (1)
c. Do you think this map shows a summer or winter anticyclone? Give one piece of evidence from the map to support your answer. (3)
d. Copy and complete the table in Resource B using the information from the map. (15)
e. Describe and explain the pattern of weather across the British Isles at this time. (4)

Resource B

WEATHER STATION	TEMPERATURE (°C)	WIND DIRECTION	WIND SPEED	CLOUD COVER	PRESSURE
X					
Y					
Z					

3. Study Resource C, a map of the weather chart for Western Europe and answer the questions which follow.

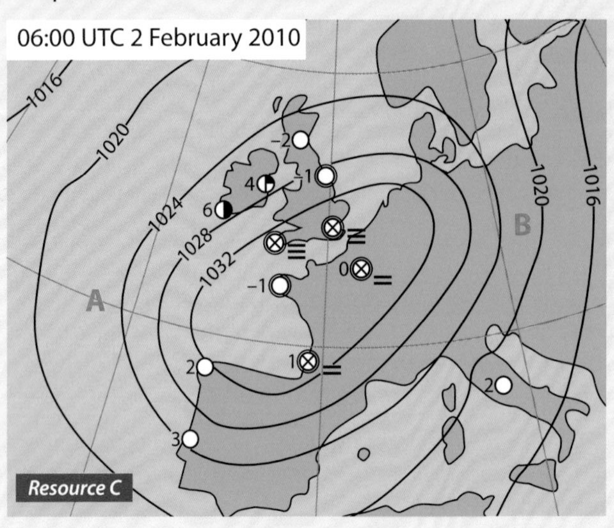

a. What would you expect the wind direction to be at places A and B on the map? (2)
b. Explain why the wind direction is different at these two places. (2)
c. Describe the weather conditions on the east of Ireland at this time. (3)
d. Describe and explain the weather conditions for London and the south of England at this time. (3)
e. Name four places on the map where fog and mist might have been recorded. (2)
f. What evidence is there on the map to support the fact that the map is for February and not a summer month? (3)

Structured notes

Copy and complete the table in Resource A to make a comparison between the different weather conditions that you would expect during depressions and anticyclones over the British Isles.

Resource A	DEPRESSIONS	WINTER ANTICYCLONES	SUMMER ANTICYCLONES
Pressure			
Temperature			
Cloud cover and cloud type			
Wind speed and wind direction			
Precipitation			
Duration			

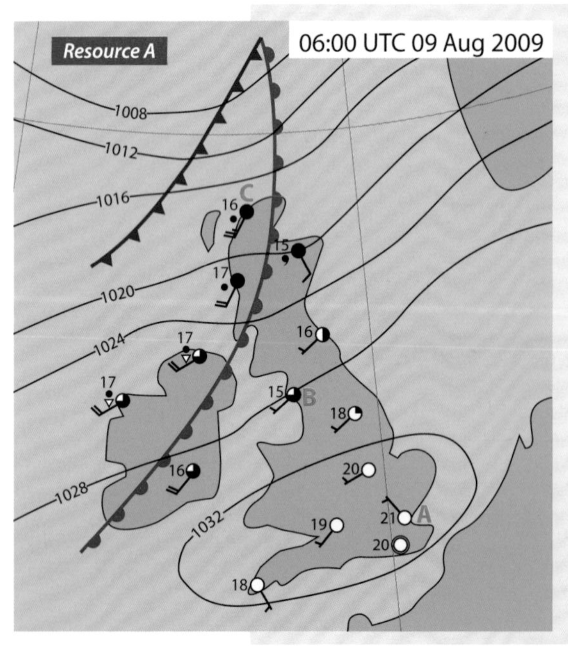

Test yourself

Study the map in Resource A and answer the questions that follow in as much detail as possible.

a. Describe and explain which parts of the British Isles have
   - The highest pressure (2)
   - The lowest pressure (2)
b. Describe and explain where the warmest parts of the UK are. (3)
c. Describe and explain where the coldest parts of the UK are. (3)
d. Where are the winds the strongest? Where are the winds the lightest? (2)
e. Compare and contrast the amount of cloud cover in different parts of the UK. (4)
f. Where might it be raining? (2)
g. Describe the weather being experienced at places A, B and C. (2 marks for each = 6)
h. Write a weather forecast for what you think the weather might be like on 10 August at each of the places A, B and C. (6 marks for each = 18)

Interpret synoptic charts and satellite images and understand the limitations of forecasting (range and accuracy)

A weather forecast is taken to try and predict what weather is going to influence a place. Weather forecasters use satellite images along with other information to help to work out what the weather in a particular location is likely to be over the next few days.

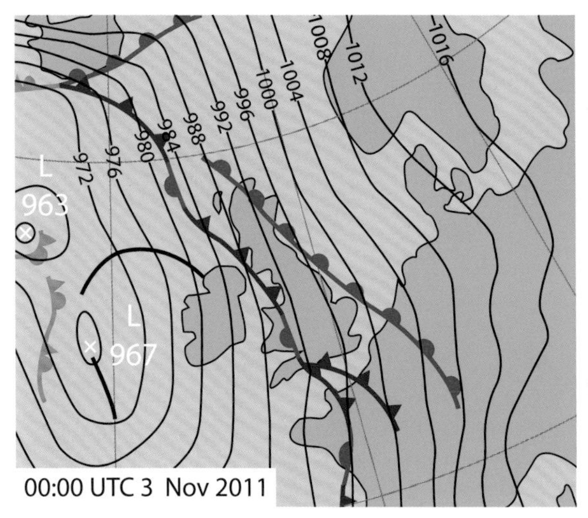

Figure 93

Synoptic chart and satellite image for 3 November 2011

The image on the right shows the infrared satellite image of the British Isles taken at 0000 on 3 November 2011. The corresponding synoptic chart is shown on the left.

When you take a look at the satellite it is pretty obvious that a band of cloud is moving from the west to the east of the British Isles. But it is only when you take a look at the synoptic chart that you note that this is linked to a depression, which is to the north west of the UK (and just south of Iceland). A warm front is in the process of passing to the east of the coast of Scotland and England, with a cold and sometimes occluded front having passed over Ireland and on its way to the western coast of Scotland and England. Both fronts would bring with them some wind and rainy conditions.

The weather forecasts that the Met Office use are governed by different formulae and weather models that help to put together an estimate of how the weather will behave in a particular place. Although the overall picture of UK weather is sometimes marginally accurate, often the weather forecasters get more local predictions, such as snowfalls, wrong. Also, weather forecasts are generally only valid for 48 hours in advance. Beyond that, the accuracy of the prediction can be seriously compromised.

A typical depression synoptic chart and satellite image

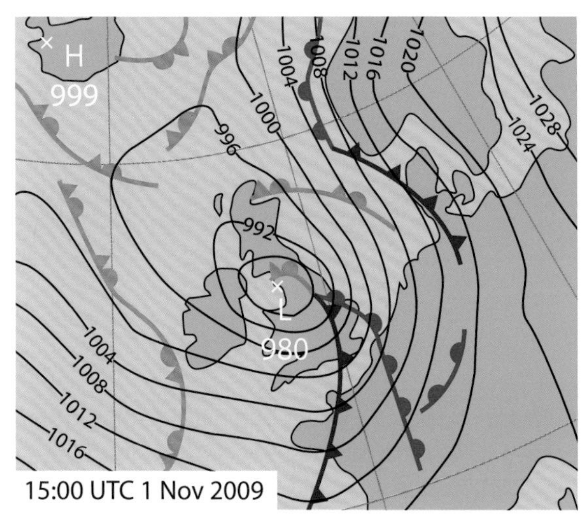

Figure 94

Synoptic chart and satellite image for a depression for 1 November 2009

The key feature of any depression is that there is usually a swirling mass of clouds that bends round to the left on a satellite image. On a synoptic chart you should try to identify the lowest pressure and look for evidence of any fronts.

A typical anticyclone satellite image

Figure 95

Satellite image for an anticyclone for 20 January 2011

Source: NASA/GSFC, Rapid Response.

The most important feature of any anticyclone is that there is usually a lack of cloud on the map over the British Isles. If there is any cloud it is likely to be high cirrus cloud. On a synoptic chart you should try to identify the highest pressure and look for evidence of calm weather conditions.

Test yourself

Use the satellite image of the anticyclone in Figure 95 and draw a synoptic chart that you might expect to see. (6)

The limitations of forecasting

Range

The UKMET unified model can help to predict the weather for up to six days in the future and other models can be used to estimate the weather more long term. However, many scientists note that the atmosphere is very unpredictable due to its changeable nature.

Accuracy

No weather forecast model can 100% accurately predict the weather. There are far too many different aspects that come under consideration. The surfaces that the weather passes over, the temperature, the microclimate and the size of a settlement can all influence the local weather in one way or another and make the weather forecast less accurate.

Test yourself

Study the satellite image and answer the questions which follow.

Source: NEODAAS/University of Dundee.

The image was taken over NW Europe at 14.00 on 14 February 1996.

a. What weather system is working over Belfast? (1)
b. What weather system is working over Milan? (1)
c. Describe and explain the weather conditions that you would expect to experience in Belfast at this time. (3)
d. Compare and contrast the weather with Milan at this time. (3)
e. What do you think the thick band of white to the left of the picture represents? (2)
f. Describe some of the changes that you might expect the weather over Belfast to experience over the next 1 to 2 days. (5)

Questions a, c and f are adapted from CCEA Higher Tier Summer 1999 Past Paper

Evaluate the effects (positive and negative) of depressions and anticyclones on the economy and people

The analysis chart and satellite image for the depression in Figure 94 are linked to a deep depression that hit the UK on 1 November 2009. The Met Office noted that this depression developed and deepened close to the UK, and predicted that this would bring heavy rain and strong winds with gales in coastal areas. They commented on the storm's impact:

"In the event, heavy rain spread north-east across the UK during the early hours of Sunday morning firstly affecting Wales and Northern Ireland and then moving on to other parts of England and eventually into Scotland. Although not a prolonged spell of rain in Wales, it was heavy while it lasted and resulted in localised flooding. Fire and Rescue services took a few hundred calls at the height of the downpour and 13 people had to be rescued from homes and vehicles due to rising flood waters in South Wales."

Source: 'Deep depression brings heavy rain on 1 November 2009', The Met Office, http://www.metoffice.gov.uk/news/releases/archive/2009/depression-brings-heavy-rain

There is no doubt that depressions and anticyclones can each have profound effects on people and the economy when they move into their airspace. However, due to our temperate climate in the British Isles it is very unlikely that we will experience truly 'extreme' weather. Hurricanes, severe flooding, long cold spells and extreme heat waves are rare here. However, our changing weather is something that can present both positive and negative influences on our environment and us.

Figure 96

The effects that weather systems can have on people and the economy

		EFFECTS ON PEOPLE	EFFECTS ON THE ECONOMY
Depressions	**Positive**	• Depressions can bring fast changing weather so people do not experience the same weather all of the time. • They bring warmer weather in the winter (and more cloud cover). • They can bring water during the summer months when there might be drought conditions.	• Depressions bring water, which is needed for the crops that farmers are trying to cultivate across the British Isles. • Clouds allow temperatures to remain above freezing, which means that animals can remain outside for longer and increases the length of the growing season.
	Negative	• Depressions bring long periods of 'bad' weather and excessive rain, which either limits outdoor activities or makes them less enjoyable. • Often one depression is followed by another, extending these periods further.	• As depressions pass, the high wind speeds can cause damage to crops, which will cost farmers money. • Strong winds can cause the cancellation of flights and ferries, disrupting transport. This disruption can also affect supermarkets and other businesses, if stock arrival is delayed.
Summer anticyclones	**Positive**	• Summer anticyclones bring 'good' weather, which often has a positive effect on people. It encourages outdoor activities and makes them more enjoyable. It also generally improves people's mood.	• Anticyclonic weather is ideal for crops in the UK, as it allows maximum heat and light to encourage the crop to ripen. However, if the anticyclone sits for a long period of time it can bring drought to an area (see negative). • The longer the anticyclone lasts, the better this is for the local tourism industry, as more people will holiday at home in Northern Ireland and spend money in tourist resorts.
	Negative	• Summer anticyclones can bring long periods of heat and dry weather which can increase the chances of drought, hosepipe bans and possibly some soil erosion.	• If an anticyclone stays for a long period of time, it can bring drought. Drought can spoil crops, causing farmers to lose money. • Irrigation can be a solution. However, it is expensive and can damage the soil if not managed carefully.
Winter Anticyclones	**Positive**	• Winter anticyclones can bring nice, clear, crisp days in the winter, providing a break from the wind and rain in a depression. • They can bring freezing temperatures, which can help to remove bugs and bacteria in the soil that otherwise may have damaged crops.	• In winter anticyclones generally only hinder the economy (see negative).
	Negative	• People with respiratory problems like asthma find it more difficult to breathe in the very cold weather. • Increased amounts of frost and ice can cause a hazard for older people, as they can slip and fall on the ice or suffer when heating fails.	• Temperature inversions can cause heavy fog, which can restrict transport, especially aircraft. • Ice and winter conditions on the roads can cause traffic accidents, which can slow down the movement of goods during the winter. • Older people might find it difficult and expensive to heat their homes in particularly cold spells of weather.

Evaluate

The word evaluate is often used in exam papers. In order to evaluate anything in a Geography exam you need to make sure that you look at both the positive and negative features before you actually come to a decision. In this instance you need to be able to explain the different ways that each of the weather systems can operate to bring contrasting types of weather. In each case, the extreme weather associated with a depression or an anticyclone will cause an impact on people and the economy. Make sure to look for both positive and negative impacts – you should be able to add to the list shown in Figure 96.

Structured notes

Read Figure 96 again and see if you can think of any other impacts and effects that depressions, summer anticyclones and winter anticyclones might have on people and the economy. You may want to consider:

a. the impact on sports players.
b. the impact on healthcare managers.
c. the impact on events organisers.

Check your learning

Now that you have studied Part 2: Weather Systems Affecting the British Isles, return to page 69 and answer the Key Questions for this section.

PART 3: THE CAUSES AND CONSEQUENCES OF CLIMATE CHANGE

a

What is the difference between the greenhouse effect and global warming?

b

What are the causes of climate change?
Natural climatic cycles
Volcanic activity
Human activity, including motor vehicle pollutants and the burning of fossil fuels

c

Evaluate the effects (actual and potential) of climate change on the environment, society and economy
Case Study: the UK (one case study from a MEDC to illustrate the effects of climate change)

d

Evaluate the sustainability of strategies to deal with climate change
International agreements (for example, Kyoto protocol)
The use of alternative sources of energy (wind power, solar power and biofuels)
Strategies to cut the use of private cars (investing in public transport, and congestion charging)
Strategies to slow the rate of deforestation in tropical rainforest areas by encouraging sustainable practices

Identify the issues and analyse the challenges associated with securing international co-operation to deal with climate change

Key questions

1. State the meaning of the term global warming.
2. What is the difference between the greenhouse effect and global warming?
3. State fully one positive and one negative effect that global warming might have for people.
4. Describe one solution to global warming.
5. Evaluate which strategy for dealing with climate change might be the most effective.
6. Explain why there needs to be international co-operation to solve the problem of climate change.

Key words

Global warming
Greenhouse effect
Fossil fuels
Deforestation
Fossil fuels
Renewable energy
International co-operation

"It is evident in the world around us that very dramatic changes are taking place".

Al Gore, in 'An Inconvenient Truth'

"The Stern Report said that global warming is very likely to intensify the water cycle and increase the risk of floods. It is an accepted part of the Stern recommendations that we have to do more."

Rt Hon Gordon Brown, 25 July 2007, during Prime Minister's Question Time

The difference between the greenhouse effect and global warming

Over the last 30 years scientists have become increasingly convinced that the climate of our world is starting to change. They point to evidence of change in the weather, changes to world weather phenomena such as El Nino, the jet stream and the melting rates of glaciers and the polar ice caps.

It has become increasingly difficult for most people to be able to follow and understand the different arguments behind climate change, and to be able to make a decision as to what is really happening. Two phrases have become important in any discussion of climate change: 'global warming' and the 'greenhouse effect' – but what do they mean?

The greenhouse effect

The concept of the 'greenhouse effect' is actually very old, discovered initially in 1824 by Joseph Fourier. It is a process where thermal radiation from the surface of the earth is 'bounced' back again due to 'greenhouse gases' (water vapour, carbon dioxide, methane, Nitrous oxide, CFCs and ozone). As a result, the temperature that surrounds the atmosphere will be made a little higher than if the sun was the only source of radiation.

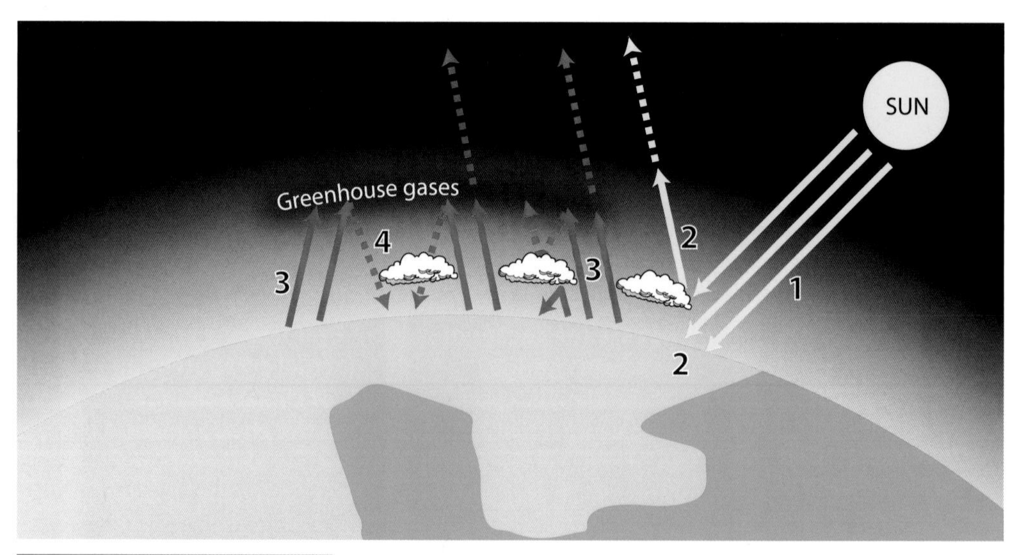

Figure 97 The greenhouse effect

How greenhouse gases help keep the earth a little warmer than normal:

1. The sun gives off energy in the form of visible light and ultra-violet (UV) radiation, which travel towards the earth.
2. Some of the energy is absorbed by the atmosphere, some is absorbed by the earth and some is reflected by clouds back into space. The rest of the energy helps to heat the surface of the earth.
3. Some of the heat energy from the earth is reflected (by seas and lakes) back into space.
4. The greenhouse gases that are in the atmosphere trap this infra-red (IR) radiation and the heat is reflected back towards the earth.

Obviously, the more greenhouse gases that exist in the atmosphere, the more heat radiation is going to be trapped and will warm up the air around our planet. Many people argue that the increase in greenhouse gases is a direct result of different aspects of human activity.

Global warming

Global warming is the name given to the process where the average temperature of the earth is rising. The greenhouse effect is one reason for the rise in global temperature but there are also other causes which might cause the global temperature to change.

The IPCC (the Intergovernmental Panel on Climate Change) indicated in the Fourth assessment report (AR4), in 2007, that during the twenty-first century the global surface temperature is likely to rise between 1.1°C to a possible 6.4°C.

At present, one of the biggest debates in science is linked to what is going on with our climate. Is the global temperature increasing as much as some scientists are predicting? Is this increase in temperature a natural phenomenon or is it something that has been caused and amplified by human activity?

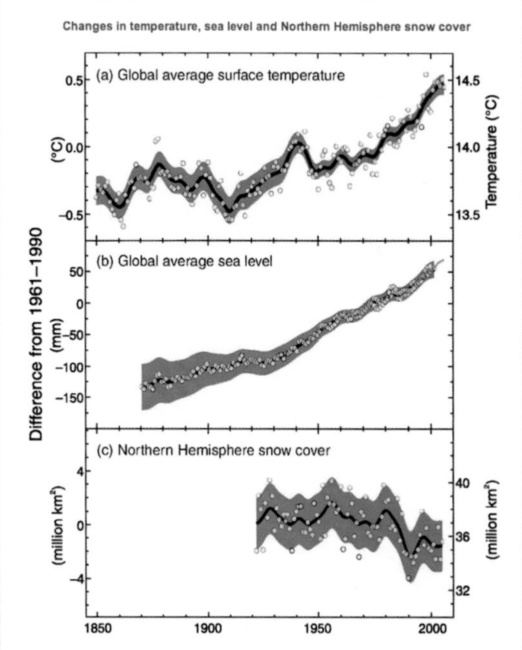

Figure 1.1. Observed changes in (a) global average surface temperature; (b) global average sea level from tide gauge (blue) and satellite (red) data; and (c) Northern Hemisphere snow cover for March-April. All differences are relative to corresponding averages for the period 1961-1990. Smoothed curves represent decadal averaged values while circles show yearly values. The shaded areas are the uncertainty intervals estimated from a comprehensive analysis of known uncertainties (a and b) and from the time series (c).

Figure 98

Fig 1.1 from the IPCC Fourth Assessment Report, Climate Change 2007 (AR4). Changes in temperature, sea level and Northern Hemisphere snow cover

Source: Climate Change 2007: Synthesis Report. Contribution of Working Groups I, II and III to the Fourth Assessment Report of the Intergovernmental Panel on Climate Change, Figure 1.1. IPCC, Geneva, Switzerland.

Test yourself

1. What are the key features of the greenhouse effect? (3)
2. Many scientists argue that humans have increased the quantities of some of the gases in the atmosphere. What are these gases and how have they added to the problem? (3)
3. Use Figure 98 from the IPCC Fourth Assessment to answer the following questions.
   a. Describe the global temperature trends from 1850 to the present. (3)
   b. What has been happening to the global temperature from the 1920s to the present? (3)
   c. Why do you think the global sea level has been increasing over the time noted on the graph? (2)
   d. Describe and explain the trend for snow cover in the northern hemisphere. (4)
4. How does the process of the greenhouse effect cause global warming? (2)
5. If the world did not experience the greenhouse effect, what would the impact on the climate be? (4)

What are the causes of climate change?

Natural causes of climate change

Natural climatic cycles

Slow changes to the earth's orbit can cause some small but important changes to the strength of the seasons over a long period of time.

The earth's orbit varies a little between being circular and more elongated every 100,000 years. This change is called the earth's 'eccentricity' and has been linked to some of the earth's glacial cycles.

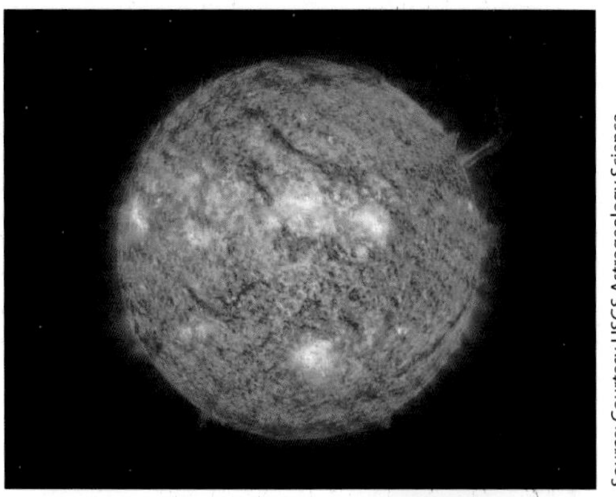

Source: Courtesy USGS Astrogeology Science Center, http://astrogeology.usgs.gov

Figure 99

Solar activity which has a big effect on our climate

The sun can also affect temperature changes. As the source for the earth's climate system, the sun's energy is not always constant. Scientists reckon that the warming of the first half of the twentieth century was down to an increase in the output of solar energy. Interestingly, using studies by a Russian Astrophysics team, WeatherAction have noted that solar activity in the last 50 years has been at its highest for the past several thousand years. They also suggest that the minimum of the cycle of solar activity will occur between 2021 and 2026, when we will face global cooling rather than global warming, leading to a deep freeze around 2050.

Some scientists also reckon that the earth goes though periods of relative warmth followed by period of relative cooling. Historical evidence shows that clothing worn across the UK in 1066 suggests that the temperatures were warmer than today (called the Medieval warming period from AD950–1300), whilst in the Middle Ages many believe that there was a period of cooling (called the Little Ice Age from AD1350–1850).

Figure 100

The volcanic eruption at Mt Pinatubo sending aerosols into the atmosphere

Volcanic activity (sulphates and sulphur dioxide)

When volcanoes erupt they send huge amounts of sulphate gas, ash and dust (called aerosols) high into the atmosphere. When the tiny particles of sulphur mix with water vapour in the atmosphere this can produce tiny droplets of sulphuric acid, which can reflect sunlight back into space. It also creates more sulphur dioxide.

The discharge from major eruptions can reach high into the stratosphere (about 19 miles high), where the aerosols can spread quickly around the world. Massive volcanic eruptions can actually cause the earth to cool down. The 1991 Pinatubo eruption is estimated to have cooled the global surface temperature by nearly 1°C. One of the coldest periods during the last 200 years occurred in the year following the Tambora eruption in 1815 (it also killed about 62,000 people).

Human causes of climate change

As the world population is increasing at a rapid rate, the number of people breathing out carbon dioxide and expelling methane is also increasing. The more people alive, the more food that farmers need to produce to feed these people. To do this, more and more natural woodland is turned into farmland and nutrient rich fertilisers are added to the soil, which can also produce an increased amount of nitrous oxide. Cattle numbers also increase, which in turn increases the amount of methane released, as cows exhale methane as a by-product of their grassy diet. Methane is the second most significant cause of greenhouse warming.

Data from US Environmental Protection Agency

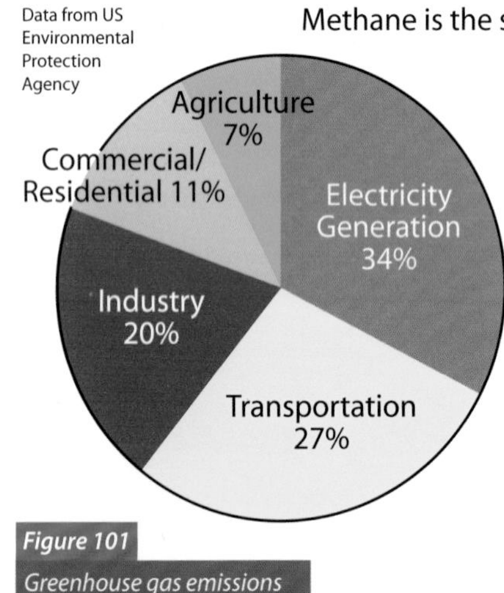

Figure 101

Greenhouse gas emissions from the USA, 2010

Burning fossil fuels (carbon dioxide)

The economy of the world runs on carbon. Carbon is the main fuel in 'fossil fuels' (coal, oil and natural gas), which are used as the main energy source all over the world. Burning them generates power for electricity, heat and light. When the carbon is burned in the air, carbon dioxide is created as a by-product. Many scientists agree that carbon dioxide makes a bigger contribution to the increase in the greenhouse effect than any of the other gases.

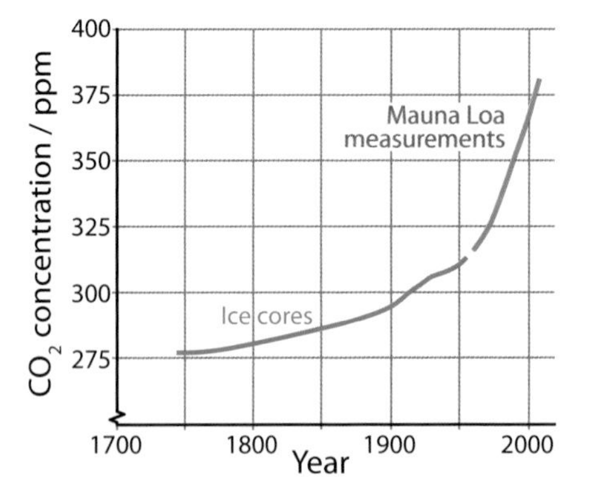

Figure 102

Atmospheric concentration of carbon dioxide (1744–2005)

Source: Neftel, A, H Friedli, E Moore, H Lotscher, H Oeschger, U Siegenthaler, and B Stauffer. 1994. Historical carbon dioxide record from the Siple Station ice core. pp 11–14. In TA Boden, DP Kaiser, RJ Sepanski, and FW Stoss (eds) Trends '93: A Compendium of Data on Global Change. ORNL/CDIAC-65. Carbon Dioxide Information Analysis Center, Oak Ridge National Laboratory, Oak Ridge, Tenn. USA and Keeling, CD and TP Whorf. 2006. Atmospheric CO2 records from sites in the SIO air sampling network. In Trends: A Compendium of Data on Global Change. Carbon Dioxide Information Analysis Center, Oak Ridge National Laboratory, US Department of Energy, Oak Ridge, Tenn., USA

There are many separate studies into carbon dioxide levels across the world which clearly indicate that the amount of CO_2 in the atmosphere has increased at an exponential rate since the 1950s.

Motor vehicle pollutants (nitrous oxide and hydrocarbons)

Most cars are generally known for the burning of oil, which contributes to increasing CO_2 emissions and makes up about 20% of all carbon emissions within the EU. However, cars and other forms of transport are also responsible for the production of hydrocarbons, nitrogen oxide and nitrogen dioxide (NO_X). A 2005 US EPA (Environmental Protection Agency) study showed that the largest emissions of NO_X came from road vehicles.

GREENHOUSE GAS	PRE-INDUSTRIAL CONCENTRATION	PRESENT CONCENTRATION	ATMOSPHERIC LIFETIME (IN YEARS)	PRINCIPAL SOURCES	GREENHOUSE WARMING POTENTIAL
Carbon Dioxide (CO_2)	280 ppm*	368 ppm	50–200	Fossil fuels Deforestation Making cement	1
Methane (CH_4)	700 ppb*	1,750 ppb	12–15	Fossil fuels Livestock Agriculture	21
Nitrous Oxide (N_2O)	275 ppb	316 ppb	120	Fertilisers Industry Aircraft & cars	310
Hydroflurocarbons	0	0.1 ppb	Up to 100	Refrigeration Aerosols	140–11700
Perflurocarbons	0	0.1 ppb	Up to 50,000	Aluminium production Electronic industry Training shoe soles	6500–9200

* ppm = parts per million, ppb = parts per billion

Source: table based on data from the NI Environmental Agency

Figure 103
Concentration of greenhouse gases in Northern Ireland

Test yourself

1. Using Figure 101, describe and explain the breakdown of greenhouse gas emissions from the USA in 2010. (4)
2. Study the graph in Figure 102, which shows the concentration of carbon dioxide in the atmosphere. Describe the changes to the concentration between:
   a. 1750 and 1850. (2)
   b. 1850 and 1950. (2)
   c. 1950 and the present day. (2)
3. The UK Department of Energy and Climate Change noted the following facts about climate change in 2011. How do they help arguments that climate change is more than just a natural phenomenon?
   a. The earth's surface has warmed by about 0.75 °C on average since around 1900 and by about 0.4 °C since the 1970s.
   b. More than 30 billion tonnes of CO_2 are emitted globally each year by burning fossil fuels. (6)

Evaluate the effects (actual and potential) of climate change on the environment, society and the economy

Currently one of the most contentious aspects of geographical science is the discussion of what the actual and potential effects of climate change might be. However, the majority of scientists, whether they recognise that human activity has accelerated the process of change or not, do accept that there are some real, observable effects that have been caused through a change to the climate.

Effects of climate change on the environment

Actual

1. **Increasing temperatures** – We know that the global temperature of the earth has risen by 0.75 °C in the last century.
2. **Increasing rainfall** – Rainfall patterns are changing. Wet places are getting wetter and dry areas are getting drier. However, seasons are starting to change as well, for example, summer rainfall in the British Isles is decreasing whilst in winter it is increasing.

Figure 104
Global-average temperature 1850–2009
Warmest decade 2000–2009

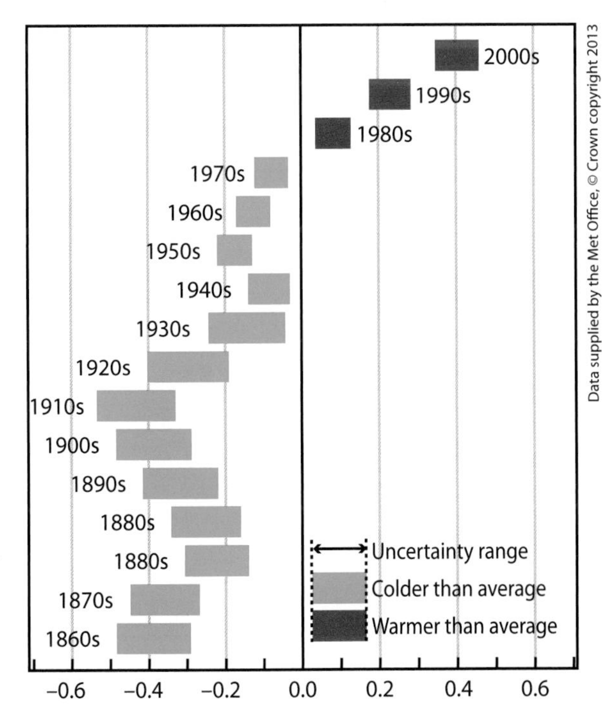

3. **Changing seasons** – In the UK spring is starting earlier and autumn and winter are starting later. Nature experts note that animal behaviour is starting to change, for example, butterflies appear earlier and birds are changing their migration patterns. Trees are also coming into leaf earlier and ecosystems seem to be changing.
4. **Glacier and icecap melt** – The warmer global temperatures have meant that polar icecaps and glaciers in areas of high land are melting at a fast rate. Some evidence suggests that sea-ice in the Arctic has retreated by nearly 20%. The Greenland and Antarctic ice sheets, which store the majority of the earth's fresh water, have both stated to shrink. Glaciers are also retreating, as has been observed in the Alps, the Rockies, the Himalayas, Africa and Alaska. NASA and the USGS have estimated that 85% of Kilimanjaro's glacier volume has been lost with about one quarter of the ice present in 1993 now gone.
5. **Sea level changes** – Since 1900, sea levels around the UK have risen by about 10 cm, and 17 cm globally. As water is released from storage in ice, the amount of water in the world's oceans will continue to rise.

Figure 105
Glacier retreat on Mount Kilimanjaro

Kilimanjaro, 1993

Kilimanjaro, 2000

Source: NASA

Potential

1. **Plants and animals** – As changes occur to the climate, some species of animals and plants might be unable to adapt. Large areas of the Amazon rainforest could be lost due to drought or uncontrolled fire under extreme change. There is also some evidence to support claims that polar bears in the Arctic are now more at risk, as they have to swim further between flows of ice to hunt for food in the summer. As they have to use up more energy, they are losing weight and body fat, and fewer are surviving.
2. **Sea level rise** – Some estimate that the release of water from ice and snow could raise the sea levels around the world by up to 5 m. Even a rise of 1 m could flood 25% of Bangladesh and other low lying areas.

Figure 106
People begin to return home following a flood in Pakistan

Source: Department for International Development

Effects of climate change on society

Actual

1. **Food supply** – The exact impact on crops and food supply will vary in each region but as climates change, the crops that were farmed in particular places will need to be replaced with new crops.
2. **Healthcare** – There will be serious health consequences for millions of people who do not have the ability to adapt to the changes of the climate. There will be an increase in malnutrition and disorders for child growth and development. More people will die due to extreme weather events, such as drought, floods, storms, heatwaves and fires.
3. **Death rates in LEDCs** – The Global Humanitarian Forum, in a report called 'Climate Change: The Anatomy of a silent

crisis', wrote that already 300,000 people each year are seriously affected by climate change at a total economic cost of £80.87 billion.

Potential

1. **Availability of water** – It is likely that there will be less water available for drinking or irrigation as rain will be more variable with a higher chance of drought conditions.
2. **Disease** – More of the world's population will be at risk from insect-borne diseases such as Malaria and Dengue fever, and water-borne diseases such as Cholera. People with respiratory problems will be most vulnerable and there could be an increase in Tuberculosis and Pneumonia.
3. **Weather** – Weather patterns across the UK may change, causing better summers and milder winters.
4. **Mortality rates in LEDCs** – The Global Humanitarian Forum noted in 2009 that nearly 500,000 people die annually due to climate change and 660 million people are seriously affected one way or another each year. 99% of these deaths are in LEDCs. The total cost to the global economy caused by these deaths is £219.96 billion per year.

Barbara Stocking, Oxfam GB Director

"Climate change is a human crisis which threatens to overwhelm the humanitarian system and turn back the clock on development. It is also a gross injustice – poor people in developing countries bear over 90% of the burden – through death, disease, destitution and financial loss – yet are least responsible for creating the problem. Despite this, funding from rich countries to help the poor is not even 1 percent of what is needed."

Effects of climate change on the economy

Actual

1. **Farming** – The traditional patterns of farming are being disrupted. It is becoming more and more difficult for farmers in marginal areas (across Africa and South America) to continue to farm the land as soil erosion increasingly affects the fields.
2. **Flooding** – The amount of damage that is caused due to the flooding of low lying coastal and river areas is increasing. The more water that is flowing across the surface of the planet, the more industrial and economic areas will be at threat from flooding.
3. **Travel** – Airlines have started to introduce 'optional' and sometimes compulsory levies on travellers to help to reduce the 'carbon footprint' of their travel plans. As awareness rises about the impact of climate change, people have to pay more to compensate for any pollution that might be generated.
4. **Green taxes/green economy** – Increasingly, countries have to produce strategies which tackle climate issues. In the UK, the government introduced its Carbon Plan in March 2011. There is a slow but obvious shift towards an economy which is less based on fossil fuels and the consumption of raw materials.

Source: ©iStockphoto.com

Figure 107
Actual impact of climate change on farming can create soil erosion

Potential

1. **GDP** – The costs of climate change could be huge. The Stern Report notes some of the factors that might affect the economics of climate change. It estimates that not taking action could cost between 5 and 20% of global Gross Domestic Product (GDP) every year.
2. **Farming** – In certain places crops will be able to be grown further north than ever before, and in others grain harvests will be improved due to warmer temperatures. However, in some places extreme drought may reduce harvests.
3. **Flooding** – The impact of flooding on cities and farmers will continue to worsen. As the climate continues to change, more and more flooding will cause problems for MEDCs and LEDCS alike.
4. **Tourism** – Travel is becoming increasingly expensive. This could mean that people are less likely to travel overseas for holidays in the future. International travel could be reduced as concerns about the environmental impact of aircraft rise.

CASE STUDY

Case Study: the UK

(one case study from a MEDC to illustrate the effects of climate change)

A number of the environmental, climatic and economic changes already mentioned are likely to impact the UK. Many scientists agree that the climate of the UK will change throughout the twenty-first century, as winters become milder and summers become drier.

The map below suggests some of the key changes that could impact the UK between now and 2080.

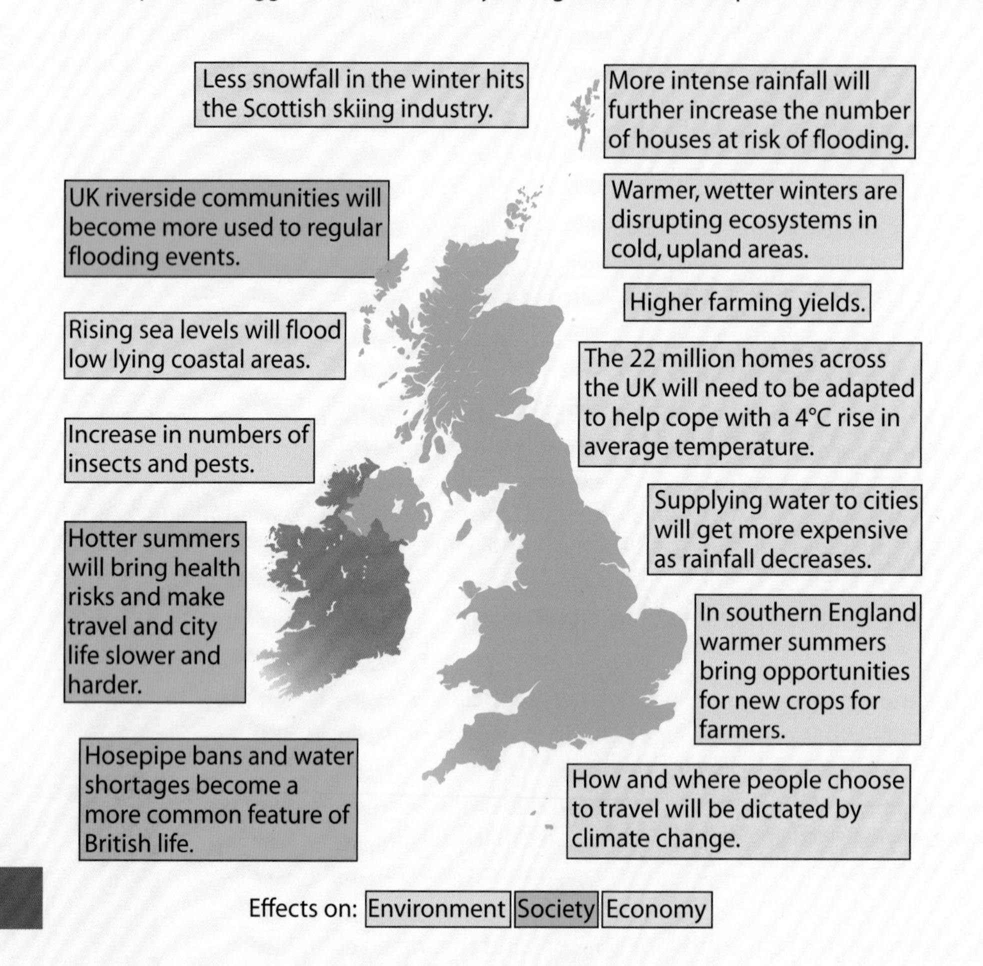

Figure 108
The effect of climate change on the UK

Evaluate the effects of climate change in the UK

The continued impact of climate change upon the UK is likely to force people to adapt to cope with these changes. This means that the environment, economy and society will need to respond to the different ways that aspects of the climate will change.

EFFECT OF CLIMATE CHANGE	HOW THIS MIGHT IMPACT THE UK IN THE FUTURE		
	ENVIRONMENT	SOCIETY	ECONOMY
Sea level rises	Coastal areas will be flooded. Their ecosystems will not have enough time to adapt to climate changes and some species will become extinct. For example, the salt marshes of Norfolk.	Homes in low-lying areas will be flooded if new coastal defences are not built. For example, Belfast and areas around Strangford Lough could be affected.	Flood defences can be extremely expensive to build and maintain (over £20 million of government spending per year). Insurance companies will have to pay out an increasing amount of money in compensation, which will push the price of home insurance up further.

EFFECT OF CLIMATE CHANGE	HOW THIS MIGHT IMPACT THE UK IN THE FUTURE		
	ENVIRONMENT	SOCIETY	ECONOMY
Warmer temperatures	As temperatures rise, both winters and summers will be warmer. Rare arctic plants and alpines will find it more difficult to survive (in upland areas of Scotland).	There could be an increase in the influx of insects. Cold winters kill off insect larvae and keep insect numbers manageable. Warmer winters could allow more swarms of insects to gather. Some suggest that mosquitoes which carry Malaria might be able to survive in the warmer climates. UK houses are not designed for higher temperatures. People may want to invest in air conditioning (which also uses up more energy, further increasing energy demands). High summer temperatures are already raising London Underground temperatures and Transport for London is spending more money on air conditioning units. This in turn will lead to increased travel prices to cover these costs.	The warmer, drier conditions will suit some crops. For example, wheat, maize and fruit will grow better and faster in the new climate. As temperatures and weather improve, more people will want to stay in the UK for holidays rather than going overseas. However, industries that rely on winter snowfalls, such as skiing in Scotland will be wiped out.
Extreme weather (reduced rainfall)	New species of plants will be able to establish themselves. Grapes could be grown in the south of England and lower rainfall levels would be good for fruit growth. However, trees might find it difficult to survive, as they need a constant supply of water.	Droughts will lead to water shortages. People will have to monitor and pay more for any water that they use and hose pipe bans will become more common. The dry conditions will also cause more forest fires and lives, animals and property will be at greater risk.	Arable crop farmers will have to invest a lot more money into expensive irrigation systems. Water will be an increasingly valuable resource and there will be conflict between a priority for use between water for drinking and farming.

Climate Change in Northern Ireland

The Northern Ireland Environment Agency notes in their resource 'Climate Change, Impact, 2009':

"Northern Ireland enjoys a very mild climate. It is buffered by the huge Atlantic Ocean – since water heats up and cools down more slowly than land does, so our climate is fairly constant, and there are no great extremes of weather. Long periods of snow and sub-zero temperatures, or sizzling heatwaves are uncommon".

Source: © Crown Copyright 2009, Climate Change, Impact (2009) CLM 002, http://www.doeni.gov.uk/niea/02_impact.pdf

Some of the predicted changes which climate change might bring to Northern Ireland by 2080 are:

1. **Temperature** – The average annual temperature is predicted to increase. Summer and autumn will warm up most and there will be more record-breaking temperatures.
2. **Precipitation** – Winters will become wetter and summers drier.
3. **Cloud cover** – There will be slightly less clouds.
4. **Humidity** – Relative humidity will decrease slightly.
5. **Wind speed** – Wind speeds will decrease slightly, especially on the east coast.
6. **Snowfalls** – There will be a big decrease of winter snowfalls.

Test yourself

1. "The effects of climate change will impact LEDCs more than MEDCs". Evaluate the extent to which you agree or disagree with this statement. (8)
2. Some scientists argue that there is as much good with climate change as there is bad.
    a. What are some of the positive impacts of climate change? (6)
    b. What are some of the negative impacts of climate change? (6)

REVISION TECHNIQUES

Using a concept map

A concept map is a useful technique which can build upon your knowledge and understanding of the things that link different ideas together. It is a great method of organising ideas graphically and looking at the relationships between the different concepts in an issue.

1. Make a copy of the concept map in Resource A. Make sure that your concept map is big and covers the whole page – you are going to need as much space as possible.

Resource A

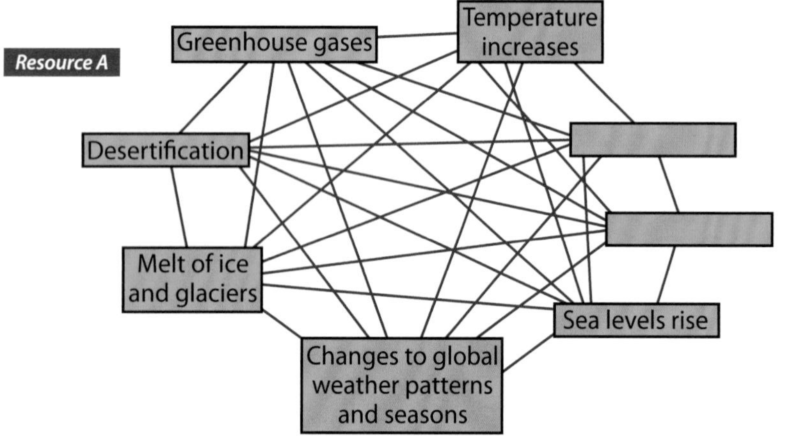

2. See if you can come up with another two words or phrases that might fit into the diagram.
3. Each line on the map represents the links between two concepts. See if you can write up an explanation between the different concepts.
4. Are any of the links between concepts two way – can they work both one way and the other way?
5. Which of the links between concepts do you think might be the most important to our understanding of climate change?

A concept map is a really useful tool for testing how many of the key facts and figures you know.

Evaluate the sustainability of strategies to deal with climate change

Source: John LeGear

Figure 109

One of the globes on display in Chicago in 2007 to urge action against global warming

Whether you agree that climate change has been accelerated by the way that we live or not, there can be no doubt that if we act more responsibly things can be different. If we use our resources effectively and minimise the amount of pollution produced, this will have positive effects, not just on the climate but also on the sustainability of our resources and our health. In recent years there have been a number of strategies used to try and encourage people to take action and start dealing with climate change.

Each of these strategies works at a slightly different level. At one extreme we have the government leaders of the world, meeting at various climate change conferences, who came up with agreements such as the Kyoto Protocol. At other levels we have strategies that encourage the use of alternative sources of energy, plus those which aim to cut down the use of private cars. Finally, there are strategies that aim to tackle the issue surrounding the increase of carbon dioxide, through policies which slow deforestation and encourage more sustainable logging practices.

Source: Peter Blanchard

Figure 110

Climate change rally in Ottawa, Canada

Strategy 1: International Agreements (such as the Kyoto Protocol)

International strategy meetings specifically designed to discuss climate change only really began in 1979. However, it was not until the second World Climate Conference in 1990 that things started to gain momentum. This was the first time that the IPCC (Intergovernmental Panel on Climate Change) was able to identify the specific causes and effects of global warming. They also started to discuss the responses that might be taken in relation to the issue. Further global meetings were called at the Earth Summit in Rio de Janeiro in 1992 and at Kyoto in 1997. Since then there have been yearly meetings of the United Nations Framework on Climate Change (UNFCCC) at a meeting that it calls COP (or Conferences of the Parties). The most recent conferences at the time of writing took place in Durban, South Africa in December 2011 and in Doha, Qatar in December 2012.

One of the most famous Conferences took place in Kyoto, Japan in December 1997. This meeting (COP3) was known as 'The Kyoto Protocol on Climate Change'. After some extensive and delicate negotiations, the conference adopted the 'Kyoto Protocol', which set out the greenhouse gas emissions reduction obligation for many of the MEDCs around the world. Many of the world's richer countries agreed to legally binding reductions in greenhouse gas (carbon dioxide, methane, sulphur dioxide, nitrous oxide and sulphur hexafluoride) emissions of between 6–8% by the year 2012.

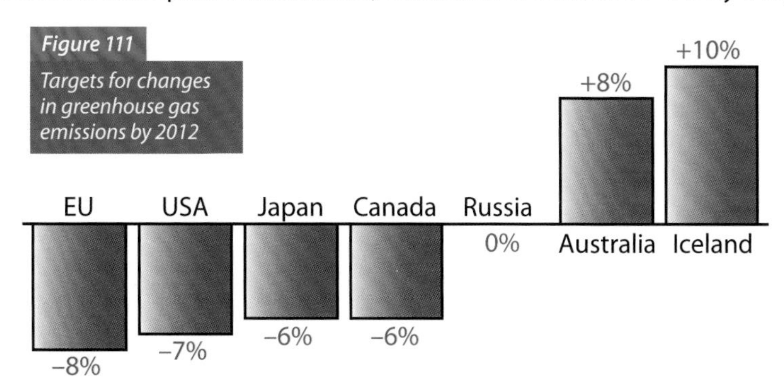

Figure 111

Targets for changes in greenhouse gas emissions by 2012

There was some disagreement within the international community. Many of the LEDCs were only beginning to develop and their economies and potential industrialisation were very reliant on burning fossil fuels. They worried that the obligation to reduce emissions could cripple their economies. They also believed that it should be the MEDCs who foot the bill for any reduction in emissions, as they had been producing them over the last two centuries.

Another aspect of the treaty was the 'carbon credit' allowance system. This would allow a high emission producing country (such as the USA or one within the EU) to purchase an allowance of emissions from a less emission producing country. This was a highly controversial measure, which many felt allowed the richer nations an opportunity to escape punishment for failing to make targets.

The USA failed to sign up fully to the agreement. They were worried about the impact that any acceptance of the problem of climate change might have on their economy. The USA produces over 15 tonnes of greenhouse gases per person per year. In subsequent COP meetings USA, China and India have all withdrawn their support from different aspects of the agreements made.

It can be very difficult to find a compromise that suits both the MEDC and LEDC world.

With thanks to Iain Magill (a GCSE Geography student)

Strategy 2: The use of alternative energy sources (wind power, solar power and biofuels)

Lord Nicholas Stern, former World Bank chief and author of the key report on the costs of climate change in the UK has spoken out (in December 2011) about how rich countries waste money and disadvantage renewable energy by giving away tax breaks, loans and other subsidies to the fossil fuel industry. He reckons that if governments were to stop these tax breaks, we could raise about £6.4 billion a year, which could go towards helping the poor prepare themselves to cope with climate change.

Figure 112

Wind farm on the Colin Mountain in Co Antrim

At present our economy has been based on the availability of fossil fuels. However, as their link with climate change becomes more apparent, and as the fossil fuels become more expensive due to the dwindling supply, changes need to be made to make sure that alternative sources of energy become available.

Wind power

Wind power is a renewable energy that is created using the force of the wind to turn the sails on a turbine to generate power. The biggest advances in wind technology have come from Denmark and California.

ADVANTAGES	DISADVANTAGES
1. Wind energy uses a naturally occurring, free source of energy. 2. Some people see wind farms as potential tourist attractions. 3. No greenhouse gases are produced. 4. They are sustainable and can last for a very long time, producing cheap, clean energy.	1. Wind can be unpredictable – some days will be calm and others will be too windy for the turbines to turn. 2. Wind turbines can cause noise pollution and many people see them as eyesores. 3. It can cost a lot of money to build turbines (over £600,000) and it can take some time to reclaim the initial outlay.

Figure 113
Solar panels reducing carbon footprint and energy bills

Solar power

Solar energy is renewable and created using the sun's energy to produce power. The sun's heat and light is converted into electricity using the energy of speeding photons within a solar panel.

ADVANTAGES	DISADVANTAGES
1. Solar energy uses a naturally occurring, free source of energy. 2. There is no pollution produced. 3. It can be used anywhere, not just in hot countries. 4. It is a very cheap and clean way of producing electricity. 5. It can easily be added to modern homes.	1. Solar energy cannot work at night. 2. It can be very expensive to build solar power stations. 3. It can be unreliable if you do not live in a sunny country. 4. Solar panels can be expensive to install and can be broken easily. 5. Although production of energy can be simple, the storage of energy for the night can be difficult, as electricity does not store easily. 6. It is not always appropriate when large amounts of energy are required.

Source: Sam Beebe, Ecotrust

Figure 114
Biofuel from Castor bean oil (in Costa Rica)

Biofuels

Biofuels are also known as Biogas or Biomass. This is another renewable energy that can be created using fermented animal or plant waste. Power is generated as the biological materials rot and create chemical energy (usually in association with methane, hydrogen and carbon monoxide gases). The gas produced can be used as a fuel. In recent years some fuel has been produced from the natural oils found in some crops.

ADVANTAGES	DISADVANTAGES
1. Biofuel energy uses a naturally occurring, free source of energy. 2. It helps to get rid of waste products on farms, rubbish dumps, etc. 3. People can produce and use their own biofuel on a very small scale. 4. Burning waste to produce gas can use up rubbish that might otherwise take up space in a landfill.	1. It can be expensive to set up. 2. It can be very difficult to control the amount of gas produced and to use it effectively. 3. Biogas contains greenhouse gases that can cause climate change. 4. Some of the contaminants in the gas can actually cause problems for engines and machinery. 5. Production of biofuel oil can involve a huge amount of land, which means less land in agriculture being cultivated for food.

With thanks to Aaron Coates and Ruth Fleck (GCSE Geography students)

Strategy 3: Cutting down the use of private cars (investing in public transport and congestion charging)

Private transport has been on the rise over the last 30 years. In MEDC countries (such as the UK) many households have more than one car, whilst in many of the newly industrialising LEDCs (such as China and India) people want to spend their newly-won disposable income on their own motorised transport. However, personal travel accounts for 13% of the UK's total greenhouse gas emissions. It is also a major factor in issues involving air pollution and congestion. The type of car that someone owns and the fuel that it consumes can make a difference to the environment.

The government uses a number of strategies that attempt (notionally) to cut the use of private cars on the road.

Source: Alex Graves

Figure 115

Increased congestion in Hyderabad, India

1. Investing in public transport

The UK Government is investing in the public transport network (such as trains, underground and bus services) to encourage commuters to use public transport rather than cars. In many cases the transport links are integrated together to allow people a smooth transition from one transport type to another. For example, people might be encouraged to cycle to the local bus station, park their bike there and hop on a bus, which will take them to the local train station, where a train will take them the rest of the way. Some estimates claim that the 'carbon footprint' of a diesel bus filled with 50 people is equal to 2 cars filled with 6 people. Public transport is seen as a cheap, fast (through the use of dedicated bus lanes), reliable and environmentally responsible mode of transport.

Ballymena firm Wrightbus have introduced a new range of hybrid buses that operate both diesel and electric engines. These buses have been deliberately engineered to reduce greenhouse gas emissions in London.

Source: Oxyman

Figure 116

An environmentally friendly Wrightbus vehicle in service in London

Figure 117

Congestion charging zone around Central London

Source: The roads shown on this map are © OpenStreetMap contributors and licensed under the Creative Commons Attribution-ShareAlike 2.0 license

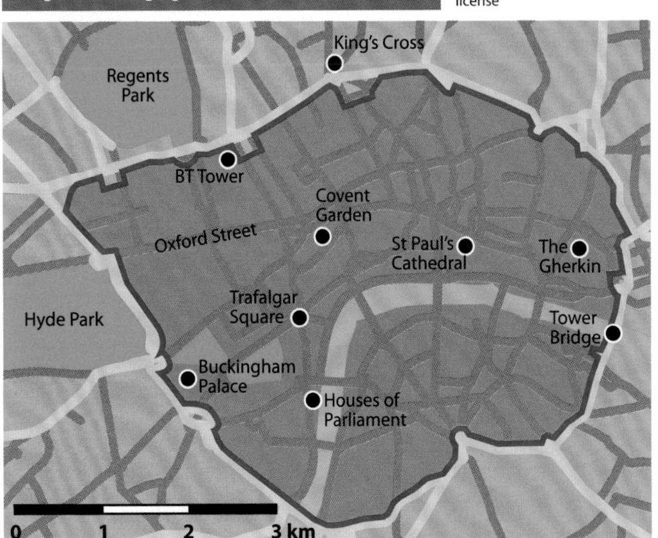

2. Congestion charging

It should be noted that most congestion charging schemes were not originally intended as a direct method to approach the issues of climate change. Transport for London explained that they started to consider congestion charging in London as a way of managing the amount of traffic in the city. In fact, TfL went as far as to suggest that they did not expect congestion charging to significantly affect the air quality. Any reduction in greenhouses gases has been by accident rather than by design. However, it is worth noting that in London nitrous oxide fell by 13.4% between 2003 and 2007, particulates by 24% and carbon dioxide by 3%.

In London drivers pay £10 each day that they enter the congestion charging zone in central London. The effect has been that traffic levels are now 15% lower and there has been a 6% increase in bus passengers. In 2009/2010 the city of London raised £148 million through charges, which by law has to be invested in improving the transport in London. Other cities, including Belfast, are also starting to consider measures such as this, though cynics say that they are more interested in the revenue that charging brings rather than any environmental benefits.

Figure 118 *Congestion charging zone, Old Street, London*

Source: Nevilley

Source: Elliot Brown

Figure 119

Car sharing can encourage people to share journeys

3. Car sharing

Some car sharing schemes have been put in place across the UK to encourage people to share travel. Car sharing is when two or more people travel together for all or part of a trip. This can help to reduce travel costs and carbon emissions, and might even allow increased security. Some estimates suggest that on average cars in the UK each carry 1.6 people, with around 38 million empty seats. Car sharing organisations have even been set up, such as NCS (National CarShare), who aim to put people who wish to car share in contact with each other. In Bradford a section of the M606 has been converted to have a car-share lane, where people with more than one person on board have priority.

Strategy 4: Slowing the rate of deforestation in tropical rainforests (and encouraging sustainable practices)

Deforestation (the cutting down of trees on a large scale) has long been linked with climate change. It can cause climate change in three ways:

1. If trees are removed, there will be fewer trees available for evapotranspiration and therefore there will be less water in the air.
2. The main method of removal in the tropical rainforests is to burn the trees. As the forest is burned, much of the stored carbon is released into the atmosphere. This combines with the oxygen and there is an increased amount of carbon dioxide created (a greenhouse gas).
3. Finally, the tropical rainforest is sometimes referred to as 'the lungs of the earth', as they will usually take in carbon dioxide and give back oxygen. Fewer trees mean that less carbon dioxide will be recycled.

Source: ©iStockphoto.com

Figure 120

The devastation brought to Tasmania due to deforestation

Many people now feel that it is important to slow down the rate of deforestation so that climate change is not accelerated. More and more forests are being cut down across the globe to make room for agriculture and the development of towns and cities.

Malaysia is an example of a country that is trying to develop but also trying to manage the way that trees are cut down, so that forestry is more sustainable and causes less of an impact on climate change. The Malaysian government has attempted to make sure that only sustainable methods of logging are used. Clearfelling (which results in everything being removed from the area) is no longer allowed and instead new logging methods are encouraged:

1. Selective cutting

This is when single or grouped trees are selected for cutting. This means that a natural regeneration of the forest can take place quickly, though it can be quite expensive.

2. Integrated cutting

This is when several different types of tree are removed at the same time. It is economical as loggers remove all that is useful when working in an area. The down side is that the good trees are taken and the poor quality trees are left, making regeneration of the area quite difficult.

3. Strip cutting

This is quite similar to clearfelling but this method often means that trees are cut down along the path of contours in the land. This method of logging can help to stop soil erosion but still destroys the ecosystem of vast strips of land.

With thanks to Lucy Frampton (GCSE Geography student)

Structured notes

Using the information about **all** of the four strategies to deal with climate change start to evaluate the positive and negative impacts of **each** strategy. You might find it useful to draw a table similar to the one below to help you organise your information.

STRATEGIES	POSITIVE IMPACTS	NEGATIVE IMPACTS
Strategy 1: International agreements (such as Kyoto)		
Strategy 2: The uses of alternative energy sources (wind power, solar power and biofuels)		
Strategy 3: Cutting down the use of private cars (investing in public transport and congestion charging)		
Strategy 4: Slowing the rate of deforestation in tropical rainforests (and encouraging sustainable practices)		

Test yourself

1. How sustainable do you think the four individual strategies are? Write a comment on each. (2 marks for each = 8)
2. How does each of the strategies specifically deal with reducing the impact of climate change? Give reasons to support your answer.
   (2 marks for each = 8)
3. Which of the strategies is the most sustainable and which is the least sustainable? Give reasons to support your answer.
   (3 marks for each = 6)
4. Many of the strategies listed can change very quickly. Carry out some research on the new developments in each of these areas. What has changed? [8]

Identify the issues and analyse the challenges associated with securing international co-operation to deal with climate change

With approximately 200 independently run nations across the world, it can be very difficult for any reasonable agreement to be made in relation to making sure that countries actually co-operate to tackle climate change.

The Independent on 9 December 2011 wrote:

"The chance of a binding new climate deal involving the world's biggest greenhouse-gas emitters, China, the US and India, looks increasingly unlikely as the UN climate conference in Durban draws to a close today. In an outcome that would dash the hopes of thousands of people and many countries who feel themselves threatened by global warming, an international treaty that would make the planet's biggest polluters cut back by a definite timetable on their carbon emissions is looking like an impossible dream."

Michael McCarthy, 'Another climate summit, another chance goes up in smoke', *The Independent*, 9 December 2011

Chris Huhne, the former UK Government Energy and Climate Change Secretary, even went as far as to say that the traditional view of developing and developed countries needed to be modernised so that emerging economies like China can take on more responsibility for climate change. For example, Singapore and Korea are considered to be LEDCs, despite being richer than MEDCs such as Romania and Bulgaria.

The only existing climate change treaty in place, the Kyoto Protocol, commits just the MEDCs to cutting carbon, which makes the distinction and international agreement on such things important.

The Kyoto Protocol, 1997

Out of all of the International conferences in recent years, the only agreement that has been made became known is the Kyoto Protocol (or agreement) in 1997. Many countries signed up to reduce emissions of greenhouse gases by 5.2% and are taking steps to introduce measures to slow down global warming. Some of the key points from Kyoto were:

1. The USA signed up to the agreement initially (but later did not pass the laws in their government). They were reluctant to make any cuts in case it damaged their economy. This was despite the fact that in 1996 the USA produced 21% of the world's carbon dioxide, even though they had only 4% of the world population.
2. Many of the LEDCs have been trying to improve their economies and feel that they need to be able to industrialise and burn fuels so that they can create jobs and improve their standards of living.

World Summit for Sustainable Development (Johannesburg, 2002)

At another conference, the 'World Summit for Sustainable Development' (WSSD) in Johannesburg, in 2002, some principles for sustainable development were established, including:

1. The polluter should pay (ie countries emitting pollution should pay a fine to control emissions).
2. Forests should be used as carbon 'sinks', with trees planted to help absorb carbon dioxide.
3. Emissions of greenhouse gases should be reduced to 1990 levels.
4. LEDCs should be helped to become more energy efficient.
5. There should be significant investment in renewable energy resources.
6. The world should prepare to deal with the problems caused by global warming and provide a sustainable solution.

According to the WSSD, a sustainable solution to global warming means:
"Keeping greenhouse gases at levels which will not dangerously upset the global climate and yet will allow economic developments to continue in both LEDCs and MEDCs".

Figure 121

One of many cartoons that commentate on the ineffectiveness of international co-operation

Source: © Seppo Leinonen, seppo.net

Why is it difficult to get international agreement?

1. Some governments and scientists do not believe that climate change is any more than a natural process.
2. Many countries find it difficult to set aside their own national interest, given the challenges being faced by their people on economic and social levels. Reducing greenhouse gases may cost a country a lot of money and damage their economy. This is unpopular, particularly in times of financial crisis and depression.
3. Some governments do not see climate change as a priority. They feel it is more important to improve standards of living and spend money on schools or hospitals.

4. Introducing strict environmental laws will cost industries a lot of money and this could lead to a loss of jobs.
5. It is a voluntary agreement. What will happen if countries fail to meet targets?

Test yourself

1. What do you think the main issues might be for countries when they try to make a deal about the climate? (4)
2. What do you think might cause the main differences of opinion that various countries have towards climate change? (4)
3. Carry out some research into the climate change stance taken by one country. Why do you think they could argue that dealing with climate change and cutting emissions might be a good thing? (7)
4. What organisations might be able to help bring the world together towards an agreement? (4)

Check your learning

Now that you have studied Part 3: The Causes and Consequences of Climate Change, return to page 85 and answer the Key Questions for this section.

ORGANISE YOUR REVISION

Draw a mind map to summarise Unit 1B 'Our Changing Weather and Climate'.

EXAM PRACTICE QUESTIONS

Some of these questions are from previous CCEA GCSE examination papers and others have been written in the same format to give you practice at answering 'exam style' questions.

Try to answer the questions with as much detail as possible. Also consider the number of marks that each question receives, as this will give you a good indication of the amount of depth that your answer needs.

Resource A

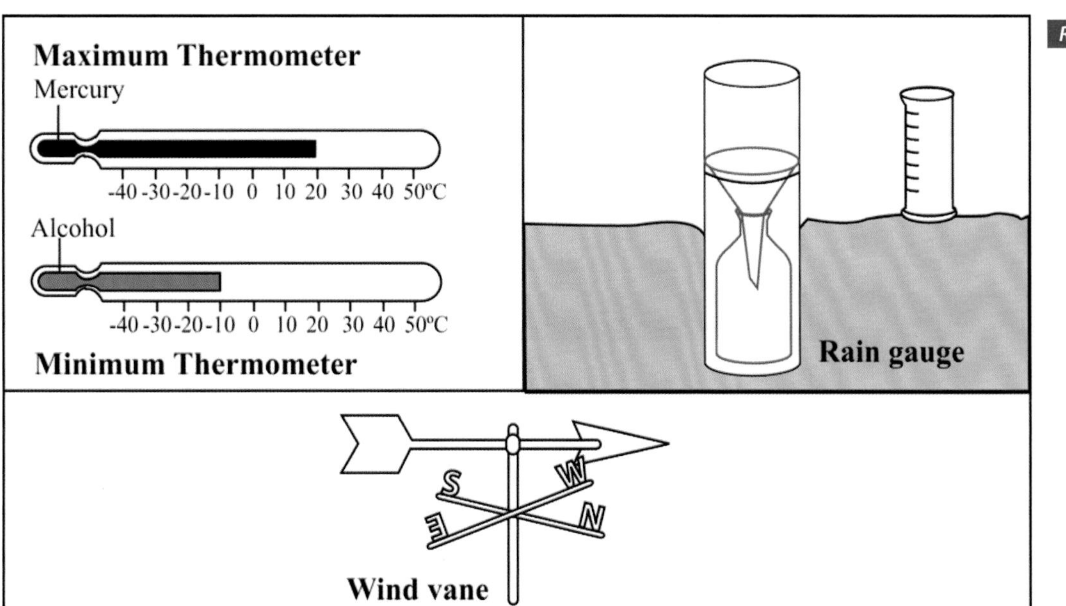

Resource B

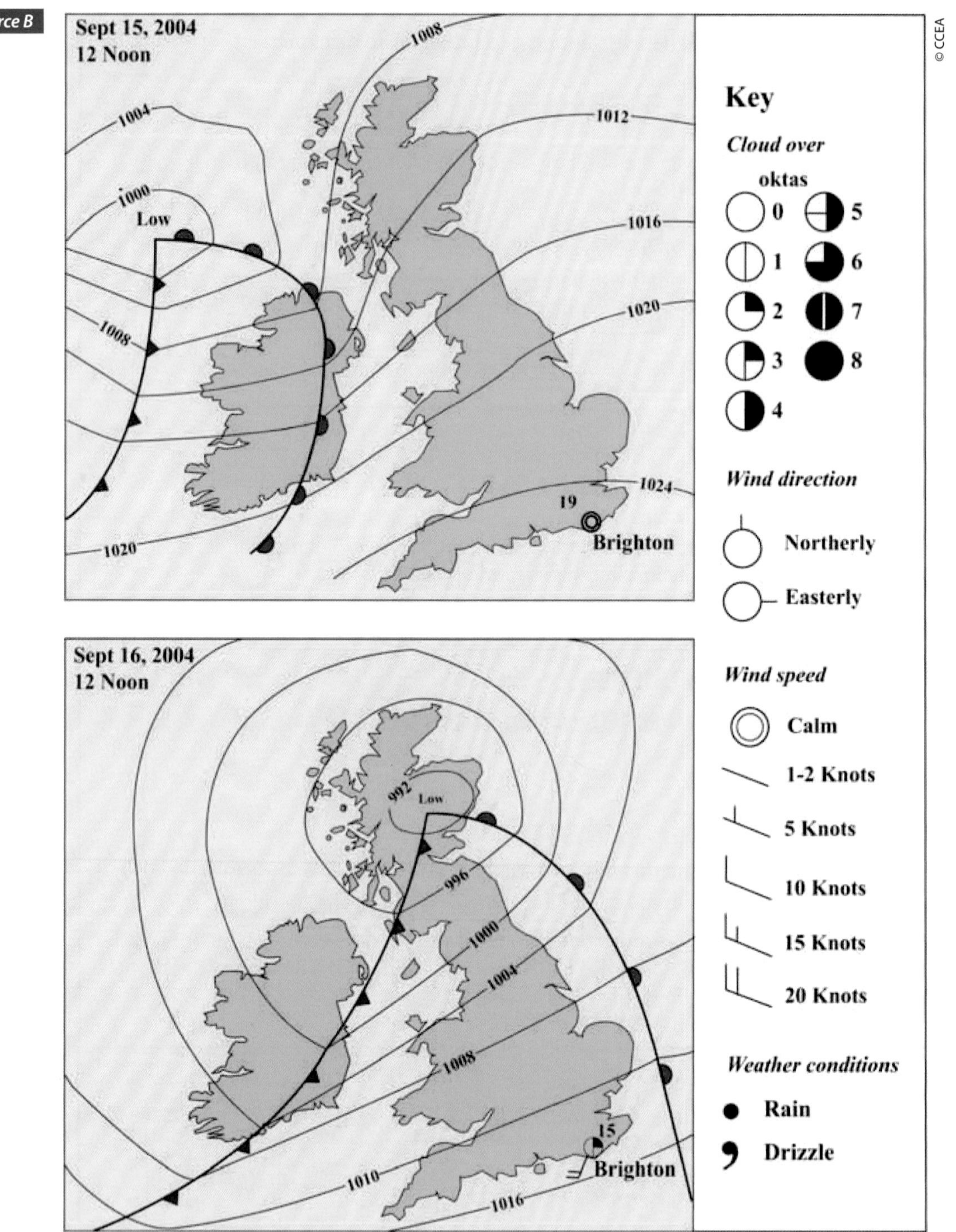

Resource C

FORECAST	YES OR NO
Wind speed will increase	
Temperatures will increase	**NO**
Cloud cover will increase	
Rain will occur	

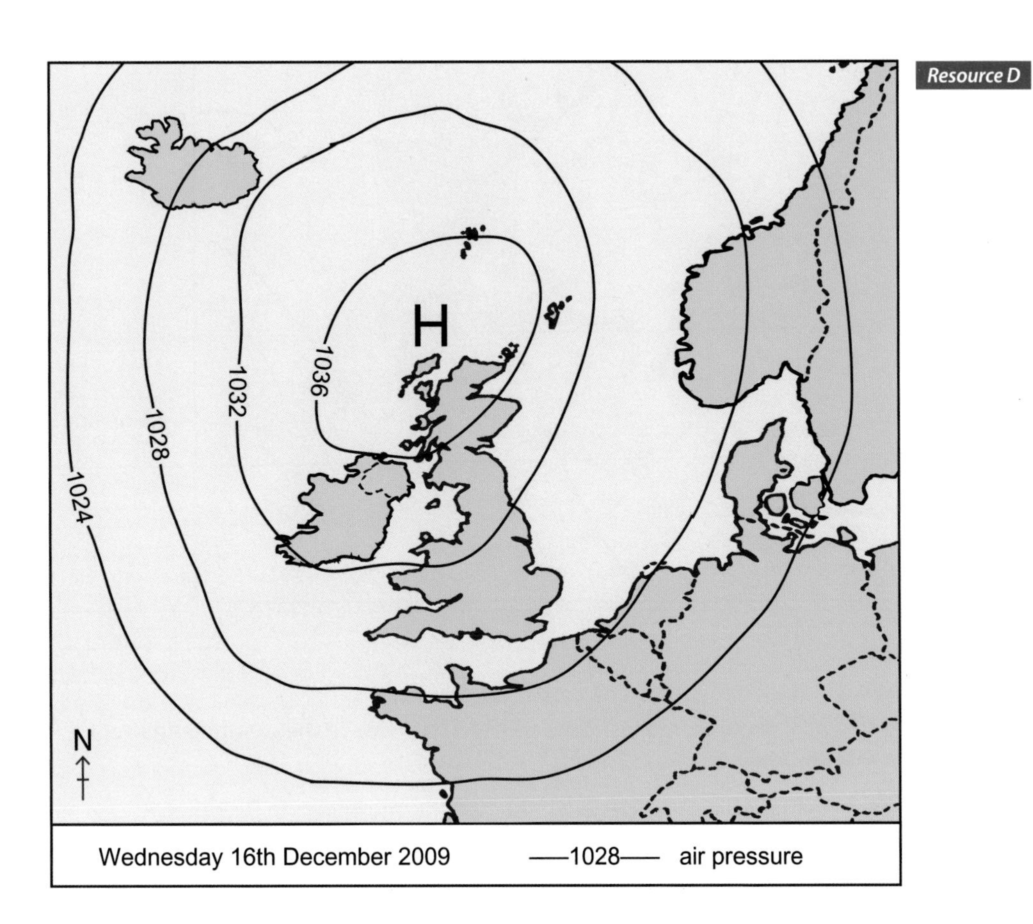

Resource D

Forecasters have warned that the weather in Britain is set to turn cold this weekend.

Temperatures will fall to as low as –5°C in London and –2°C in Edinburgh by Sunday.

Over the next few days the country will experience dry, calm and cold conditions.

Resource E

NATURAL CAUSE	CAUSE OF CLIMATE CHANGE	HUMAN CAUSE
	Vehicle emissions →	
	Cycles of climatic change	
	Volcanic eruptions	
	Burning fossil fuels	

Resource F

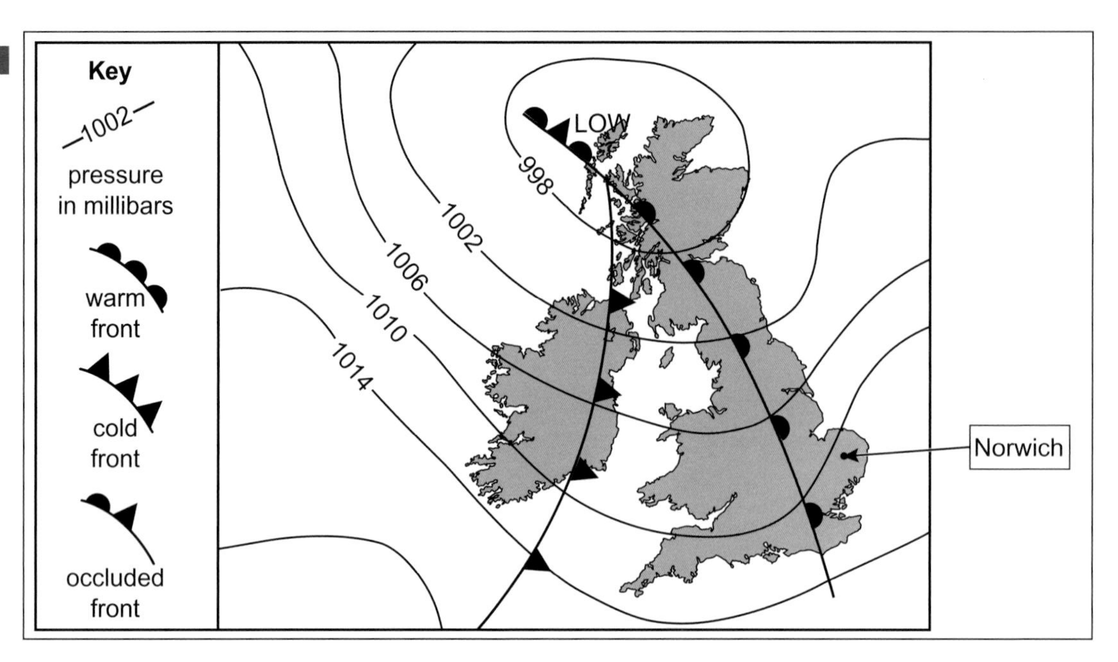

Foundation Tier

1 Study Resource A which shows some instruments used to measure the weather. Answer the questions which follow.

a. State the difference in temperature between the two thermometers in Resource A. Choose your answer from the list below. (1)

20°C
30°C
40°C

b. Explain why temperature readings are usually taken in the shade. (2)

c. Outline **one** factor that would need to be taken into account when locating either a rain gauge or a wind vane instrument. (2)

2. Study Resource B which shows the weather systems over the British Isles on 15 and 16 September 2004. Answer the questions which follow.

i. Using the weather map for 15 September, choose the pressure value shown for the area near Brighton from the list below. (1)

1016 mb
1020 mb
1024 mb

ii. Using the weather map for 15 September, name the front over Ireland. (1)

iii. Using the weather map for 15 September, describe the weather at Brighton. (3)

iv. The forecast predicted that the weather could change at Brighton from 15 September to 16 September.

Make a copy of the table in Resource C and write YES or NO beside each statement to show if the forecast was accurate. One has been completed for you. (3)

3. Study Resource D which shows information about a weather system over the British Isles in December 2009.

i. What type of weather system is over the British Isles on Wednesday 16 December 2009? Choose your answer from the list below. (1)

Front
Depression
Anticyclone

ii. Which of the following terms means lines of equal air pressure on a weather map? (1)

Contour
Isotherm
Isobar

4. Make a copy of the table in Resource E and complete it by drawing arrows to show some causes of climate change. (3)
5. Explain **one** negative effect of climate change in a country that you have studied. (1 mark for the country and 3 marks for the negative effect = 4)
6. Describe **and** explain **one** sustainable method for dealing with the issue of climate change. (6)

Questions and resources from CCEA Past Papers and Specimen Papers
June 2010 (Q3, 4, 5; Resources D, E), Specimen 2010 (Q1, 2, 6; Resources A, Bi and ii, C)

Higher Tier

1. Study Resource A which shows some of the instruments used to measure the weather.
   i. State the meaning of the term **weather.** (2)
   ii. State the difference in temperature on the thermometers. (2)
2. Study Resource B which shows the weather systems over the British Isles on 15 and 16 September 2004.
   i. Name the front to the west of Ireland in the weather map for 15 September. (1)
   ii. Describe how the position of this front has changed from 15 September to 16 September. (3)
   iii. Explain why the weather at Brighton on 15 September was warm and sunny. (3)
3. Study Resource F which shows information about a weather system.
   i. What is a synoptic chart? (2)
   ii. State the name of the weather system located over the British Isles. (1)
   iii. The temperature in Norwich will change as this weather system passes. Describe **and** explain how the temperature will change. (2 marks for description and 3 marks for explanation = 5)
4. State **two** effects of climate change (**one** positive and **one** negative). (2)
5. Explain **one** sustainable strategy which could be used to deal with climate change. (5)

Questions and resources from CCEA Past Papers and Specimen Papers
June 2011 (Q3, 4, 5; Resource F), Specimen 2010 (Q1, 2; Resources B i and ii)

1C The Restless Earth

> "We learn geology the morning after the earthquake"
> **Ralph Waldo Emerson**

It is only quite recently that scientists have been able to make an educated guess about what is happening underneath the earth's crust. All over the surface of the earth there is much evidence of the processes that have helped to shape it but we still do not really know how the engine of the earth works. The theory of plate tectonics provides some understanding of what is happening and how particular areas and zones of activity have produced some quite spectacular and often disastrous events.

This theme takes a look at the rocks beneath our feet, how they were formed, how they move and how these movements can often produce cataclysmic consequences both within the UK and further afield.

The theme is divided up into four parts:

1. Basic rock types
2. Plate tectonics theory
3. Tectonic activity in the British Isles
4. Earthquakes: can they be managed?

PART 1: BASIC ROCK TYPES

The formation and characteristics of basic rock types
Igneous: basalt and granite
Sedimentary: limestone and sandstone
Metamorphic: slate and marble

Key questions

By the end of this section you will be able to answer the following questions:

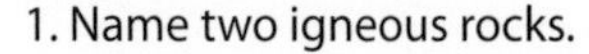

1. Name two igneous rocks.
2. Explain how limestone is formed.
3. Compare the formation of slate and limestone.
4. What are the similarities and differences between basalt and granite?
5. How are metamorphic rocks formed?
6. Describe the key characteristics of sedimentary rocks.

Key words

Igneous
Sedimentary
Metamorphic

The formation and characteristics of basic rock types

Rock is a solid, natural mass of mineral material which makes up the crust of our earth. There are three groups of rocks:

- Igneous
- Sedimentary
- Metamorphic

Igneous rocks: basalt and granite

Igneous rocks are those that have been formed by the cooling and solidifying of molten rock (magma) from underneath the earth's crust. This molten rock is called lava on the surface of the earth. Crystals are usually evident in the rock. However, if the rock cools quickly as the lava hardens on the surface, there will be little evidence of the crystals (for example, basalt). If the magma is cooled slowly underground, then a crystalline structure will be more evident (for example, granite).

Sedimentary rocks: limestone and sandstone

Sedimentary rocks are those that have been formed by the sediments that have built up over a long period of time, usually under water. The sediments are made up of weathered and eroded material that builds up in layers. As more and more material is added, pressure pushes air and water out, and the sediment gets cemented into a rock.

The mineral particles in the rock are usually made from mud, sand and clay. Limestone and sandstone are good examples of sedimentary rocks. These rocks can contain fossils. Sedimentary rocks are very easy to recognise due to the their flat structure in stratified beds.

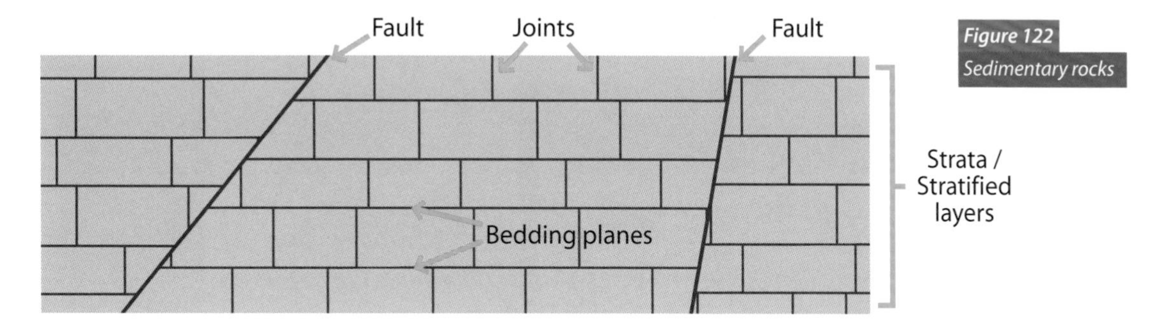

Figure 122
Sedimentary rocks

Metamorphic rocks: slate and marble

Metamorphic rocks are those rocks that have been changed from an earlier state through the addition of pressure or heat. The rocks would originally have been igneous or sedimentary. For example, marble is a metamorphic rock which was once limestone, a sedimentary rock.

Metamorphic rocks are tough and can be highly resistant to erosion, as they have undergone very hard conditions in their formation. They are often used as building materials, for example, slate.

ROCK GROUP	ROCK TYPE		FORMATION
Igneous	**Basalt**	**Granite**	Basalt: Lava erupts from volcanoes and solidifies. Granite: Magma hardens within the earth's crust.
Sedimentary	**Sandstone**	**Limestone**	Sandstone: Particles the size of grains of sand are pressed together. Limestone: Shells and skeletons of tiny creatures on the sea floor are compressed.
Metamorphic	**Marble**	**Slate**	Marble: Limestone is altered by heat or pressure. Slate: Shale, clay and volcanic ash is changed through pressure.

Check your learning

Now that you have studied Part 1: Basic Rock Types, return to page 106 and answer the Key Questions for this section.

PART 2: PLATE TECTONICS THEORY

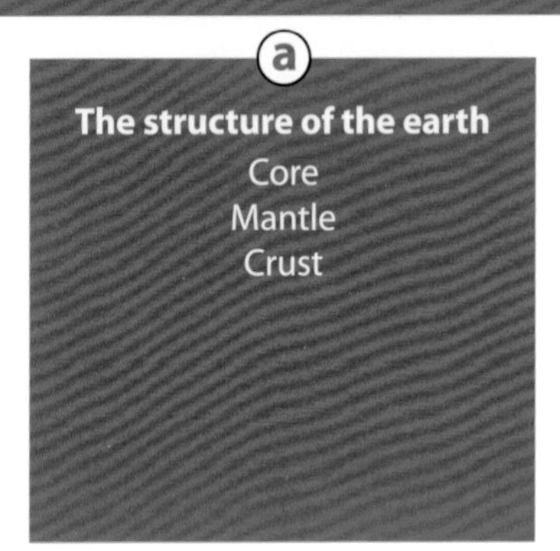

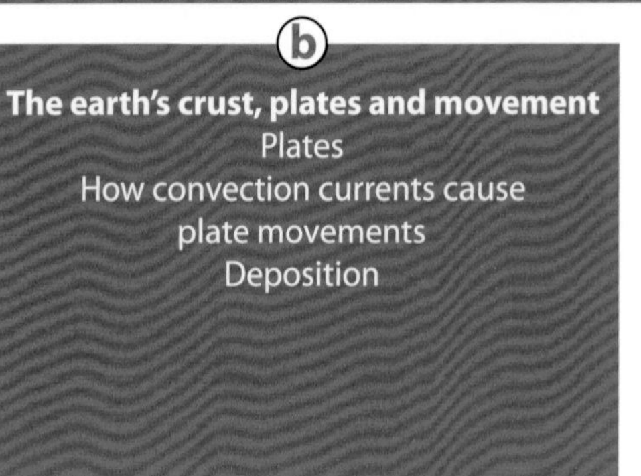

c

The processes and landforms associated with plate margins:

Constructive plate margins: mid-ocean ridges

Destructive plate margins: subduction zones and ocean trench

Collision zones: fold mountains

Conservative plate margins: fault lines

Key questions

By the end of this section you will be able to answer the following questions:

1. State the meaning of the term plate.
2. Explain why plates move.
3. Describe and explain the formation of two landforms associated with a destructive plate margin.
4. Explain how a mid-ocean ridge is formed at a constructive plate margin.

Key words

Core
Mantle
Crust
Convection current
Plate
Plate margin/boundary
Constructive
Destructive
Conservative
Collision
Mid-ocean ridge
Ocean trench
Subduction zone
Fault line

Figure 123
The structure of the earth

Crust
Mantle
Outer Core
Inner Core

The structure of the earth

The theory that underpins scientists' best educated guess about what happens under our feet is called 'the theory of plate tectonics'. This theory has been around in various forms for about 100 years. Originally, it built on the ideas behind 'continental drift' and later 'seafloor spreading'.

The ground beneath our feet is made up of a series of layers. Much of our understanding about the make up of the different layers is based on scientific experiments that measure the speed and action taken when earthquake waves move through the earth. Our understanding of what the differences are between each layer is still very limited. The distance from the edge of the crust to the centre of the earth is around 6,370 km, yet the deepest that we have been is around 3 km (in some South African gold mines).

Core

At the earth's core, the temperature will be over 6,000 °C. Scientists reckon that the material here is much more dense than material at the crust. The core is believed to be composed mostly of iron and nickel, with a liquid outer core surrounding the solid inner core. The outer core is about 1,200 km deep and the inner core is around 2,200 km deep.

Mantle

The mantle is the thickest layer of the earth (at 2,900 km). It is made up of silicate rocks that, due to the high temperatures (about 3,500 °C), remain molten and flow and move about. The mantle layers are sometimes known as the **asthenosphere**.

Crust

The crust is the outermost section of the earth. It is a thin layer, varying in different places from 5 to 70 km in depth. The crust is broken up into several pieces, which are known as plates. These plates are forced to move about due to the process of the convection currents within the mantle (as explained on page 110). The crust and some of the more solid mantle are sometimes known as the **lithosphere**.

Test yourself

Before we go into detail about the mechanics of how plates move, let us take a moment to think about what you might already know from your study of geography or science.

1. Make a copy of the concept map below which contains some of the key terms associated with tectonic activity.

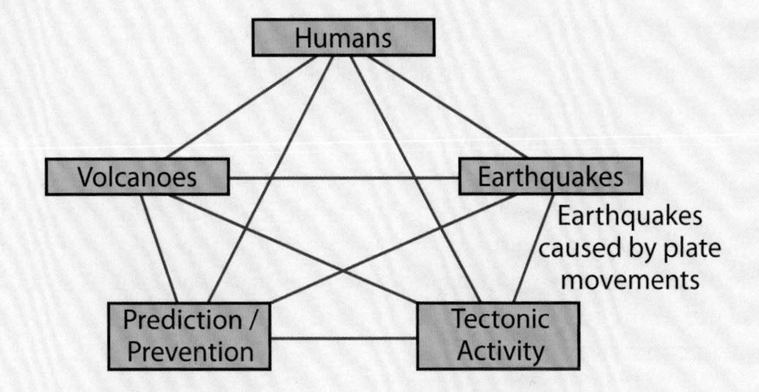

2. Write a sentence to link each of these concepts together. The first one has already been done for you. (9)
3. Try to come up with a two sentence definition of each of the five concepts. (5 × 2 marks = 10)

The earth's crust, plates and movement

The earth's surface is not static – it is moving. It only moves a small amount each year (sometimes less than 1 cm), but over many years these small movements can have a big impact. Scientists have noted that the hard crustal rock seems to sit on top of a layer of molten mantle rock. These sections of crustal rock (called plates) are moved about by the forces and currents of pressure and temperature inside the earth.

The theory of plate tectonics was originally proposed by Alfred Wegener. He was the first person to look at the jigsaw of plates and land masses from around the world, and put them together into a closely fitting land mass (that was later called Pangea). His idea was that the continents had once formed one single land mass and that something had allowed them to float away from each other, like icebergs.

Since then, it has become more obvious that a link clearly exists between plates, the location of earthquakes and the location of volcanoes. Scientists have been able to identify that the major instances of tectonic activity (either as an earthquake or a volcano) takes place in these specific zones of activity, which are located at the boundaries or margins of the plates. One famous zone of activity is called 'the Ring of Fire', which includes many active volcanoes that surround the margins of the Pacific Plate.

USGS
science for a changing world

KEY
Depth of focus
● 0–69 km
● 70–299 km
● 300–700 km
Active volcano △
Plate boundary

Source: USGS

Figure 124
The location of the major earthquakes and volcanoes from 1900

How convection currents cause plate movement

Deep inside the thick mantle layer of the earth, the extreme heat from the earth's core causes mantle rock to melt. As the rock becomes liquid, it becomes less dense than the material above it and starts to rise towards the surface. The molten rock (called magma) rises and attempts to break through cracks in the crust where possible (see Figure 125). The magma in the mantle layer will then interact with the crust (often due to friction from the moving mantle rock) causing sections of the crust to move towards or away from each other. Any excess magma, now that it is far away from its source of heat, will start to cool and solidify, and as the rock becomes more dense, the rock will sink back deeper into the mantle, where it will become heated again.

Plate types

There are two different types of plate that make up our earth's crust:

- **Continental plates** are usually thick plates that form at land masses/continental areas.
- **Oceanic plates** are usually thinner plates that form at the bottom of ocean floors.

Both types of plate are part of the seven major plates (North American, Pacific, South American, Antarctic, African, Eurasian and the Australia Plates) and the minor plates (which include the Nazca, Cocos, Caribbean, Arabian, Indian and Philippine Plates) that join to make our earth.

Figure 125
Convection currents in the mantle

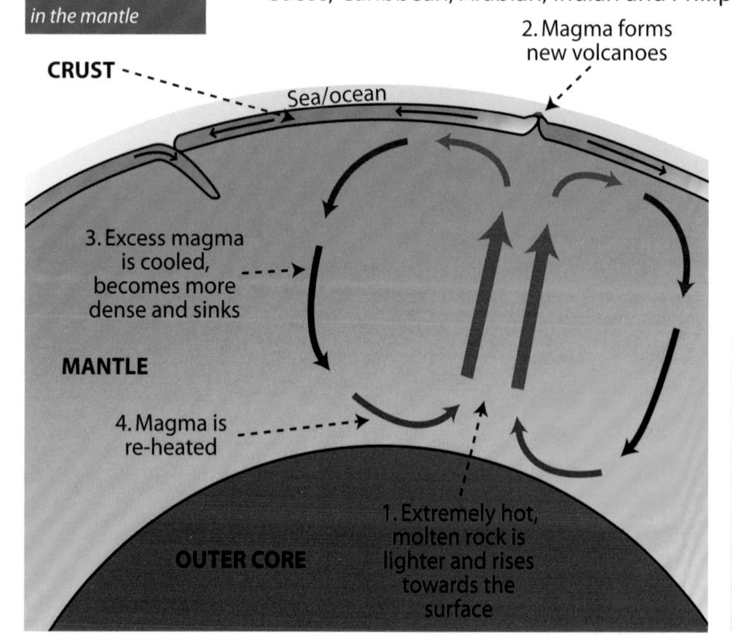

CONTINENTAL PLATES	OCEANIC PLATES
• 35–100 km (thick) • Rocks can be very old • Generally contains 'light' rocks that are less dense • Example: Granite	• 6–10 km (thin) • Rocks can be very young • Generally contains 'heavy' rocks that are more dense • Example: Basalt

Test yourself

1. Which of the plates in Figure 126 are continental plates and which are oceanic? (10)
2. Describe how convection currents help to move plates. (4)
3. What are five key differences between continental and oceanic plates? (5)

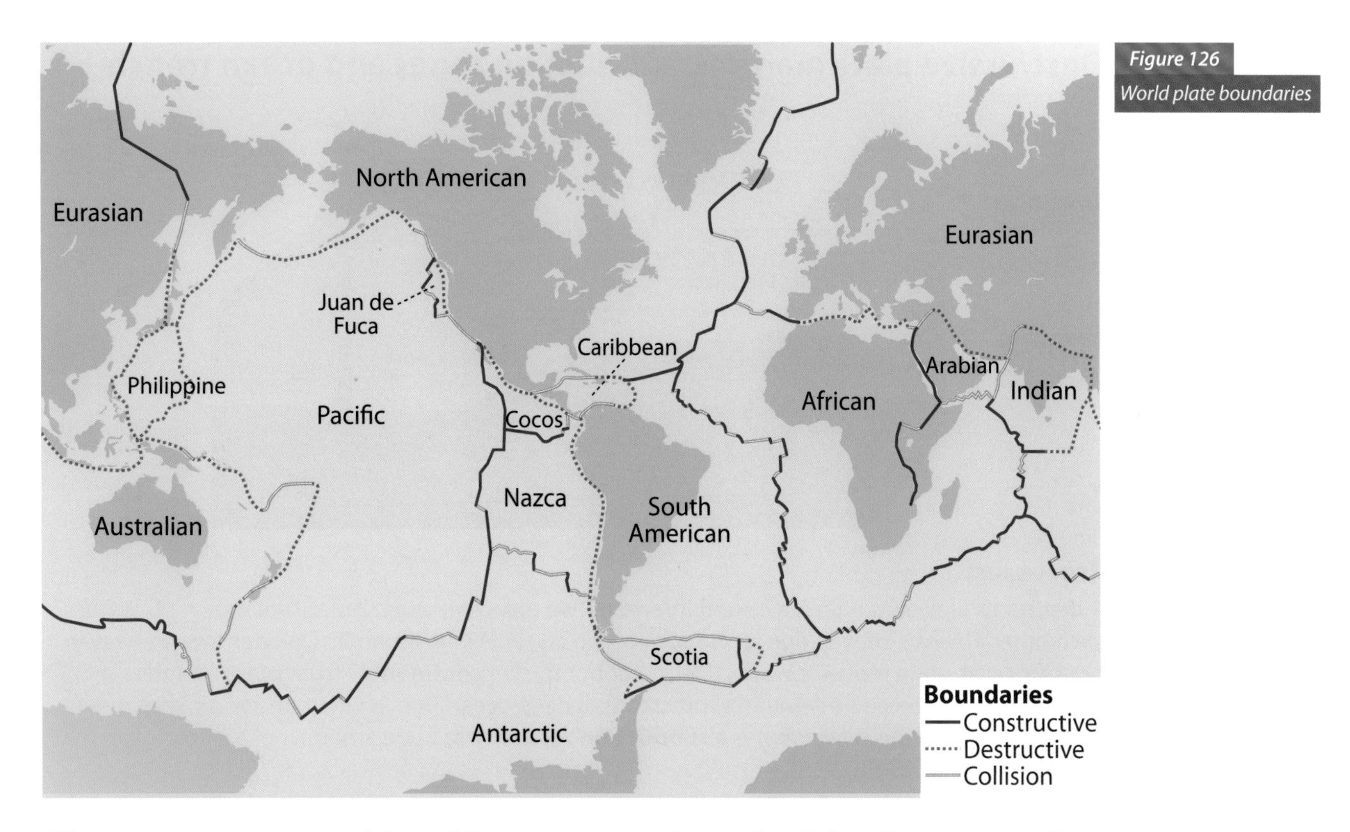

Figure 126
World plate boundaries

The processes and landforms associated with plate margins

Plates rarely move at consistent speeds each year. Some plates move regularly, whilst others have not moved in centuries. The amount of tension at each plate boundary is dependent on the amount of force that is being pressed by the surrounding plates.

Constructive plate margins: mid-ocean ridges

Figure 127
Constructive plate margin

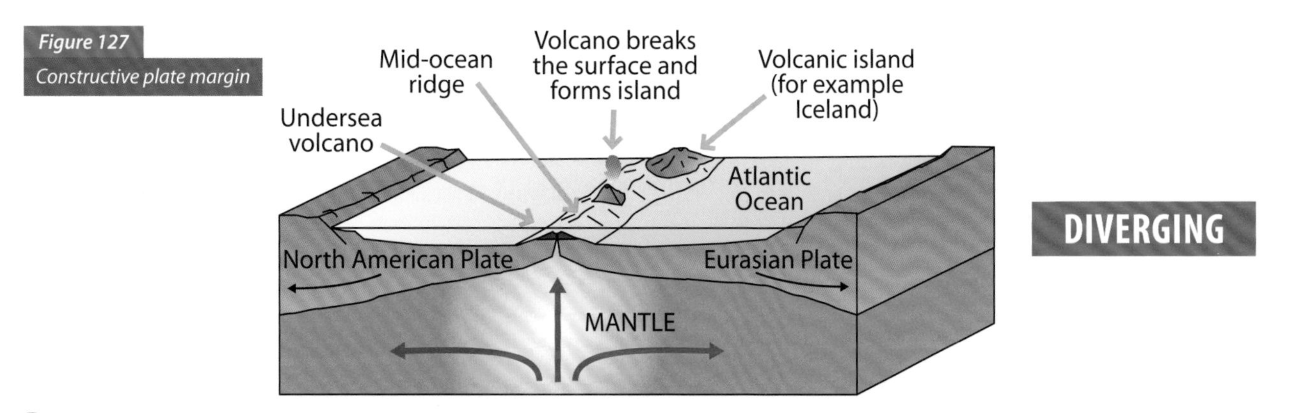

Processes:

A **constructive** plate margin is a boundary where two **plates** are moving away from each other. They are moved as the convection currents push the plates further away.

This type of movement can usually only take place under the oceans. As the two plates move away from each other, they create a gap in the seabed (which is called a mid-ocean ridge) and this gap is filled with fresh magma material from deep within the earth. Undersea shield volcanoes will be formed, some of which might continue to grow with every eruption until they break the surface and form small islands. In some cases these islands will continue to grow and will be stretched with the movement of the plates away from each other (for example, the island of Iceland is increasing in size by 2.5 cm per year). When the plates move, this can sometimes trigger some small earth tremors/earthquakes.

Landform: Mid-ocean ridge
Example: Iceland

Destructive plate margins: subduction zones and ocean trench

Figure 128
Destructive plate margin

CONVERGING

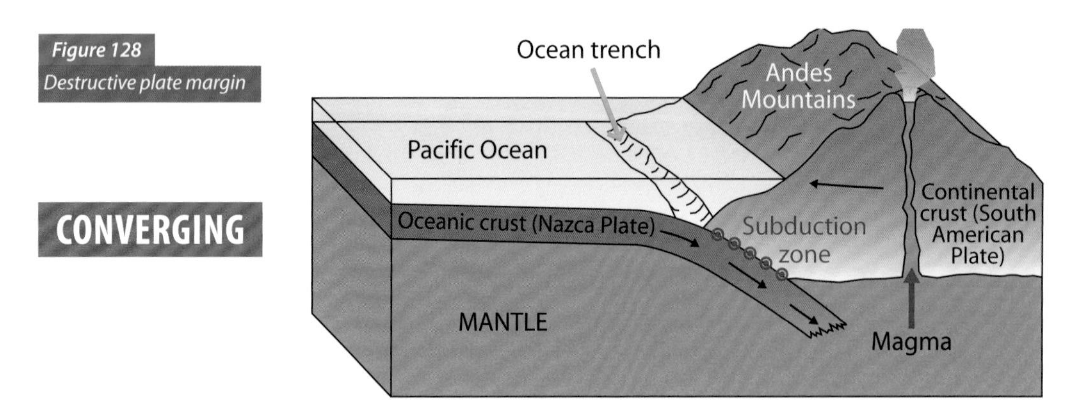

Processes:

A **destructive** plate margin is a boundary where two different types of plate are moving towards each other. They are moved due to the convection currents in the mantle. One denser and heavier **oceanic crust plate** moves towards the lighter but thicker **continental crust plate** and the **oceanic plate** is forced underneath. This creates a deep ocean trench where the two plates meet. The movement of this plate creates a **subduction** zone, where huge amounts of friction mean that earthquakes can be felt.

As the oceanic plate continues to sink, the plate is heated by the surrounding mantle and begins to melt. The increased amount of molten material (or **magma**) creates more pressure on the magma chamber of any local dormant or active volcanoes. Magma starts to rise and makes its way to the surface quickly, causing violent volcanic eruptions and **composite cone volcanoes** to be formed (for example, Nevado del Ruiz, in Colombia).

In some cases the mountain areas are pushed further up as the continental crust collides with the oceanic crust and **fold mountains** are created (for example, the Andes).

Landform: Subduction zone and ocean trench
Example: Nevado del Ruiz, Colombia and the Andes Mountains

Collision zones: fold mountains

Figure 129
Collision zones

COLLIDING

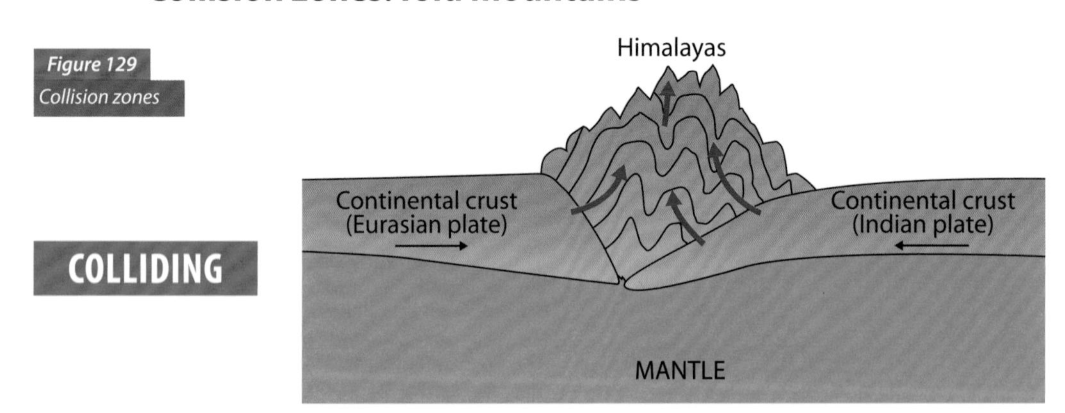

Processes:

A collision zone (plate margin) is a boundary where two **continental crust plates** are forced by convection currents in the mantle to move towards each other. The continental crust cannot be pushed down into the mantle, so the plates push each other upwards. Sometimes violent earthquakes can signal plate movement.

A good example of where this process can be seen in action is where the Indian Plate is moving into the Eurasian Plate. The resulting battle between the two plates has created an uplift of the mountains in this area, creating the highest range in the world, the Himalayas.

Usually fold mountains are made from sedimentary rock. They start when rivers deposit sediment in a depression between two large plates, in what is known as a geosyncline. As more and more sediment is laid down on the seabed, pressure is put on it and it forms sedimentary rock. The plates then move towards each other and the rock is crumpled upwards to form the folded mountains. When the rock is folded upwards it is called an **anticline** and when it is folded downwards it is a **syncline.**

Landform: Fold mountains
Example: Himalayan mountains

Figure 130
Formation of fold mountains

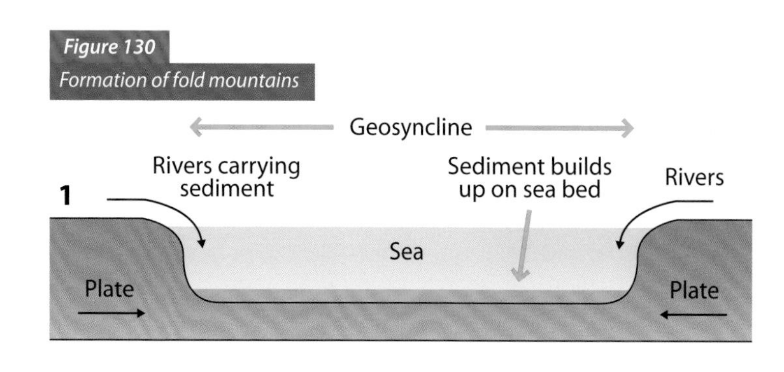

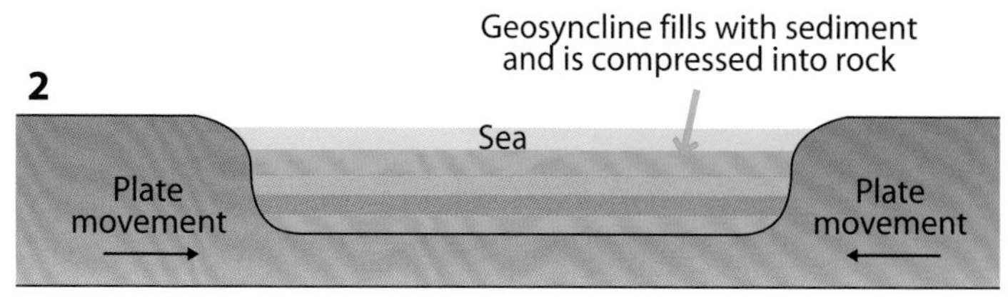

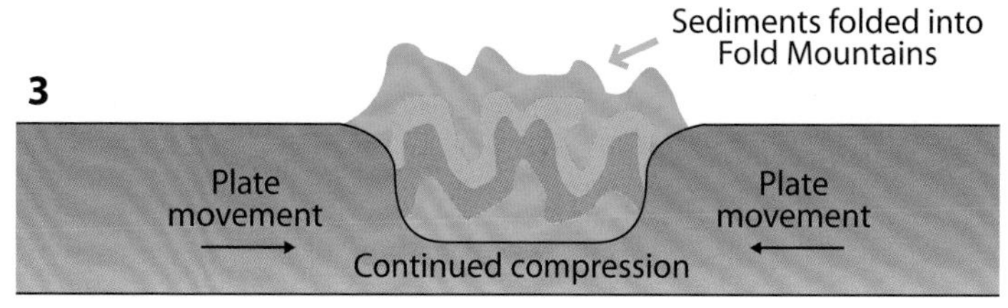

Figure 131
Evidence for fold mountains

Conservative plate margins: fault lines

Processes:

A **conservative** plate margin is a boundary where two **plates** are slipping past, rather than moving towards or away from each other. No new crust is being created and crust is not being destroyed.

In some cases the plates might be moving in two different directions (one to the north and one to the south), or the plates could be moving in the same direction, but at different rates. The plates do not pass each other smoothly. There is a huge amount of friction between the two plates, which can build up over time. When the pressure of the friction is released to allow the plate to move, the plate will be 'jerked' forward and the momentum can cause an earthquake.

One example of a place where this is happening is along the San Andreas Fault, which is a **fault line** that separates the North American Plate and the Pacific Plate. Both plates move northwards, however, the North American Plate is moving at a faster speed.

Landform: Fault lines
Example: San Andreas Fault, California

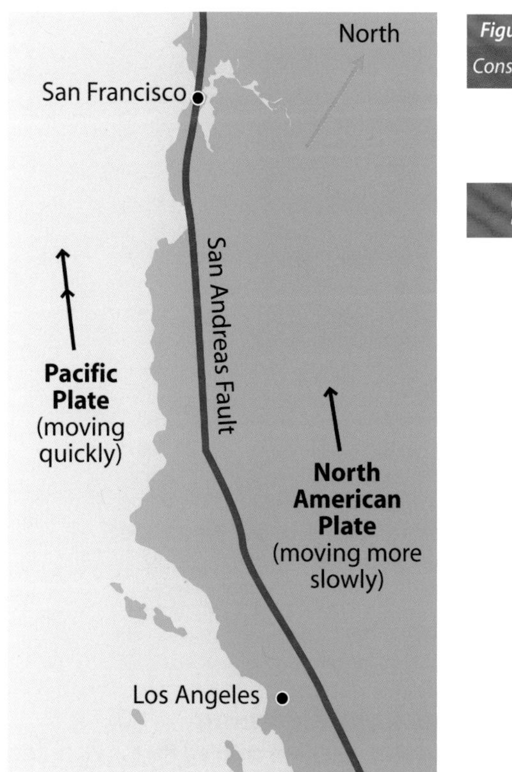

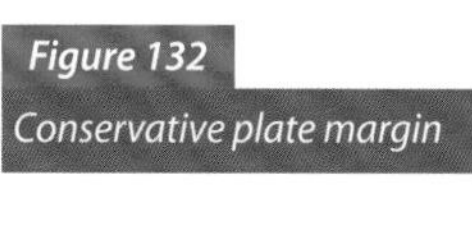
Figure 132
Conservative plate margin

U.S. Geological Survey/photo by Scott Haefner

Figure 133

Looking southeast along the San Andreas fault in the Carrizo Plain, California

Structured notes

1. Draw a diagram for each of the different plate boundaries.
2. Add the following labels to your diagram:

(a) **Constructive**
- Oceanic crust
- Eurasian Plate
- North American Plate
- Magma
- Convection currents
- Mid Atlantic ridge
- Shield volcano (for example, Iceland)
- Earth tremors

(b) **Destructive**
- Oceanic crust
- Continental crust
- Nazca Plate
- South American Plate
- Fold mountains (for example, Andes)
- Composite cone volcano (for example, Nevado del Ruiz)
- Subduction zone
- Deep ocean trench

(c) **Collision**
- Continental crust
- Indo-Australian Plate
- Eurasian Plate
- Collision zone
- Fold mountains (for example, Himalayas)

(d) **Conservative**
- Pacific Plate
- North American Plate
- Fault line (for example, San Andreas Fault)
- Plates move at different speeds

3. Below each of the diagrams write a paragraph to explain what is happening at this margin using the following words and phrases:

(a) **Constructive**
- Convection currents
- Move apart, not smooth
- Non-violent volcanic eruptions
- Small earthquakes
- New land

(b) **Destructive**
- Friction
- Jerking movement
- Violent earthquakes
- Melted oceanic crust
- Violent volcanic eruptions
- High fold mountains

(c) **Collision**
- Forced upwards
- Violent earthquakes
- No volcanic activity
- Fold mountains
- Mountains continue to grow

(d) **Conservative**
- Friction
- Different speeds of movement
- Plates get stuck
- Sudden jerking movement
- Violent earthquakes
- No volcanoes

4. Write a definition for the following words/phrases:
- Plate
- Zone of activity
- The 'Ring of Fire'

Check your learning

Now that you have studied Part 2: Plate Tectonics Theory, return to page 108 and answer the Key Questions for this section.

PART 3: TECTONIC ACTIVITY IN THE BRITISH ISLES

(a)

a. Identify landscape features created by tectonic activity and explain their formation
Lava Plateau (for example, Antrim Plateau)
Basalt Columns (for example, Giant's Causeway)
Volcanic Plugs (for example, Slemish Mountain)

(b)

Describe and explain the causes and Impact of an earthquake in the British Isles
Case Study: Market Rasen, 2008

Key questions

By the end of this section you will be able to answer the following questions:

1. Describe and explain the formation of basalt columns.
2. Show how a lava plateau can be formed.
3. Explain how volcanic plugs like Slemish Mountain can be found within Northern Ireland.
4. Explain the causes and the main impacts of an earthquake which has happened in the British Isles?

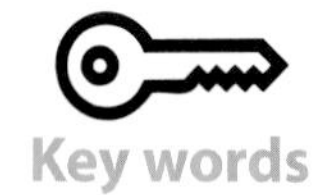

Key words

Volcano
Volcanic plug
Lava plateau

Identify landscape features created by tectonic activity and explain their formation

The British Isles as a whole and the landscape of Northern Ireland in particular (Figure 134) have been shaped by the tectonic processes underneath the surface of the earth. These features give us a glimpse of the different ways that our landscape was put together during a very different time. At one stage Ireland sat on the edge of the Eurasian Plate but continued activity at this constructive plate boundary gradually pushed Ireland further away from the North American plate. The newly created volcanic material on the seabed then pushed Ireland even further away from the zone of activity. This means that today there is very little tectonic activity taking place within Northern Ireland, and the volcanic features are extinct.

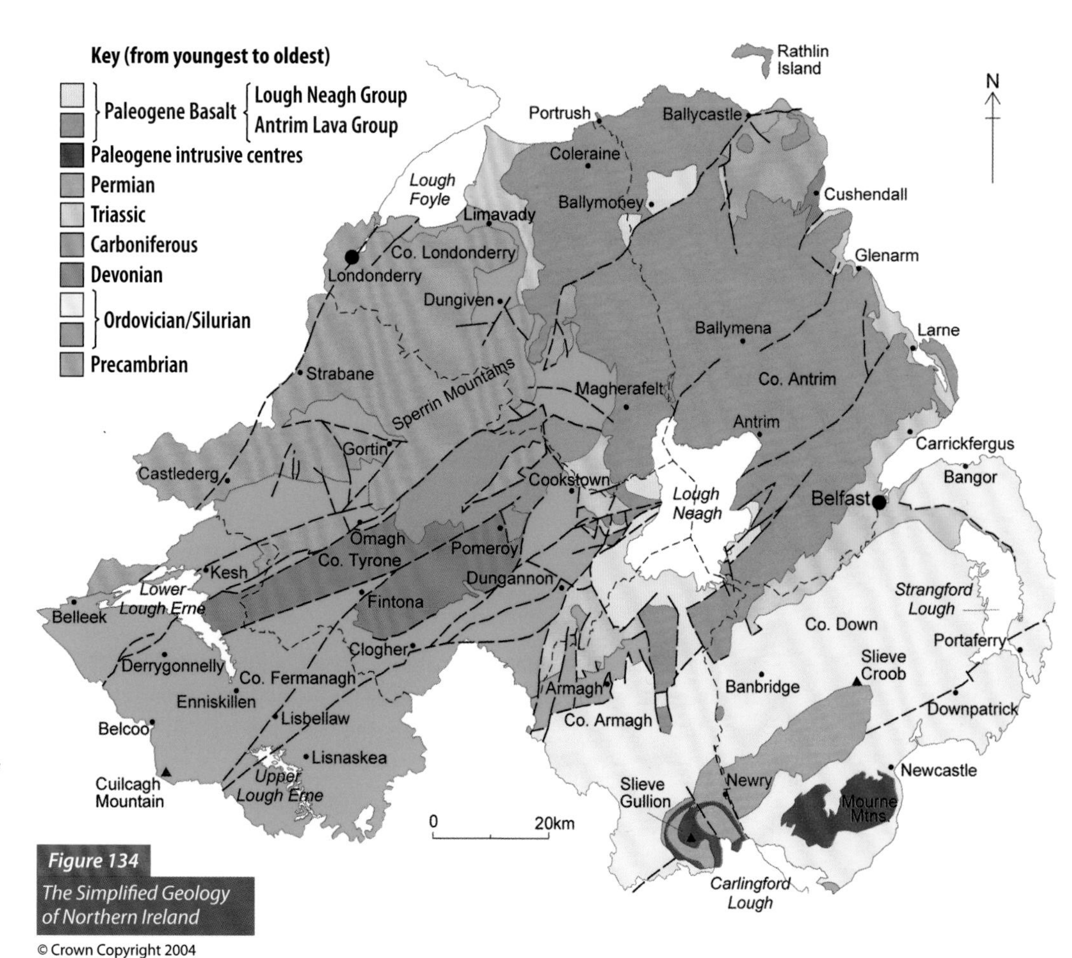

Figure 134
The Simplified Geology of Northern Ireland

Each of the following landscapes was formed at a time when there was a lot more going on underneath the surface of the earth. As Ireland sat at the edge of a constructive plate margin, this meant that volcanoes and earthquakes were common features.

Lava plateau (for example, Antrim Plateau)

Figure 134 shows the simplified geological features that cover Northern Ireland. Most of the tectonic features that we study are found in what geologists call the Permian to Palaeogene age, within what is called the 'Antrim lava group'.

The Antrim lava group is what has formed the flat lava plateau known as the Antrim Plateau. The plateau was formed by a series of basalt lava flows that built up on top of each other, creating an extensive extrusive feature. (Extrusive means that the feature is formed above the earth's surface, when magma flows out onto the surface as lava and igneous rock remains.)

Figure 135
The formation of the Antrim Plateau

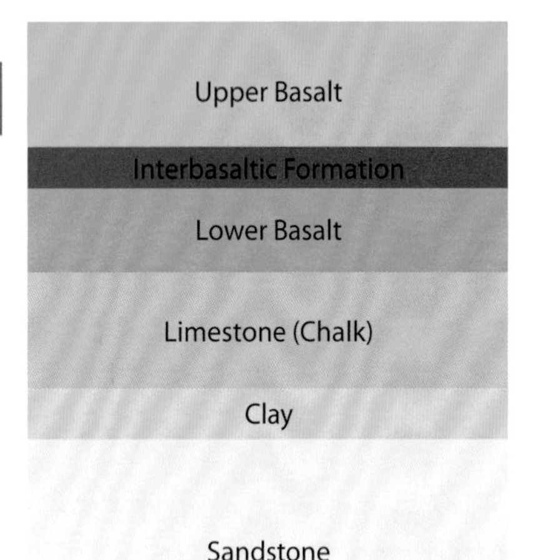

1. Before the activity that created the volcanic features, the bedrock was made up of old layers of sandstone, which had been covered with clay and limestone (chalk).
2. The first stages of volcanic activity were explosive and produced huge amounts of dust and fragmented rock. This became what is known as the lower basalt. Following this, further volcanic activity brought more lava flows, and the lava came through cracks and fissures in the ground (which can today be seen as dykes and sills). Many small ash cone volcanoes (at Carrickarede and Tardree Hill, near Antrim) spewed huge amounts of lava.

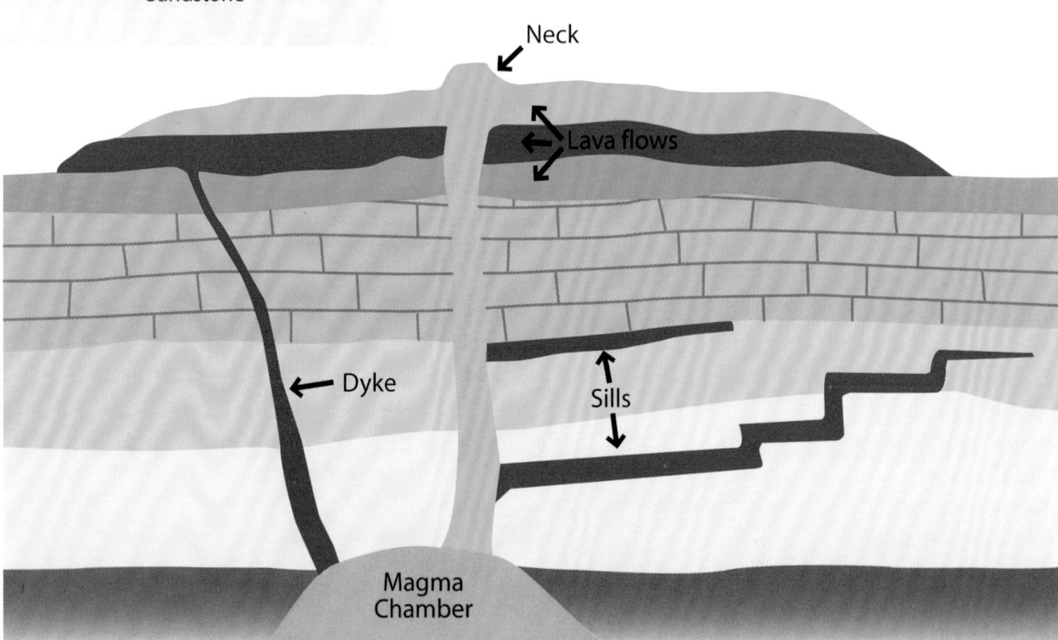

3. Following a period of relative inactivity (which allowed some of the basalt to be weathered and broken down into red, iron-rich soils) new activity (called the Interbasaltic formation) caused localised eruptions, which created features such as the Giant's Causeway (see Figure 137).
4. The final period of tectonic activity brought further lava flows to the surface through fissures that covered the land and formed the cap of the Antrim Plateau.

Three major flows of volcanic activity created the Antrim Plateau and allowed it to spread as far north as the North Coast and as far south as the edge of Belfast (to Cave Hill and Carnmoney Hill).

Basalt columns (for example, Giant's Causeway)

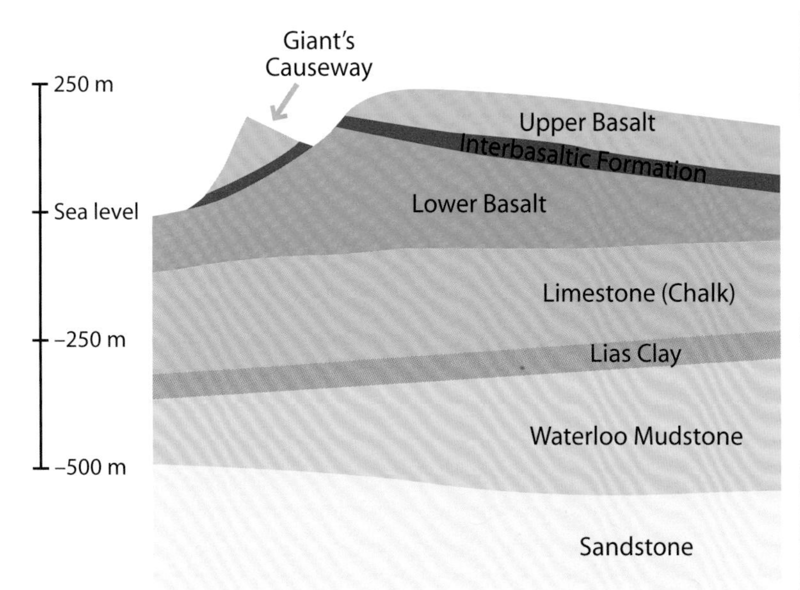

Figure 136 The rock structure near the Giant's Causeway

Figure 137 The iconic 'Middle Causeway' that forms part of the Giant's Causeway

The Giant's Causeway is a great example of a spectacular extrusive feature which is caused when the cooling conditions for basalt are perfect. Geologists believe that columns occurred when lava erupted (probably from the Carrickarede fissure) into an ancient river valley where the lava became trapped and ponded, and cooled very slowly. While the thick lava flows were cooling, some of the surface areas were flooded with water to form temporary lakes. Some of this floodwater could have been from heavy rainfall or from rivers.

As the lava started to cool cracks started to appear on the surface, similar to what happens when mud dries in the sun. The lava lost heat upwards into the atmosphere and downwards into the ground, causing the column shapes to extend many metres.

Therefore the hexagonal columns associated with the Giant's Causeway were a result of perfect cooling conditions. Tension caused by different rates of cooling and shrinking within the column split the column into regular tablets of stone, with curved joints that fitted together in a ball and socket arrangement. The hexagonal shape was the most efficient way to fill a space.

Figure 138 The hexagonal stones

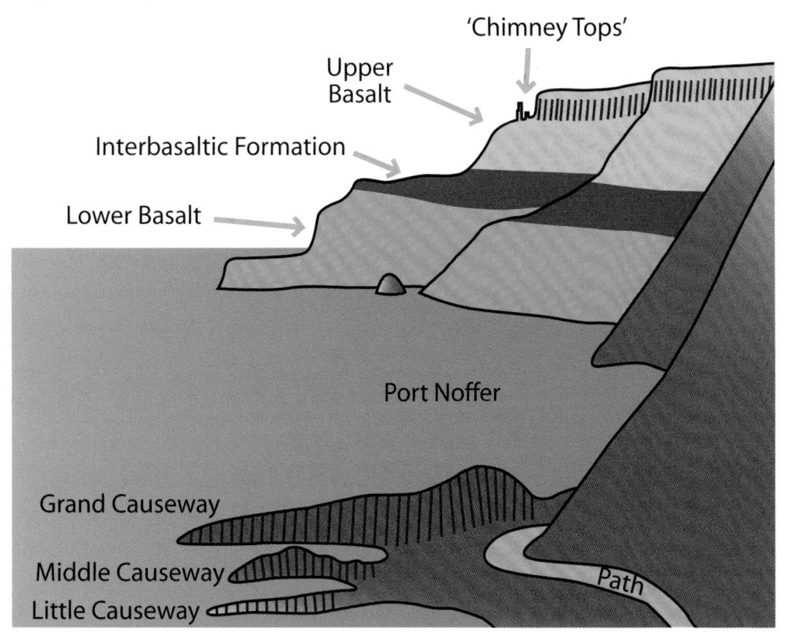

Figure 139 Sketch map (left) and photograph (below) showing the headland at Port Noffer and the different lava flows

Volcanic plugs (for example, Slemish Mountain)

There are a number of volcanic plugs scattered across the Antrim Plateau including Scawt Hill, Ballygally Head, Carrickarede and Carnmoney Hill. However, the most famous is probably Slemish Mountain.

It is one of the largest of the dolerite plugs in the Antrim Plateau. Dolerite is similar to basalt but it is cooled a little more slowly. It rises to 437 m above sea level on the south side of the Braid Valley, about 4 miles from Broughshane.

Figure 140 *The formation of Slemish*

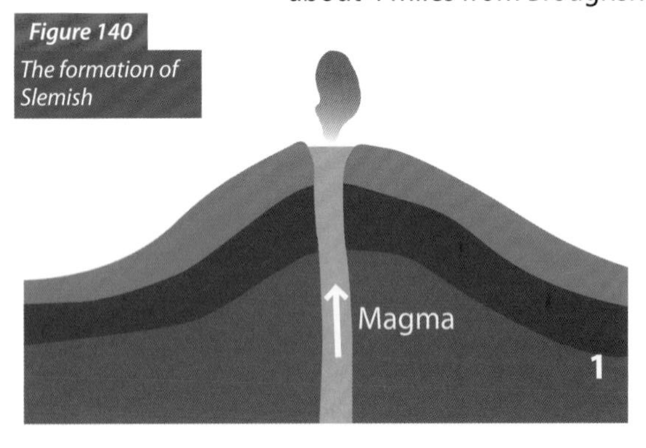

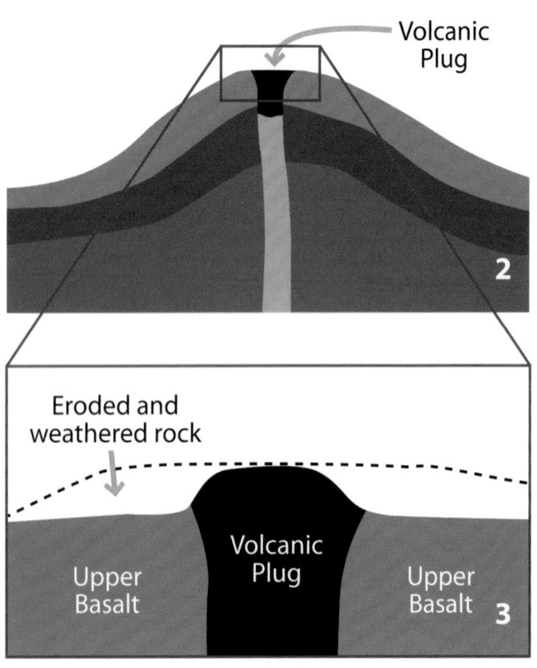

1. Volcanic plugs are intrusive features. (Intrusive means that the feature occurs below the earth's surface, when the magma does not reach the surface.) They are formed when magma is forced up from under the crust to move up inside the vent of a volcano. Magma pours through the vent of the Slemish volcano as lava and covers the surrounding area with lava, which hardens into basalt.
2. As volcanic activity ends the vent is blocked by a lump of hardened magma, which forms rock called dolerite. Dolerite is harder and more resistant to erosion than basalt.
3. Over time rain, ice and wind weathers away the softer basalt slopes of the volcano, leaving the harder dolerite exposed above the flat landscapes of the Antrim Plateau.

Figure 141 *Slemish*

Describe and explain the causes and impact of an earthquake in the British Isles

The British Isles are no longer positioned on the edge of a plate boundary, which means that there is much less happening under our feet. The earthquakes and volcanic eruptions which might have taken place to shape our landscape no longer happen. Any earthquakes which do happen are more to do with old, shallow fault lines settling in the rocks, rather than deep tectonic activity. Any earthquakes that do take place in the British Isles are usually located on the western side of the British mainland. Ireland is almost completely free of earthquakes.

The largest known British earthquake took place at Dogger Bank in 1931, when a magnitude 6.1 earthquake took place. However, the most damaging UK earthquake was in Colchester in 1884 when 1,200 buildings needed repaired, chimneys collapsed and walls cracked.

Case Study: Market Rasen, 2008

The Market Rasen earthquake, 27 February 2008 at 0056

One of the biggest earthquakes to be measured in recent years was located about 2 ½ miles north of Market Rasen in Lincolnshire, England on 27 February 2008. The earthquake was measured at a magnitude of 5.2 on the Richter scale and at a depth of 12 miles. The earthquake itself was estimated to have lasted around 10 seconds.

The Causes

The cause of this earthquake was nothing to do with the movement of major plate boundaries, as Market Rasen is nowhere near the edge of a plate boundary. However, Intraplate earthquakes can happen, and in this case the tectonic pressures from the North Atlantic ridge and the African plate caused stress which led to a sudden rupture along the strike slip fault line in the local area. Some geologists even think that the fault line that runs from Doncaster to Lincoln might be the reason for this movement.

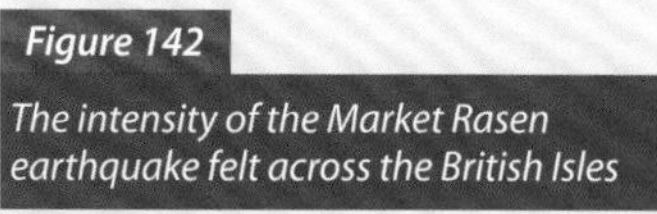

Figure 142 The intensity of the Market Rasen earthquake felt across the British Isles

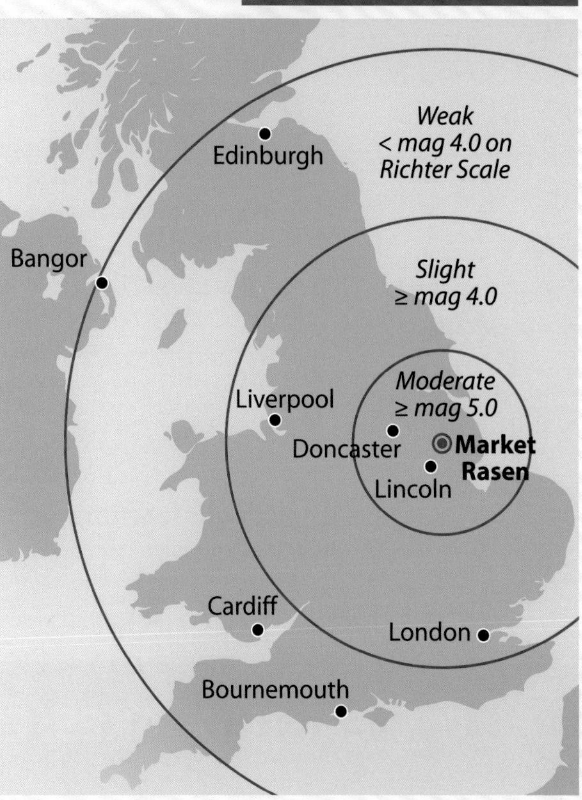

Figure 143 Damage caused at a strike slip fault (not Market Rasen)

The impact

Compared with some of the earthquakes discussed later in this book, the impact of this earthquake was extremely minor.

- Some buildings were reported to shake for up to 30 seconds.
- There were 15 small after shocks recorded (between 1 and 2 on the Richter scale) over the following weeks and months.
- It was the biggest earthquake to hit the UK in 24 years.
- There was some structural damage to local homes and businesses – roofs collapsed and chimney pots fell from a few houses. Firefighters attended three properties to make chimneys safe.
- Birds and pets became very agitated and 77 hens died in Pete Sargent's shed, which was located directly above the tremor.
- There were no deaths, but one man, 19 year old David Bates, suffered a fractured pelvis when a chimney smashed through the roof and landed on him as he lay in bed. A stone cross had fallen from St Thomas's church, damaging the church roof and dislodging tiles before smashing onto the ground.

Figure 144 The Market Rasen earthquake caused minimal damage to people and property

- The Spire at St Mary Magdalene church at Waltham, Leicestershire was also damaged, costing around £100,000 to repair.
- The estimated cost of damage to homes was around £10 million.
- The earthquake was felt as far away as Bangor, Northern Ireland and Bournemouth.
- Emergency services all over the north of England were called by people who had woken up to find their house shaking (5,000 calls in one hour).

Test yourself

1. Describe and explain how lava is formed into a plateau on the Antrim Plateau. (6)
2. Describe the processes that cause the formation of columnar basalt such as the Giant's Causeway. (4)
3. What is a volcanic plug and how is it created? (3)
4. What caused the earthquake at Market Rasen in 2008? (4)
5. What were some of the main impacts? (5)
6. Why do many people consider Market Rasen to be a minor earthquake event? (3)

Check your learning

Now that you have studied Part 3: Plate Tectonics in the British Isles, return to page 115 and answer the Key Questions for this section.

PART 4: EARTHQUAKES: CAN THEY BE MANAGED?

a

The global distribution and causes of earthquakes in relation to plate boundaries

The physical consequences of earthquakes
Liquefaction
Tsunami

The causes and impacts of earthquakes and evaluation of the management responses
MEDC Case Study:
The Great Tohoku earthquake (Japan, 11 March 2011)
LEDC Case Study:
The Haiti earthquake (Haiti, 12 January 2010)
Identify the plates involved
Describe the short and long term impacts on people and the environment
Evaluate the management response to the earthquake (including prediction and/or precautions before the event and immediate and long term strategies implemented after the event).

Key questions

By the end of this section you will be able to answer the following questions:

1. Where do earthquakes occur around the world?
2. Explain what liquefaction is.
3. Explain what a tsunami is and describe how it works.
4. Describe the cause of an earthquake in an LEDC.
5. Describe the cause of an earthquake in an MEDC.
6. Explain the differences between the impact of an earthquake in an LEDC compared with the impact in a MEDC.

Key words

Earthquake
Richter scale
Seismograph
Epicentre
Focus
Liquefaction
Tsunami

The global distribution and causes of earthquakes in relation to plate boundaries

Figure 145 *Features of an earthquake*

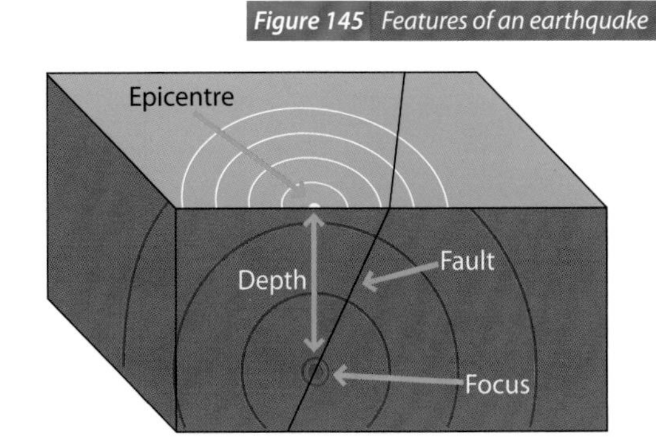

An earthquake is described as a "fault rupture that generates seismic waves". This occurs when rocks on either side of a weakness in the earth's crust (a fault) causes the ground to vibrate and shake. Earthquakes are very powerful natural hazards.

When a movement takes place deep inside the earth, the vibrations called seismic waves travel from the **focus** (the place where the earthquake originally occurs) and then from there to the surface. The **epicentre** is the place on the earth's surface which is the shortest distance from the focus. This is the place where the intensity or the magnitude of the energy released is felt the most.

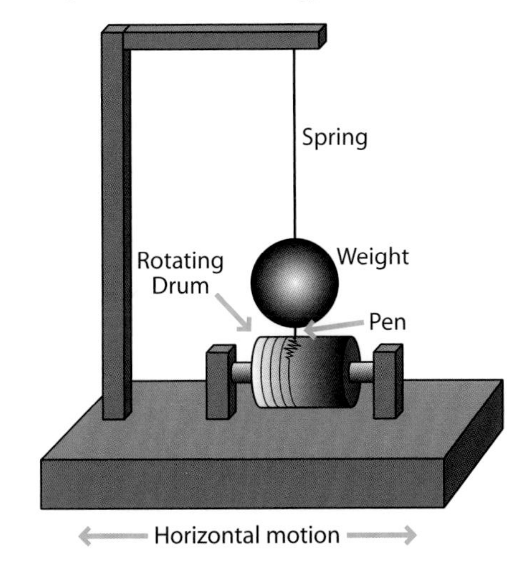

These seismic waves can be recorded using a **seismograph**. Generally this consists of a weight with a pen attached, which is suspended from a spring. During an earthquake event the base of the seismometer will move horizontally, and the motion is converted into electrical voltage and recorded onto paper.

The strength of an earthquake is usually referred to as its magnitude. A number of magnitude scales have been developed to help us understand the amount of power released during an earthquake event. However, the most common scale that is used around the world was developed in 1935 by Charles Richter. The Richter scale is a logarithmic scale, which means that an earthquake that measures a magnitude 6.0 earthquake will be 10 times greater than a 5.0 magnitude earthquake.

Figure 146 *A seismograph used to measure earthquakes*

HOW THE RICHTER SCALE CAN BE USED TO MEASURE EARTHQUAKES			
MAGNITUDE	NO PER YEAR	DESCRIPTION	EFFECTS
Less than 2.0	Continual	Micro	Rarely felt, only detected by seismographs.
2.0–2.9	1.3 million	Minor	
3.0–3.9	130,000		Sometimes vibrations felt, but rarely causes damage.
4.0–4.9	13,000	Light	Some rattling. Some shaking of indoor items. Might break windows or cause unstable objects to fall.
5.0–5.9	1,319	Moderate	Can cause major damage to poorly built buildings. Slight damage to well designed buildings. Furniture will be moved.
6.0–6.9	134	Strong	Can be destructive in urban areas.
7.0–7.9	15	Major	Can cause serious damage over a large area. Buildings can be moved from foundations. Cracks in the earth. Pipes break.
8.0–8.9	1	Great	Causes damage over a very wide area. Buildings are destroyed and few structures are left standing.
9.0–9.9	1 every 10 years		Devastating over an extremely wide area, near total destruction.
More than 10	Very rare	Massive	Never recorded but impact would be catastrophic.

Although we have seen that small earthquakes can occur at any time and any place, there is a clear pattern for the distribution of the major earthquakes that have been recorded around the world. We can identify that major earthquakes occur in zones of activity, which coincide with the plate boundaries that are spread across the surface of the earth (Figure 124, page 110).

Test yourself

1. On a copy of a blank world map, use the resource below to locate and plot the main earthquake events that occurred in 2010 and 2011. (½ mark each = 15)

Resource A

2010			2011		
Date	**Magnitude**	**Region**	**Date**	**Magnitude**	**Region**
3/1/10	7.1	Solomon Islands	1/1/11	7.0	Santiago del Estero, Argentina
12/1/10	7.0	Haiti	2/1/11	7.2	Araucania, Chile
26/2/10	7.0	Ryukyu Islands, Japan	13/1/11	7.0	Loyalty Islands
27/2/10	8.8	Bio-Bio, China	18/1/11	7.2	SW Pakistan
27/2/10	7.4	Bio-Bio, China	9/3/11	7.3	NE Honshu, Japan
11/3/10	7.0	Libertador, Chile	11/3/11	9.0	NE Honshu, Japan
4/4/10	7.2	Baja California, Mexico	11/3/11	7.9	NE Honshu, Japan
6/4/10	7.8	N Sumatra, Indonesia	11/3/11	7.7	NE Honshu, Japan
9/5/10	7.2	N Sumatra, Indonesia	7/4/11	7.1	NE Honshu, Japan
27/5/10	7.1	Vanuatu	24/6/11	7.3	Fox Islands, Aleutian Islands
12/6/10	7.5	Nicobar Islands, India	6/7/11	7.6	Kermadec Islands
16/6/10	7.0	Papua, Indonesia	10/7/11	7.0	NE Honshu, Japan
18/7/10	7.3	New Britain, Papua new Guinea	20/8/11	7.2	Vanuatu
23/7/10	7.3	Mindanao, Philippines	20/8/11	7.1	Vanuatu
23/7/10	7.3	Mindanao, Philippines	24/8/11	7.0	N Peru
23/7/10	7.6	Mindanao, Philippines	3/9/11	7.0	Vanuatu
4/8/10	7.4	Mindanao, Philippines	15/9/11	7.3	Fiji region
10/8/10	7.3	Vanuatu	21/10/11	7.4	Kermadec Islands
12/8/10	7.1	Ecuador	23/10/11	7.1	E Turkey
3/9/10	7.0	South Island, New Zealand	14/12/11	7.1	E New Guinea, Papua New Guinea
29/9/10	7.0	S Papua, Indonesia			
25/10/10	7.8	Kepulauan, Indonesia			
21/12/10	7.4	Bonin Islands, Japan			
25/12/10	7.3	Vanuatu			

Source: Figures USGS, http://earthquake.usgs.gov/earthquakes/eqarchives/year/2011/

2. How would you describe the location of the major earthquakes (more than magnitude 7.0)? (5)
3. How does the pattern of earthquake activity compare to the distribution pattern shown on the map in Figure 124 on page 110? (4)

The physical consequences of earthquakes

Thankfully the majority of earthquakes that occur around the world are too small to cause any damage. In addition, many take place deep in the Pacific Ocean or in very remote areas where there are very few people living.

It is often said that it is not actually earthquakes that kill people, rather it is the intensive shaking of the ground and its affect on buildings that caused damage.

Two impacts that occur when an earthquake strikes are liquefaction and tsunamis. However, their side effects on people are quite different.

Liquefaction

Liquefaction occurs when an earthquake hits an area and shakes the wet soil. The shaking causes the water within the soil to start to rise to the surface, and this process turns solid soil and rock into a liquid mud. Buildings will start to sink and tip over as the support for the foundations is waterlogged and cannot maintain the weight of the buildings. For example, in 1964, in the city of Niigata in Japan, an earthquake of magnitude 7.5 caused some reinforced concrete buildings to sink into the soil.

NOAA National Geophysical Data Center

Figure 147
Apartment houses in Niigata, Japan damaged due to liquefaction in 1964

Tsunami

A tsunami is a large wave which is created when an underwater earthquake sends shockwaves through the water, causing a surge of water to move towards the coastline. Often the energy transferred due to a tsunami can travel for thousands of miles across the oceans.

Figure 148 *Tsunami warning sign*

Figure 149 *How a tsunami happens*

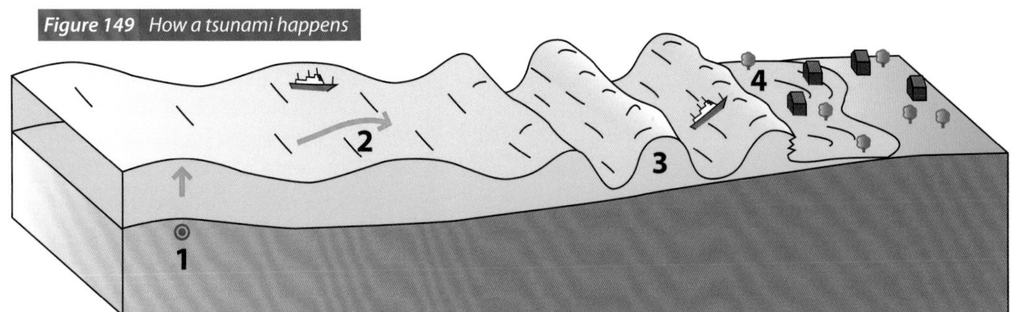

1. A rupture in the sea floor pushes water upwards and starts the waves moving.
2. The waves move rapidly across the deep ocean, reaching speeds of up to 500 km/h.
3. As the waves near land, they slow to 45 km/h but get squeezed upwards by the sloping beach and the waves starts to increase in height.
4. The waves climb to 10–40 metres in height and move inland, destroying everything in their way.

The causes and impacts of earthquakes and evaluation of the management responses

MEDC Case Study: The Great Tohoku earthquake (Japan, 11 March 2011)

CASE STUDY

The US Geological Survey (USGS) reported that in 2011 there were 20 magnitude 7.0 or larger earthquakes. Of the estimated 21,400 people who died due to earthquakes in 2011, over 20,352 of these were as a result of the earthquake which took place at Great Tohoku (northeast Honshu), Japan on 11 March. It is estimated that over 5,314 people were injured and 130,927 were displaced with 332,395 buildings, 2,126 roads, 56 bridges and 26 railways being destroyed or damaged by the earthquake or the tsunami which followed.

U.S. Marine Corps photo by Lance Cpl. Garry Welch/Released

Figure 150
A large ferry is stranded inland following the tsunami at Aichi, Japan

Causes

The Great Tohoku earthquake was a magnitude 9.0 undersea earthquake, which took place just off the coast of Japan at 14:46 local time on Friday 11 March 2011. The epicentre of the earthquake was approximately 43 miles east of Tohoku, at a focus depth of about 20 miles. The earthquake lasted for six minutes and also caused a massive tsunami, which reached heights of 40.5 m in Miyako and travelled about six miles inland at Sendai.

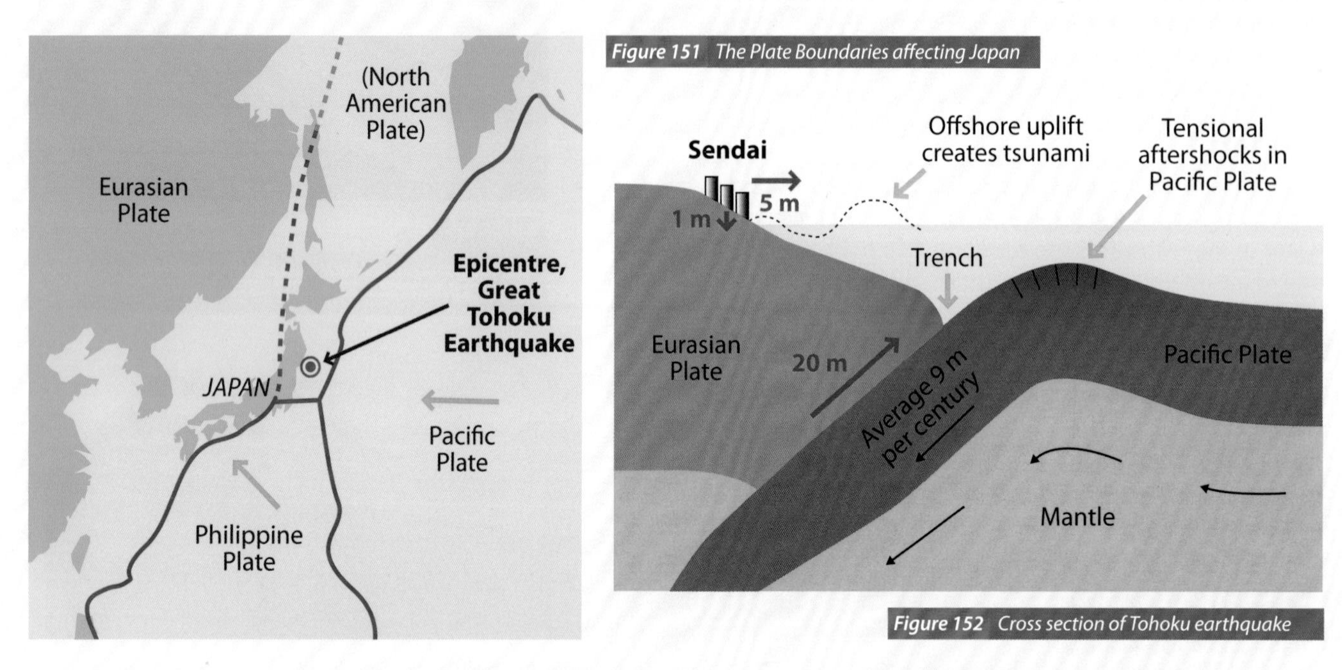

Figure 151 The Plate Boundaries affecting Japan

Figure 152 Cross section of Tohoku earthquake

Japan lies in one of the most active tectonic regions in the world. Both the Philippine Plate and the Pacific plate are moving towards the Eurasian/North American Plate at a fairly fast rate (about 8 cm per year). This is a classic destructive plate boundary, where a subduction zone has formed as the thin oceanic plates are moving in to clash with the thick continental plate. A huge amount of friction was built up over time, which led to old, cold oceanic crust from the Pacific plate being subducted beneath north Honshu and leading to what are known as megathrust earthquakes.

The amount of energy released during the earthquake was estimated as being equivalent to 600 million times the energy of the Hiroshima bomb (and nearly double that of the 9.1 magnitude, 2004 Indian Ocean earthquake and tsunami).

Figure 153 Aerial view of damage to Wakuya, Northern Japan following the earthquake

U.S. Navy photo by Mass Communication Specialist 3rd Class Alexander Tidd/Released

Short term impacts (on people and the environment)

IMPACTS ON PEOPLE	IMPACTS ON THE ENVIRONMENT
Death and injury – An estimated 20,352 died, 5,314 people were injured and 130,927 were displaced. **Nuclear crisis** – Although the Fukushima nuclear plant was protected with a 5 m high tsunami barrier, a 9 m high wave came ashore and flooded the plant's generators and electrical wiring. People lost energy immediately and this took some time to restore. **Defences ineffective** – Japan had spent billions of dollars building anti-tsunami defences at heights of 12 m. However, the tsunami just washed over them, rendering them totally ineffective. The flood waters in some cases moved six miles inland, destroying houses, factories, roads and other buildings. **Damage** – 332,395 buildings, 2,126 roads, 56 bridges and 26 railways were destroyed or damaged by the earthquake or the tsunami which followed. 300 hospitals were damaged and 11 were totally destroyed. An estimated 23,000 cars and trucks were damaged or destroyed. Ports were closed for nearly three weeks. One dam ruptured and another six were found to have cracks. **Power supplies** – Around 4.4 million households in north east Japan were left without electricity. Power blackouts were experienced for around three months in many areas.	**Fore and aftershocks** – The earthquake was preceded by a number of foreshocks (one measuring 7.2 on 9 March) and hundreds of aftershocks. Three aftershocks measuring more than magnitude 6 took place on the same day and another three measuring more than magnitude 7. Some scientists estimate that over 800 earthquakes of magnitude 4.5 or more have been recorded since the main quake. Each quake was caused by plate movement, which created further tears and fissures in the ground, and damage across Japan. **Tsunami** – The earthquake triggered a tsunami (which hit about 30 minutes following the earthquake) up to 40 m high in Japan. It devastated entire towns and resulted in the loss of thousands of lives. However, waves also travelled to the east. Tsunami warnings were issued right across the Pacific Ocean and were felt in Alaska and Chile (over 11,000 miles away but measured 2 m tsunami waves). Flooding caused huge amounts of damage. Pollution carried in the water settled far inland and sources of clean drinking water were affected in some cases for up to a year. **Land subsidence** – some coastal areas in Japan experienced land subsidence as the earthquake dropped the beachfronts in some places by more than 50 cm. This made the areas much more susceptible to flooding.

Long term impacts (on people and the environment)

IMPACTS ON PEOPLE	IMPACTS ON THE ENVIRONMENT
Economy – The World Bank estimates that the economic cost of this event was US$235 billion, making this the most expensive natural disaster in world history. They reckon it will take Japan five years to rebuild. **Tsunami** – A Japanese government study found that only 58% of people in the coastal areas that were affected by the tsunami heeded the tsunami warnings immediately after the earthquake and headed for higher ground. Only 5% of those who attempted to evacuate were caught in the tsunami, whereas the water hit 49% of those who did not heed the warning. **Further casualties** – Three members of the Japan ground self-defence force died whilst conducting relief operations. The Japanese government estimate that a further 922 people have died as a result of the harsh living conditions since the earthquake. **Nuclear energy** – The damage caused by the earthquake resulted in the meltdown of seven reactors, which led to the evacuation of the local area and a 12 mile exclusion zone was set up (affecting 200,000 local residents). Radiation levels at one point were over eight times normal levels. Radioactive water and leaks contaminated local areas, with radioactive soil and hotspots being found as far south as Tokyo. Food products were also contaminated, including some fish and beef. The amount of damage to nuclear safety raised safety concerns both in Japan and around the world regarding nuclear energy. Protests about the use of nuclear power in Japan were organised in Tokyo on 27 March 2011 and all Japanese nuclear reactors were taken off-line until June 2012. **Transport** – Japan's transport network suffered huge disruptions. Sections of the Tohoku Expressway were damaged, Sendai airport was hit by the tsunami wave, four trains were derailed and over 1,100 sections of rail line needed to be repaired. **Rebuilding** – A panel called the 'Japan move forward committee' proposed that young adults and teenagers could help rebuild parts of Japan devastated by the earthquake.	**Landmass movement** – The quake moved parts of North East Japan 2.4 m closer to North America, making parts of the Japanese island landmass wider than before. **Coastline movement** – A 250 mile stretch of coastline dropped by 0.6 m, allowing the tsunami to travel further inland. **Plate movement** – Some people estimate that the Pacific plate has slipped westwards by between 20 and 40 m. **Seabed movement** – The seabed near the epicentre shifted by 24 m and the seabed off the coast of the Miyagi province has moved by 3 m. **Earth axis shift** – The earthquake shifted the earth's axis by between 10 and 25 cm, shortening the day by 1.8 microseconds. **Liquefaction** – Liquefaction occurred in many of the parts of Tokyo built on reclaimed land. Around 30 homes were destroyed and 1,046 buildings were damaged by this process alone. **Aftershocks** – Japan has experienced over 900 aftershocks since the earthquake, some of which were over magnitude 7, killed people, caused damage to electricity supplies and damage to the Higashidori nuclear power plant. **Antarctica** – In Antarctica the seismic waves from the earthquake were reported to have caused some massive slabs of ice (one about the size of Manhattan Island) to fall from the Sulzberger Ice Shelf.

Figure 154

Land subsidence and liquefaction caused by the earthquake

Evaluation of the management response

As a country with a long tectonic past, the Japanese have learnt how to live with the threat of earthquakes. They have spent many billions of dollars on attempting to keep buildings and people safe in the event of both earthquakes and tsunamis. In this case, both the earthquake and the tsunami were too powerful for the preparations and very quickly the Japanese emergency services were stretched beyond breaking point.

Prediction and precautions BEFORE the earthquake

1. **Earthquake prediction** – Japan has spent over £70 million trying to predict earthquakes before they happen. They use lasers to measure possible earth movements. The scientists thought they could predict the location of the next Japanese quake but believed that Tohoku was not in the danger zone.
2. **National Disaster Prevention Day** – every year on 1 September (the anniversary of the Great Kanto earthquake, 1923) the Japanese government holds earthquake and tsunami drills to make sure that the emergency and rescue services are ready for any disaster.
3. **Earthquake-proof buildings** – Japan spends billions of pounds on buildings that are designed to be resistant to the effects of earthquakes. Buildings built since 1981 often use a range of techniques which make them more flexible and less affected by liquefaction: for example shock absorbers, flexible steel frames and deep foundations.
4. **Early warning systems** – Japan also has the world's most sophisticated earthquake and tsunami early warning systems. The tsunami warning system was set up in 1952 with 300 sensors. Tsunami safety has been a focus for coastal city planning throughout the nation. Hundreds of earthquake and tsunami proof shelters have been built and some cities have built tsunami walls and floodgates to stop waves from travelling inland through river systems. In Tokyo, the Earthquake Early Warning system, which fetches data from over 1,000 seismographs across Japan, sent out warnings of impending strong shaking to millions during the Tohoku earthquake. The Japanese Metrological Agency believe that this slight warning saved many lives.

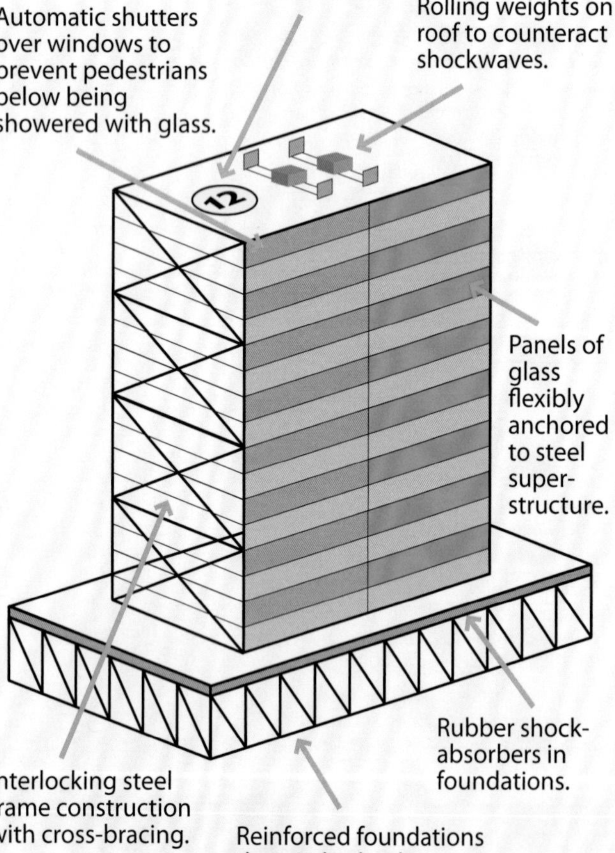

Figure 155
Various strategies for making buildings 'earthquake-proof'

The big question is, how effective have these measures been in the face of a massive earthquake and tsunami? Most of the buildings are built to withstand tremors up to magnitude 7 and tsunami waves of 12 m, and anything more than this would put the preparations under severe pressure. The magnitude 9 earthquake and waves of 40 m severely tested even this, the most prepared of nations.

Immediate and long term strategies AFTER the earthquake

1. **Refugees** – The earthquake created over 300,000 refugees and resulted in shortages of food, water, shelter, medicine and fuel for survivors. By the end of July 2011, 87,000 people were still evacuated and roughly 46,000 were living in temporary housing. The Red Cross reported that by August 2011, 52,358 prefabricated houses had been completed in the three worst affected areas (Fukushima, Miyagi and Iwate). Many of the temporary shelters were built by the army very quickly following the earthquake, which helped people survive the cold nights that followed. The spread of disease and starvation were not major problems. Even with huge amounts of surface flood water there were not huge problems with water-borne diseases such as cholera.

2. **Rebuilding** – The Japanese government set up a Reconstruction Design Council who were determined to bring 'Hope beyond the despair'. The government announced that more than 23 trillion Yen would be made available over the next 10 years to aid rebuilding programmes. Sendai airport, which was badly damaged by the tsunami, was rebuilt and reopened in June 2011. New buildings have to conform to rigorous standards (last updated in 2007) and must be earthquake proof. Small buildings must have frames made of concrete and need to be reinforced. Houses are not to be made of wood.

Figure 156

Prefabricated housing going up at Rikuzentakata

3. **Evaluating the emergency response** – Many Japanese schools are designated as evacuation centres in times of disaster, but the reality is that many are poorly equipped. However, since the March 2011 earthquake, new measures are being taken to upgrade the centres to help save and sustain life in the next big emergency.
4. **Tsunami barriers** – New 18 m high barriers are being built around some of the remaining nuclear power stations.
5. **Social factors** – Many festivals went on as normal during the summer months in Japan in an effort to raise spirits. The Japanese women's football team were surprisingly successful in the 2011 World Cup tournament and Tokyo hopes that a bid for the 2020 Olympics will help to serve as a symbol of Japan's recovery from the 11 March earthquake and tsunami.
6. **Economic factors** – Even when some industrialists (such as carmakers Toyota and Honda) wanted to restart production in their factories following the earthquake, they faced supply problems as the factories that made components for their vehicles were severely damaged. Some analysts predicted that this would lead to production capacity shortfalls for at least 18 months. In addition, some also had to endure restrictions on energy use whilst the Japanese power supply issues were sorted out. The Japanese Ministry of Health, Labour and Welfare reported that the number of people filing for unemployment benefit reached nearly 164,285 in August 2011 in the three worst affected regions.
7. **Aid** – Aid organisations in Japan and worldwide responded to the disaster. The Japanese Red Cross reported over $1 billion in donations (many from overseas). They also distributed over 30,000 emergency relief kits and 14,000 sleeping kits in the evacuation centres in Iwate, Miyagi, Kukushima, Yamagata and Akita. As people moved to temporary and pre-fabricated buildings, they had no basic equipment to resume everyday life and official requests were that the Japanese Red Cross would provide household appliance packages to the displaced. They have provided 110,000 packages (each containing a washing machine, refrigerator, TV, rice cooker, microwave oven and hot water dispenser. Google provided a special crisis response team which provided key information for people who were caught up in the disaster and allowed people to use a person finder service to help them find displaced friends and relatives.

EMERGENCY RELIEF KIT CONTENTS	SLEEPING KIT CONTENTS
25 utility items including towels cup portable radio utensils papers and pen toothbrush plasters flashlight	mattress pillow ear-plugs eye mask

Figure 157

Contents of the Japanese Red Cross relief and sleeping kits

Figure 158
An eyewitness account of life during the aftermath of the Great Tohoku earthquake

Yuki Kumagaya and his wife Teruko were forced to move into a temporary home in Iwate following the Tsunami. "We were pleasantly surprised to see the electronic household sets already installed as we came through the door of the prefabricated house" says Yuki on 11 April. Yuki remembers the tsunami caused by the Chile earthquake in 1960 which reached the coastline of Tohoku. "The recent tsunami was nothing compared to what we had back then" says Yuki. The only thing the couple were able to take with them was an emergency bag with the documents for their pension and insurance when the tsunami came.

Source: 'Japan Earthquake and Tsunami Six Months On', Irish Red Cross, 8 September 2011, http://www.redcross.ie/news/publications/japan-earthquake-and-tsunami-six-months-on/

Test yourself

1. Describe the causes of the Great Tohoku earthquake. (3)
2. What is a tsunami? (2)
3. Identify and explain some of the short term impacts of the earthquake on the local people. (4)
4. Identify and explain some of the short term impacts of the earthquake on the local environment (4)
5. Describe the long term impacts of the earthquake on people. (3)
6. Evaluate the effectiveness of prediction and precaution measures before the earthquake. (4)
7. Identify and evaluate one long term strategy used following the earthquake. (5)

CASE STUDY

LEDC Case Study:
The Haiti earthquake (Haiti, 12 January 2010)

The US Geological Survey (USGS) reported that in 2010 there were 22 magnitude 7.0 or larger earthquakes. Of the estimated 227,000 people who died due to earthquakes in 2010, over 222,570 of these were as a result of the Haiti event on 12 January. According to official estimates (from the United Nations High Commissioner for Refugees – UNHCR) this earthquake injured 300,000 people, displaced 1.3 million, and left 97,294 houses destroyed and 188,383 damaged in the south of the country.

Causes

The Haiti earthquake was a magnitude 7.0 earthquake that took place near the town of Logane, around 16 miles west of Port-au-Prince, the capital city of Haiti, at 16:53 local time on Tuesday 12 January 2010. The epicentre of the earthquake occurred inland, at a focus depth of about 8 miles, on a blind thrust fault that made up part of the Enriquillo-Plantain Garden fault system (a strike-slip fault system). The USGS estimated that around 3.5 million people living in the local areas experienced intense shaking, which would usually lead to very heavy damage, even to earthquake-resistant structures. The damage caused was more severe than for other similarly sized quakes due to the shallow depth of the earthquake.

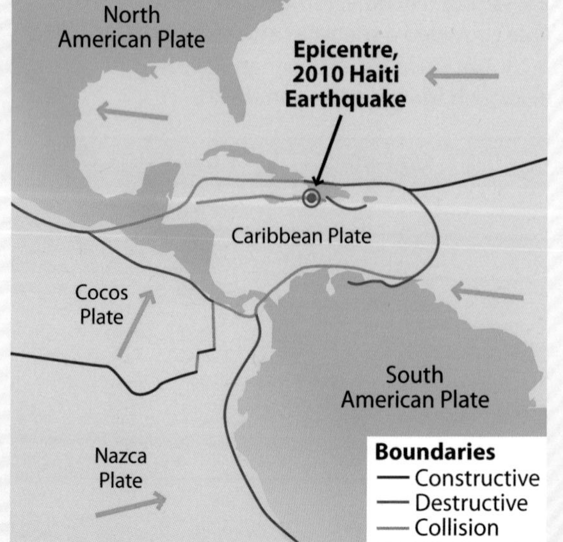

Figure 159
Plate map of the Central America area

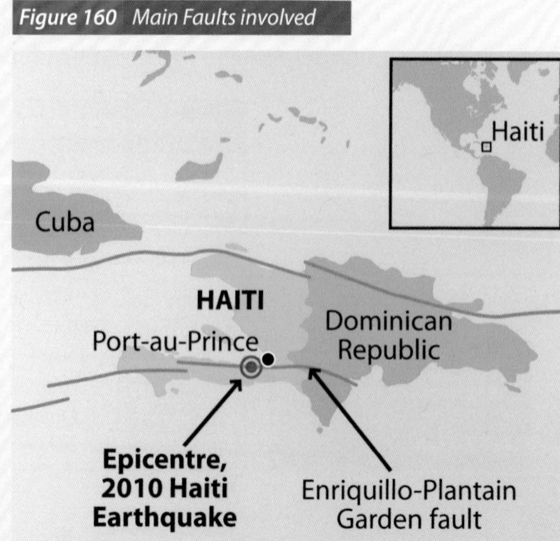

Figure 160 *Main Faults involved*

The earthquake in Haiti was caused by movement at the boundary of the Caribbean and North American Plates. In this area the Caribbean Plate is moving eastward at around 20 mm per year, in a strike-slip (conservative) motion. The movement is not smooth and takes the form of a number of jolts, which can produce earthquakes. In the area where this earthquake struck there had been little movement or activity for 250 years. This earthquake is reckoned to have caused a slip of about 1.8 m.

Short term impacts (on people and the environment)

U. S. Geological Survey/photo by Walter Mooney

Figure 161
Haiti earthquake reduces buildings to rubble

Figure 162
Massive destruction causes roadslides and makes rescue operations difficult in Port-au-Prince

IMPACTS ON PEOPLE	IMPACTS ON THE ENVIRONMENT
Location – The earthquake took place very close to the capital city of Haiti, Port-au-Prince, a very densely populated part of the country. **Death and injury** – An estimated 222,570 died, 300,000 were injured and 1.3 million were displaced. 97,294 houses were destroyed and 188,383 were damaged in the south of the country. Around 3.5 million people were affected by the quake. **Damage** – A catastrophic amount of damage was reported. Many hospitals collapsed, the airport and port were seriously damaged, roads were blocked with debris or the surfaces broken, communications were severely interrupted, radio stations went off the air for about a week, the National Palace was severely damaged and the Prison Civile was destroyed, allowing 4,000 inmates to escape. Many of the municipal buildings, the UN offices and the World Bank in Haiti were destroyed. Petrol reserves were decimated as the earthquake caused leaks to pipes and stores. Much drained away or was set on fire. The education system was also severely damaged, as more than 1,300 schools were destroyed. **Poverty** – Haiti was already an exceptionally poor LEDC before this earthquake hit. It was listed as 145th out of 169 countries in the UN Human Development Index and is the poorest country in the western hemisphere. More than 70% of the people in Haiti live on less than $US 2 per day, only half of the people living in Port-au-Prince had access to latrines and only a third had access to tap water. Many poor people lived in shanty towns, such as Canape Vert in Port-au-Prince. Residents would build houses with whatever was available nearby, such as wood, mud, cheap concrete and bricks. These constructions were unsteady and collapsed easily during the earthquake.	**Aftershocks** – Over the next month, 52 aftershocks measuring more than magnitude 4.5 or more were recorded in the area, with three reaching magnitude 5.9. This caused further damage and hampered rescue efforts. **Mini-tsunami** – A 'mini-tsunami' was reported in the small fishing town of Petit Paradis. Three people were swept out to sea and died, as boats and debris were swept out into the ocean. **Damage** – The earthquake was very shallow which increased the intensity of the shaking. Shallow focus earthquakes often cause more damage to natural ecosystems, buildings and property. **Location** – Much of Port-au-Prince's residential areas were built on hillsides that surround the city and many homes tumbled down the slopes due to the force of the earthquake. Haiti has experienced huge amounts of deforestation in recent years and the earthquake led to landslides, as the soil and rock was released by the earthquake.

Long term impacts (on people and the environment)

Figure 163
Port-au-Prince before and after the earthquake

IMPACTS ON PEOPLE	IMPACTS ON THE ENVIRONMENT
Industry and employment – The clothing industry accounts for more than two-thirds of Haiti's exports and it was hit severely by the earthquake. An estimated one in every five jobs in Haiti was lost. **Homelessness** – Many people continued to sleep in the streets, on pavements or in cars after the earthquake, as they were worried that structures would not be able to withstand aftershocks. **Aid** – The impact of the earthquake meant that it took nearly a week for relief efforts to be organised properly and the government handed over control of the airport and ports to the US to hasten and ease flight operations. Very quickly, even the most simple of medical supplies began to run out (such as antibiotics). **Health risks** – Many bodies remained unburied for long periods of time and in the heat and humidity, the corpses started to smell and decompose, causing serious health risks to survivors. **Crime** – The slow distribution of resources led to sporadic violence and looting in Port-au-Prince. **Rebuilding** – Over £3.42 billion was pledged internationally to help the reconstruction of Haiti. It was overseen by the International Haiti Reconstruction Commission, headed by Bill Clinton and Haitian Prime Minister Jean-Max Bellerive.	**Physical** – This earthquake had very few impacts on the physical environment. Most of the impacts were caused by the proximity of the earthquake to a major urban area. **Relocation** – In April, due to the potential threat of mudslides and flooding from the upcoming rainy season, the Haitian government began operations to move thousands of refugees to a more secure location north of the capital. **Out migration** – In the months that followed, many people started to return to the rural areas where they had lived and farmed prior to migrating into Haiti before the earthquake.

Evaluation of the management response

Haiti was not prepared for this disaster. Many of the short and long term impacts continue to this day, as the government of Haiti and the associated aid agencies have struggled to improve the conditions on the ground in the country. Homelessness, crime, gang cultures, disease and poverty all continue to be the main issues for someone living in Haiti today.

Prediction and precautions BEFORE the earthquake

1. **Earthquake prediction** – Haiti did not have an organised system for trying to predict or warn people about earthquakes or potential natural disasters. It does have a basic national Centre for Meteorology (in conjunction with the National Hurricane Centre it can help provide details on Tropical Cyclones) but does not have a similar organisation for seismology.

2. **Buildings** – Many of the buildings in Haiti were temporary constructions and not built to withstand the most basic damage.
3. **Emergency services** – The emergency services were inadequate and unprepared for any form of rescue following the disaster. Many more people would have died if international aid organisations and governments had not stepped in when they did.

Could this earthquake have been predicted?

To date no earthquake event has been predicted to the exact time and place. However, some geologists in 2007 and 2008 had identified that a possible magnitude 7.2 earthquake could take place in the Enriquillo-Plantain Garden Fault zone, as it was at the end of a seismic cycle started from the 1692 Jamaica earthquake. Even if the earthquake had been predicted fully, though, there is the question of what exactly the Haitian government could have done to minimise the consequences of the earthquake.

Immediate and long term strategies AFTER the earthquake

Figure 164

Refugees in Port-au-Prince following the earthquake

1. **Refugees** – Over 3 million people were left homeless by the impact of this earthquake initially. Six months later, by July 2010, the number of people in relief camps of tents and plastic sheets was 1.6 million, and almost no transitional housing had been built. Most of the camps had no electricity, running water or sewage disposal, and the tents were beginning to fall apart. Crime was widespread in these camps.
2. **Aid** – Many different countries and international aid organisations mobilised quickly to send emergency aid to Haiti. Disaster relief teams, emergency field hospitals, emergency food supplies, water and materials were all supplied quickly in response to urgent requests for assistance by the Haitian authorities. The International Red Cross co-ordinated a global response and the US government, through its USAID programme, took a lead in the organisation of aid to people in Haiti. The US response grew to include 20,000 civilian and military personnel, who supported the largest urban food distribution in history (3.5 million people), emergency shelter distribution to 1.5 million people and a vaccination campaign that inoculated 1 million people. The amount of assistance that USAID provided to Haiti up to 26 August 2010 totalled nearly $US 656 million.
3. **Rebuilding** – By July 2010, 6 months after the disaster, as much as 98% of the rubble from the earthquake was still not cleared. Most of the capital city was still impassable and thousands of bodies remained in the rubble. Even a year later, aid agencies continued to report that the majority of rubble remained in situ. House building could not be started in earnest until the rubble was cleared.
4. **Evaluating the emergency response** – It took a while for emergency rescue teams to reach devastated areas quickly. British search teams were the first to reach Leogane, near the epicentre of the earthquake on 17 January. The Red Cross described the town as "severely damaged . . . the people there urgently need assistance".
5. **Cholera** – In late October 2010, an outbreak of cholera was confirmed in Haiti which required further relief efforts, including emergency supplies, blankets, buckets and disinfectant supplies. Cholera treatment facilities were established to support isolation and treatment for cases, and to prepare for the spread of the disease (which eventually had around 6,000 beds for cholera treatment). By the end of 2010, more than 3,300 had died of cholera.
6. **Food and Economic security** – assistance was given by some aid agencies to promote the agricultural sector, which would hopefully increase food productivity and increase farm incomes.

Comparing the impact of the earthquake in the MEDC and LEDC

The earthquakes described in this section were easily the two biggest earthquakes to occur in 2010 and 2011, yet their impact was very different. Even though the earthquake in Japan was a large magnitude 9.0 earthquake and created huge problems due from the tremors, aftershocks and tsunami, the Japanese people were in a much better position to deal with the consequences of the earthquake than the people in Haiti. The shallow magnitude 7.0 earthquake in Haiti shook the urban areas surrounding Port-au-Prince to its core, decimating an already poverty-stricken country further.

In the Japanese earthquake event, the Japanese government reacted swiftly and were able to mobilise a large force to move from rescue to rehabilitation quickly. The Haitian government was disorganised and had no money to allow them to react to the crisis, so were wholly reliant on aid from overseas. In effect, the US government in conjunction with the UN took on responsibility of making Haiti work for a year.

LEDCs are always going to suffer more than MEDCs in natural hazards because:

1. **Prediction** – LEDCs do not have the money to invest in the technology which helps to predict and warn people about earthquakes.
2. **Protection** – LEDCs do not have the money or the technology to plan buildings which are designed to be earthquake proof.
3. **Preparation** – LEDCs do not have the resources to cope with the rescue services, relief, aid, rebuilding or long term hazards, such as lack of water, starvation and spread of diseases. Education systems are also weaker, so people are less aware of what they should do in the event of a disaster.

Writing a Case Study Essay

Title: What were the short and long term impacts of the Great Tohoku earthquake in March 2011?

Tips: Break your response up into four main paragraphs covering the following things:

1. **Introduction**
   - Introduce where the earthquake took place (perhaps including a map).
   - Explain a few things about what caused the earthquake, and why the earthquake and tsunami happened (ie the type of plate boundary).
2. **Short term impacts**
   - Explain how each of the short term impacts affected both the people **and** the environment.
3. **Long term impacts**
   - Explain how each of the long term impacts affected both the people and the environment.
4. **Conclusion**
   - In your opinion, explain why the earthquake caused so many deaths and so much destruction in the area.
   - How do you think Japan coped with the earthquake?

Check your learning

Now that you have studied Part 4: Earthquakes: Can they be managed?, return to page 120 and answer the Key Questions for this section.

ORGANISE YOUR REVISION

Draw a mind map to summarise Unit 1C 'The Restless Earth'.

EXAM PRACTICE QUESTIONS

Some of these questions are from previous CCEA GCSE examination papers and others have been written in the same format to give you practice at answering 'exam style' questions.

Try to answer the questions with as much detail as possible. Also consider the number of marks that each question receives, as this will give you a good indication of the amount of depth that your answer needs.

Resource A

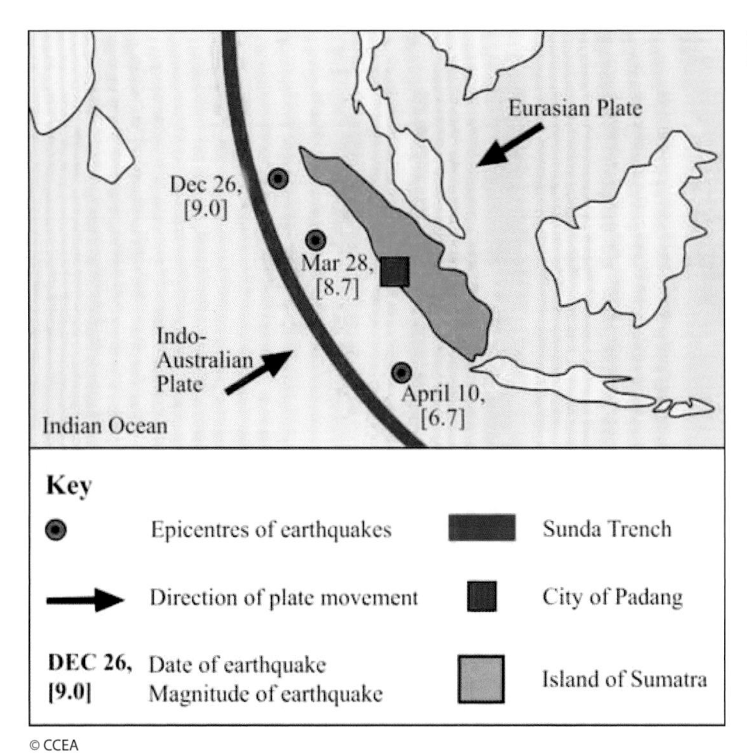

© CCEA

Resource B

City of Padang

A B C D E F

© CCEA

Key

Eurasian Plate	
Mantle	
Indian Ocean	
Indo-Australian Plate	
Sunda Trench	B
Sumatra	C

Resource C

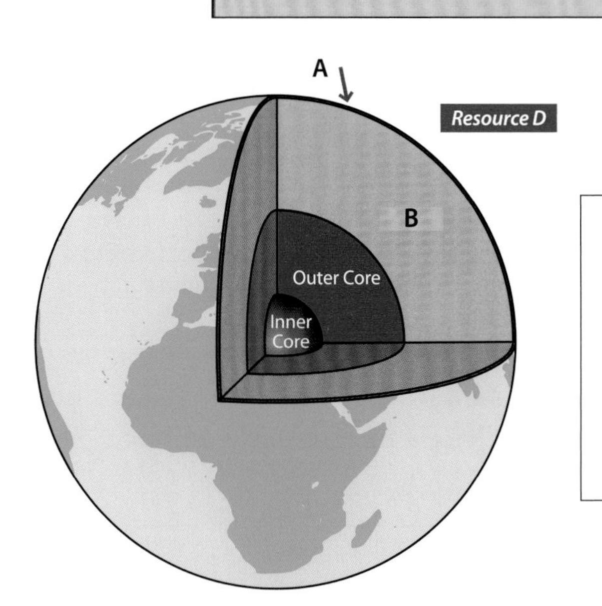

Resource D

Resource E

Why earthquakes cause many deaths in LEDCs

- ***Buildings are not earthquake proof***
- ***Poor communications***
- ***Lack of equipment to rescue survivors***
- ***Slow and poorly organised emergency services***

© CCEA

Resource F

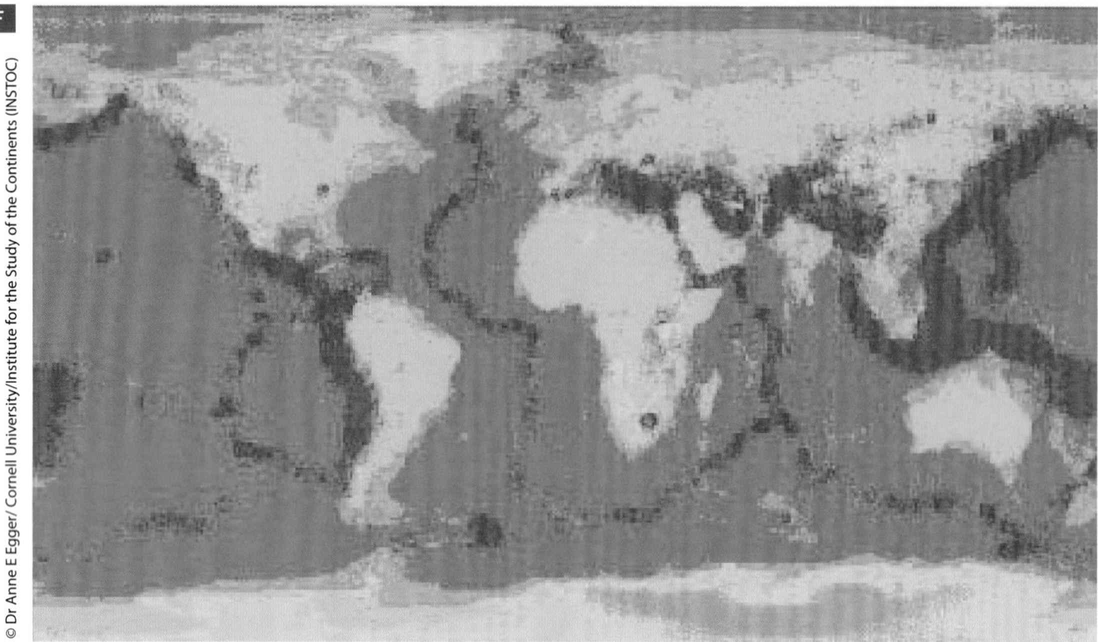

Resource G

Haiti Earthquake was not a Surprise to Some Geologists

On 12 January 2010, the LEDC country of Haiti was struck by an earthquake measuring 7 on the Richter Scale. Port-au-Prince, the capital of Haiti, is only 16km from the plate boundary where the Caribbean Plate is sliding alongside the North American Plate.
The earthquake, which had a shallow focus, resulted in almost 200 000 deaths and widespread destruction.

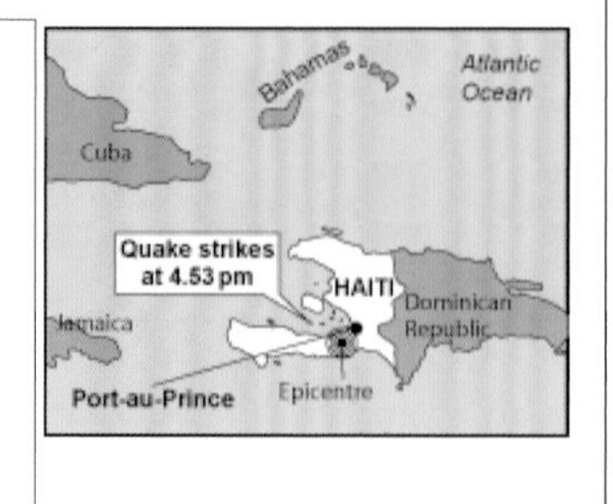

Foundation Tier

1. Study Resource A which shows information about earthquakes in 2004 off the coast of Sumatra, Indonesia. Answer the questions which follow.
   i. Name the scale used to measure the magnitude of an earthquake. (1)
   ii. Study the Resource B which shows a cross section through the Sunda trench. Make a copy of the key in Resource C and complete it by inserting the letters A to F, using Resource A to help you. (4)
   iii. Using Resources A and B to help you, explain why the city of Padang has had several earthquakes. (3)
2. Study the Resource D which shows the structure of the earth.
   i. Two layers are not named in Resource D. Write the labels for A and B. Choose your answer from the list below:
      volcano trench mantle island crust
   ii. Classify the following statements as true or false. (3)
      - The inner core is the hottest layer of the earth.
      - The surface of our planet is split into plates.
      - Plates move due to ocean currents.
      - Continents are moving very slowly.

3. The Giant's Causeway is an area with basalt columns.
   All rocks can be placed into one or other of the following categories.
   i. Write down the category to which basalt belongs. (1)
      igneous sedimentary metamorphic
   ii. Explain the formation of basalt columns. (4)
4. Study Resource E which shows some reasons why earthquakes in LEDCs kill many people.
   Choose **two** of these reasons and explain how each could lead to many deaths in LEDCs. (4)
5. Earthquakes can have severe impacts on people.
   i. Name an earthquake you have studied. (1)
   ii. State **two** ways the people prepared to cope with the earthquake before it happened. (2)
   iii. Describe **one** impact this earthquake had on the people living in the area. (3)
   iv. Describe **one** long term response put in place after the earthquake to reduce the impact of future earthquakes in this area. (3)

Higher Tier

1. Study Resource A which shows information about earthquakes in 2004 off the coast of Sumatra, Indonesia.
   i. Name the type of plate boundary shown in Resource A. (1)
   ii. Study Resource B which shows a cross section through the Sunda trench. Make a copy of the key in Resource C and complete it by inserting the letters A to F, using Resource A to help you. (4)
   iii. Using Resource A to help you, explain why earthquakes have recently occurred at this plate boundary. (3)
2. Study Resource D which shows the structure of the earth.
   i. What are the labels for A and B. (2)
   ii. Explain why plates move. (3)
3. Study Resource F which shows the global distribution of earthquakes.
   Describe the global distribution of earthquakes as shown by the map. (4)
4. Study Resource G which gives some information about the earthquake which affected Haiti in 2010.
   i. Name the group of islands to the north of Haiti. (1)
   ii. Explain why earthquakes in LEDCs tend to cause more deaths than those in MEDCs. (4)
   iii. Which type of plate margin is found near Port-au-Prince? (1)
   iv. Explain why tectonic plates move. (4)
5. (i) Explain the cause of a named earthquake you have studied. (3)
   (ii) Explain **two** strategies put in place after this earthquake to reduce loss of life in future earthquakes. (6)
6. Describe the impacts caused by an earthquake in the British Isles which you have studied. (6)

Unit 2:
Living in Our Natural World

2A People and Where They Live

2B Contrasts in World Development

2C Managing Our Resources

2A People and Where They Live

"The increase in the world's population represents our victory against death"
Julian Simon

In early 2012 the world population reached 7 billion people and continues to grow at a fast rate. Many demographers estimate that it will reach 8 billion by 2020. The population of the world exploded over 50 years ago and although the population totals in many MEDCs are now rising at a slow rate, the population totals in many LEDCs are increasing rapidly. The population of India and China combined makes up nearly half of the total world population. This population increase is getting faster all the time and it is starting to put pressure on the land, space and resources.

This theme takes a look at how and why the world population is growing so steadily. It considers the influence this is having on our settlements and in particular the impact on the growth of cities.

Theme 2A is divided up into 3 parts:

1: Population Growth, Change and Structure
2: Settlement Site, Function and Hierarchy
3: Urbanisation in LEDCs and MEDCs

PART 1: POPULATION GROWTH, CHANGE AND STRUCTURE

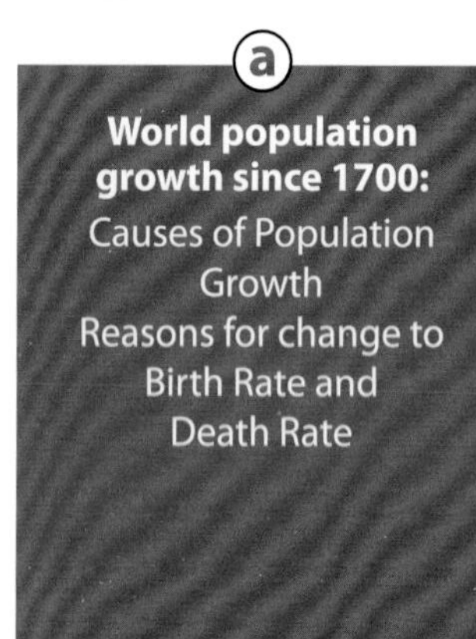

(a) World population growth since 1700:
Causes of Population Growth
Reasons for change to Birth Rate and Death Rate

(b) Skills: Using GIS to investigate in-migration within an MEDC (Northern Ireland)
Obtain migration data (the origin and numbers of people migrating into Northern Ireland)
Select and use digital graphing and mapping techniques for presentation
Analyse, interpret and evaluate data

(c) The positive and negative impacts of international migration
Case Study: International migration from Turkey to Germany
Floodplains
Numbers migrating, origins and destination
Impacts on services and the economy

(d) Population Structure
Compare and Contrast structure of MEDC and LEDC

(e) The Social and Economic implications of aged and youth dependency

Key words

Birth Rate
Death Rate
Natural Increase
Migration (immigration and emigration)
Push and Pull factors
Population structure
Dependency (youth and aged)
Geographic Information Systems (GIS)
Population Pyramid

Key questions

By the end of this section you will be able to answer the following questions:

1. Describe some of the causes of global population growth.
2. Explain how birth rates can differ between LEDCs and MEDCs.
3. Give one reason why death rates in LEDCs are being reduced.
4. Describe what a GIS is.
5. What is migration?
6. Evaluate the impact of people migrating into one European country that you have studied.
7. Describe how the population structure of an MEDC can indicate a population that is ageing.

World population growth since 1700

What causes the world population to grow?

The study of population is all about the balance that can be achieved through births and deaths. This is known as the 'population balance'. The population of a country will increase if the number of babies born is greater than the number of people who die. In the same way, the population will decrease if more people die than are born. However, migration can also have a part to play in the increase or decrease of total population.

Birth rate (or crude birth rate as it is also known) is the number of live births each year per thousand of the population in an area.

Death rate (or crude death rate) is the number of deaths each year per thousand of the population in an area.

When there is a growth of people (a higher birth rate than death rate) we say that there is a **Natural Increase.** If there is a decline in the number of people (a lower birth rate than death rate) we say that there is a **Natural Decrease.**

Test yourself

1. Take a look at the table below which shows the birth and death rates for five different countries.

Resource A

COUNTRY	BIRTH RATE	DEATH RATE
Malawi	41	13
Canada	9	8
Niger	51	14
Italy	8	11
Nepal	22	7

Source: The World Factbook, 2010. Washington, DC: Central Intelligence Agency, 2010

a. Which countries have a Natural Increase in Population? (4)
b. Which country has a Natural Decrease? (1)
c. Which country's population will remain pretty steady? (1)
d. Which country has the fastest growing population? (1)

Figure 1
Population growth is most evident in big cities, such as Moscow, Russia.

Throughout the world as a whole the birth rate is much higher than the death rate. This is the cause of what is known as the 'population explosion'. However, the balance between births and deaths varies from one country to another.

The World Population Balance organisation wrote the following on its website in 2010:

> "Current global population of over 6.8 billion is already two to three times higher than the sustainable level. Several recent studies show that Earth's resources are enough to sustain only about 2 billion people at a European standard of living.
>
> An average European consumes far more resources than any of the poorest two billion people in the world. However, Europeans use only about half the resources of Americans, on average.
>
> Currently the 6.8 billion of us are consuming about 25% more resources than Earth is producing – during any given time period. For example, in the past 12 months we have consumed the resources that it took the planet about 15 months to produce. We are consuming our resource base.
>
> Obviously, this 25% overshoot is not sustainable. Another crucial point to understand is this: the longer we overshoot and consume more resources than the sustainable level, the more the long-term 'sustainable level' actually declines!"
>
> Source: World Population Balance, www.worldpopulationbalance.org, data from www.footprintnetwork.org/atlas

It used to be fairly easy to explain why there was such a difference between birth and death rates in LEDCs. Birth rates remained high and steady as people had large families so that the children could provide free manual labour on the farm. Death rates remained high as LEDC governments had little money to invest in quality healthcare, so people would have a low life expectancy. However, in recent years this has become much harder to explain as the death rates in LEDCs have fallen rapidly, with many LEDCs having a much lower death rate than MEDCs.

Birth rates in LEDCs remain at a very high level. In fact over 165 (out of 224) countries had a crude birth rate (CBR) over 12/1000 in 2010.*

* Data from United Nations

This means that our world is currently experiencing what some people call a 'demographic time bomb'. The population of the world has expanded so rapidly that there are concerns that we might not have enough resources to sustain any further increase.

The graph below (Figure 2) shows how population has increased across various world regions.

The pattern of growth shows that before 1750 the world population was smaller than 1 billion people. There was a slow but steady growth from 1750 to 1850 but from 1900 the world population started to grow at an exponential rate. The rise was so extreme that it is often referred to as a 'Population Explosion'. Huge increases in the population of Asia and Africa continued the explosion well past 2010.

Figure 2

Population by continental region

Data from United Nations Population Division: World Population Prospects: The 1998 Revision

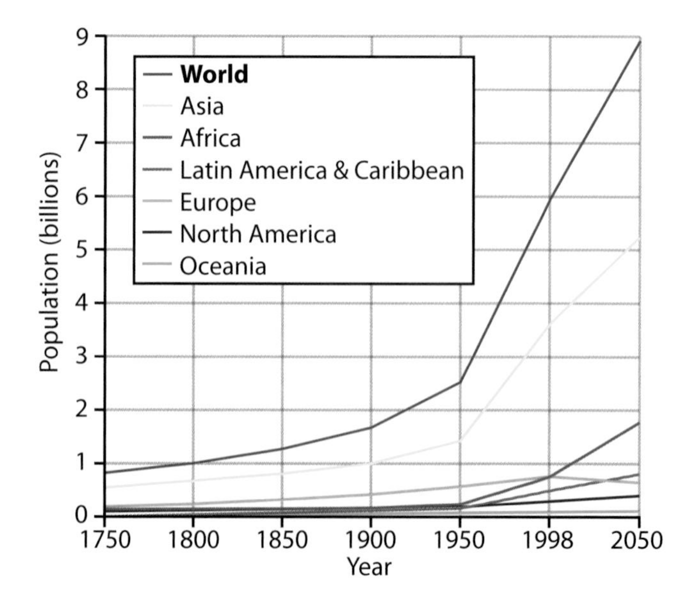

The United Nations produced the data below (Figure 3) to illustrate how it expects (or projects) the population of the world to continue to increase over the next 40 years.

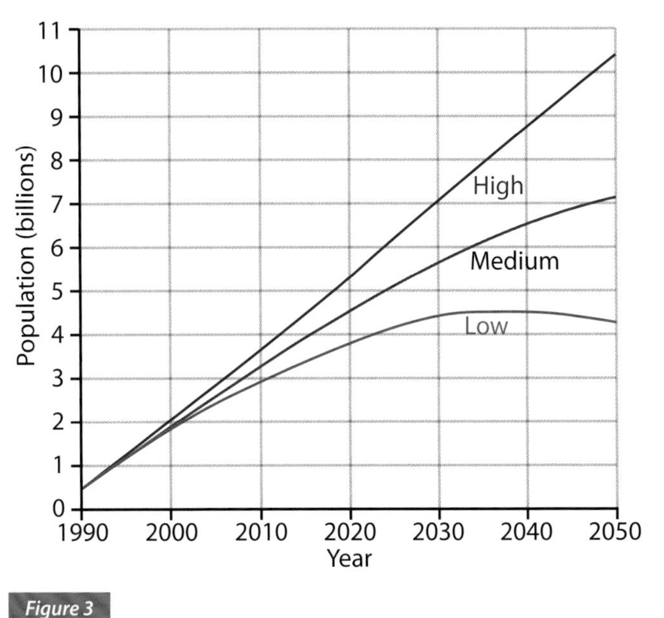

Figure 3

Population Projections until 2050 (from UNPD)

Data from United Nations Population Division: World Population Prospects: The 1998 Revision

The graph below (Figure 4) takes things a step further, showing the 10 largest countries (in terms of population).

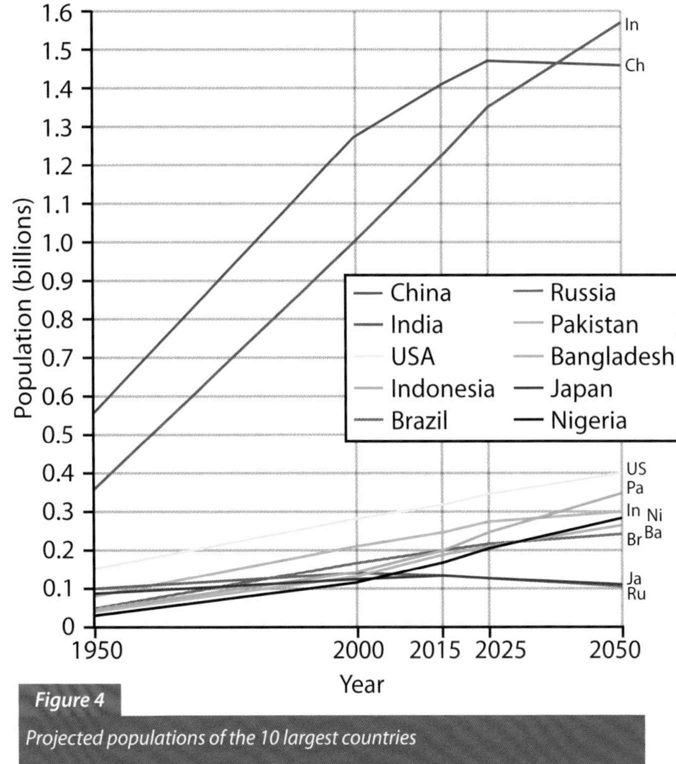

Figure 4

Projected populations of the 10 largest countries

Data from United Nations Population Division: World Population Prospects: The 1998 Revision

Test yourself

1. Which two countries have had the biggest increase in population from 1950 to 2025? (2)
2. What was the population increase of India from 1950 to 2015? (2)
3. What was the population increase of China from 1950 to 2015? (2)
4. During what year does the UN expect that India will become the most populous nation on the planet? (1)
5. Which country is the only one in the top 10 where a population decrease is noted from 2000 to 2025? Why do you think the population might be decreasing in this country? (3)

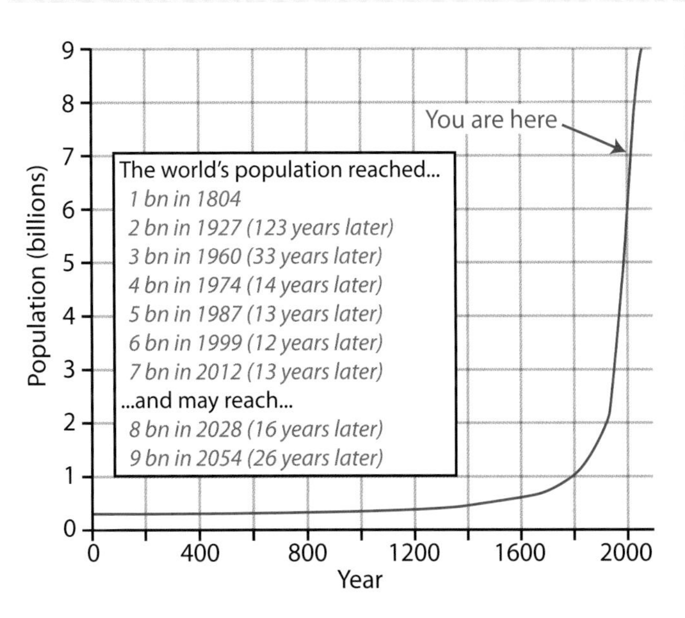

Figure 5

Graph of the Total World Population over the last 2000 years

Data from United Nations Population Division: World Population Prospects: The 1998 Revision

One of the best methods of describing population change is to talk about the amount of time it takes for the population to double in size (or doubling time). The world population reached 1 billion in 1804 and it took 123 years for the population to double to 2 billion. It only took 47 years to then double from 2 to 4 billion.

Test yourself

1. Use Figure 5 to describe the growth of the world's population:
   a. up to 1700. (2)
   b. from 1700 to 1900. (2)
   c. from 1900 to 2000. (2)
2. Using your knowledge of natural increase, birth rate and death rate, explain the changes in the growth rate shown in Figure 5. (4)
3. Make a copy of the table below and answer the questions that follow.

Resource A

COUNTRY	BIRTH RATE	DEATH RATE	NATURAL INCREASE
Australia	14	7	7
Brazil	25	8	
China	17	7	
Ethiopia	46		30
India	29	10	
Indonesia		8	16
Mexico	27	5	22
Nigeria		12	31
Russia	9	15	
Sweden		11	1
USA	15		6

Source: The World Factbook, 2010. Washington, DC: Central Intelligence Agency, 2010

a. On your copy of the table fill in the missing values. (9)
b. Which countries of the world have the highest natural increases? (2)
c. Which countries of the world have the lowest natural increases? (2)

In many countries around the world the death rates have fallen dramatically. This can be explained by improvements in medical care, medicines, hospitals and sanitation. Many of the diseases which commonly caused death, such as cholera, typhoid and smallpox have been controlled due to these improvements.

One final indicator that can help us understand population growth is **Infant Mortality.** Infant mortality rate measures the number of infants per 1,000 who die before they reach the age of one. In many MEDCs this rate is extremely low, but in many LEDC countries the rate can be very high.

In the 1950s the total population of Kenya was just over 6 million. The life expectancy was around 41 years and the infant mortality rate was around 200 deaths per 1,000 births. On average, a child would be born in a rural setting, far from any medical care and into a family with little education, money or formal employment. Around 1 in every 5 children would not even make it to their first birthday and the majority of people would not see 40 years of age.

The population of Kenya was estimated at 41 million people in 2011. Around 42% of the population are now under the age of 15, which means that Kenya displays the main characteristics of a youthful population (the average age is 19 years). Only 2.7% of the population survive beyond the age of 65 (life expectancy is 60 years). Currently the population is growing by 2.5% a year. The birth rate is 34/1000 and the death rate is 9/1000. By 2011, many more Kenyan people were living in cities (around 22%). These cities provide more opportunities for inhabitants, with access to better healthcare and to education. By 2011, the Infant Mortality Rate was 52 deaths per 1,000 live births. Together, these two factors mean that many more children can survive the early years beyond infancy.

Survival in this LEDC has changed. More people live beyond the age of one and life expectancies throughout the country have increased. Death rates have dropped but this only emphasises the difference between the high birth rate and low death rate. This creates a population that is increasing at a fast rate, with the Kenyan population estimated to double within 27 years.

The model of demographic transition (population change)

Probably the best way to demonstrate the different factors that affect the birth and death rate is to take a look at the **Demographic Transition Model (DTM)**. This model takes countries through 4 or 5 stages of growth or development and explains the main reasons for changes in birth rate and death rate throughout.

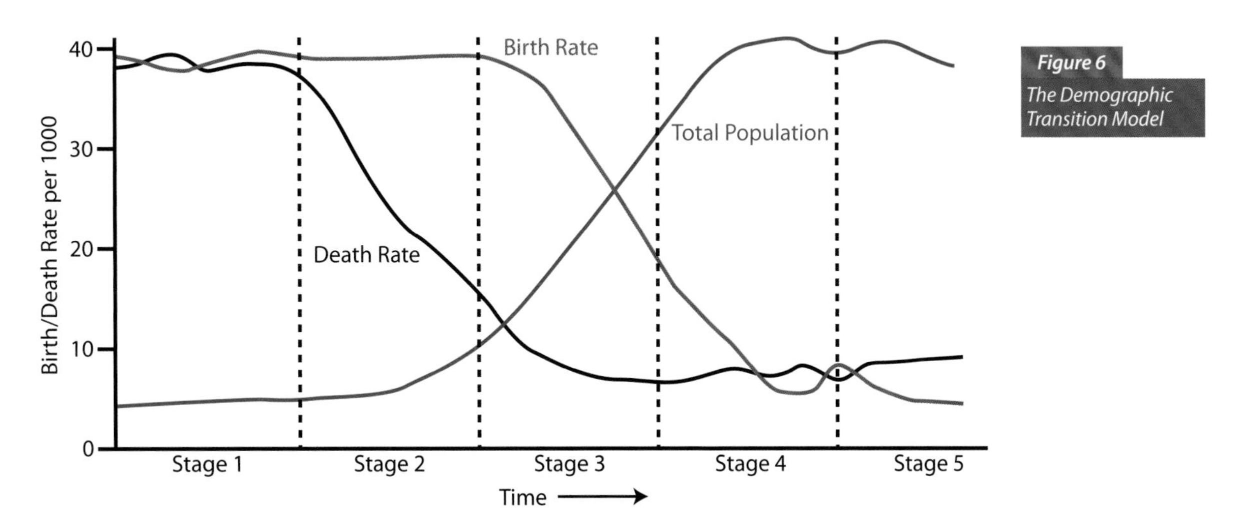

Figure 6
The Demographic Transition Model

A model is used in geography to represent reality. In this case the model is used to show the changing face of population. The DTM consists of three graphs built onto the same axis. Three line graphs are drawn with the number of births/deaths up the left hand side and the time along the bottom.

The DTM was based on the population changes that take place in MEDCs. As a result, the population changes that happened in the UK look very similar to what is shown on the DTM. Demographers (people who study population) have noticed that there are five very distinct stages that countries usually go through as their population changes. Not all countries will progress through these stages at the same speed and some countries might even miss out a stage or two as they develop quickly. As a country progresses, their death and birth rates will change and this will alter the total population within the country.

During **Stage 1**, both birth and death rates are high. This is because the population is largely undeveloped. As a country begins to develop, there is usually a focus on industrial advancement and the economic situation of people will improve. If people have more money their standard of living increases and they will start to live longer. In **Stage 2**, life expectancy increases and death rates fall. The birth rate remains high and this difference between the two population rates is what begins the population explosion in a country. Eventually, at **Stage 3**, the birth rate also begins to fall and social and economic pressures produce a reduction in the number of babies born within the country. **Stage 4** is when a form of equilibrium occurs within the population. Both the birth and death rates are low and can fluctuate. Some countries, mostly western European states, can experience a negative population growth as the birth rate slips below the death rate in **Stage 5**. This can produce a population which will eventually decline, as the population will not continue to replace itself.

Structured notes

1. Make a copy of the DTM in Figure 6.
2. Make notes on what happened to both the birth and death rates at each of the stages in the DTM. You should be able to explain why there are changes to either the birth or the death rates at each stage.

Key facts about the changes to birth and death rates in the Demographic Transition Model

The following are the main facts that can be used to explain why the birth and death rates change through time using the DTM.

Stage 1

Here both **birth and death rates fluctuate** at a high level (about 36/37 per 1,000), giving a small population growth (a youthful population).

Birth rates are high as:

- there is no birth control or family planning.
- many children will die in infancy and parents will have a large number of children to ensure that some actually reach adulthood.
- many children are needed to work on the land. Farmers have no money for workers or machinery so children are an inexpensive way of gaining labour for the farm.
- children are regarded as a sign of virility in some cultures,
- some religious beliefs (eg Roman Catholics, Muslims and Hindus) encourage large families.

Death rates are high (especially among children) due to:

- disease (bubonic plague, cholera, and even chickenpox and measles).
- famine, food shortages and poor diet.
- poor hygiene, due to a lack of piped water, sewage and basic toilet facilities.
- little medical care, with few doctors, hospitals or drugs.

Stage 2

Birth rates remain high, but **death rates fall rapidly** to about 18/19 per 1,000 people, giving a rapid population growth.

Death rates fall due to:

- improved medical care (vaccinations, hospitals, doctors, new drugs and scientific inventions).
- improved sanitation and water supplies.
- improvements in food production (both quality and quantity).
- improved transport to move food, medical care, etc.
- a decrease in child mortality.

Stage 3

Birth rates fall rapidly to around 18 per 1,000 people, while death rates continue to fall slightly (15 per 1,000 people) to give a slowly increasing population.

Birth rates fall due to:

- an increased use of family planning (contraceptives, sterilisation, abortion and government incentives).
- a lower infant mortality rate, which means that children are now surviving through to adulthood and there is less need to have as many.
- increased industrialisation and mechanisation, which means fewer labourers are needed.
- an increased desire for material possessions (cars, holidays and bigger homes) and a reduced interest in large families.
- increased equality for women in society, changing their role as solely child bearers.

Stage 4

Both birth rates (15 per 1,000) and death rates (12 per 1,000) remain low, fluctuating slightly to give a steady population.

Stage 5

Birth rates (7 per 1,000) fall below the death rates (9 per 1,000) and the population finds itself in a state of population decrease. Population is ageing and is dominated by older people.

Test yourself

A good way to assist your understanding of the demographic journey of populations is to draw out your own DTM graphs to show the actual experiences for two very different countries.

1. The table below shows the birth and death rate information for England and Wales.

Resource A

Births, Deaths and population change in England and Wales from 1700 to 2050

YEAR	BIRTH RATE PER 1,000	DEATH RATE PER 1,000	TOTAL POPULATION
1700	36.0	30.5	5.5 million
1725	35.5	34.0	
1750	39.0	30.5	6.5 million
1775	38.0	30.5	
1800	38.5	22.0	9.0 million
1825	36.0	20.5	
1850	32.5	22.5	17.9 million
1875	34.5	22.0	
1900	28.0	17.0	32.5 million
1925	20.0	12.5	
1950	16.0	11.0	43.6 million
1975	16.5	11.0	
2000	11.6	9.0	52 million
2025 est	13.0	8.0	
2050 est	11.7	7.6	75 million

Data from UK Office for National Statistics

a. Using graph paper, draw a graph with the birth and death rate values up the left hand side, the years along the bottom and total population up the right. (2 marks for axes and labels)
b. Plot all of the figures above so that you have three distinct lines in your graph. (3 marks for each line = 9 marks)
c. Give your graph a suitable title. (1)
d. Describe the changes to the death rate of England and Wales from 1700 to 2000. Can you identify any potential events that had an impact on the death rate? (3 + 3 = 6 marks)
e. Describe the changes to the birth rate of England and Wales from 1700 to 2000. Can you think of any social or economic changes that might have allowed the birth rate to fall? (3 + 3 = 6 marks)
f. Explain in detail why you think the total population of England and Wales grew so quickly from 1850 to 1900. (3)
g. Describe the potential changes to the population of England and Wales through to 2050. (2)

2. Imagine that you have to produce a report for the 'Census of India' office, giving a description of the Demographic Transition of India.

Resource B

Births, Deaths and population change in India from 1900 to 2050

YEAR	BIRTH RATE PER 1,000	DEATH RATE PER 1,000	TOTAL POPULATION
1900*	49.2	42.6	295 million*
1925	46.4	36.3	263 million
1950	41.7	22.8	361 million
1975	37.1	14.8	548 million
2000	28.5	9.2	1.08 billion
2025 est	16.2	9.4	1.46 billion
2050 est	13.2	10	1.66 billion

Data from Populstat, www.populstat.info
*1900 figure also includes Pakistan and Burma (what was known as British India).

a. Draw out the DTM graph for India using the statistics provided in the table. ($3 \times 3 = 9$ marks)

b. Describe the changes to birth rates, death rates and the increasing total population of India. Attempt to explain any reasons there might be for these changes. (2 marks for describing changes to birth rate, 2 marks for describing changes to death rate, 2 marks for describing the difference in trend between BR and DR, plus 2 marks for attempting to explain reasons for this trend = total 8 marks)

3. What differences do you notice about the shape and structure of the two DTM graphs that you have drawn out for England and Wales, and for India? (6)

4. Use the demographic Fact File comparison table below to describe and explain the main difference between births and deaths in India in 2011.
(3 marks for description and 3 marks for explanation = 6 marks)

Resource C

Demographic Fact File comparison (2011)

	INDIA	UK
Population	1.2 billion people	61.5 million people
Growth rate	1.4%	2.9%
Birth rate (2009)	22.8 per 1,000	12.7 per 1,000
Death rate (2009)	674 per 1,000	9.4 per 1,000
Life expectancy	63	79
Fertility rate	(62 for males/64 for females)	(77 for males/82 for females)
Infant mortality rate	2.68 children per woman (from 3.1 in 2001)	1.96 children per woman (from 1.63 in 2001)
	53.0 deaths per 1,000 live births	4.8 deaths per 1,000 live births

Data from United Nations Demographic Yearbook, unstats.un.org

Skills: Using GIS to investigate in-migration within an MEDC (Northern Ireland)

Many of us today have digital devices in our pocket that allow us to pinpoint our exact location or look up our location on a map. All of this technology is underpinned by the advances made using geographical information systems (or GIS).

GIS is described by the Ordnance Survey as being "a family of computer software tools that allow information to be linked to a geographical location." GIS stores information in a series of layers. Each layer contains data about a different feature for a location, in the form of maps, graphs and photography.

Today, more and more government agencies and businesses are using data captured in GIS to help them. GIS can be used to help predict which houses might be in danger of flooding, to analyse crime patterns or the spread of diseases across a country. It might even be used to help pizza delivery drivers plan their delivery routes.

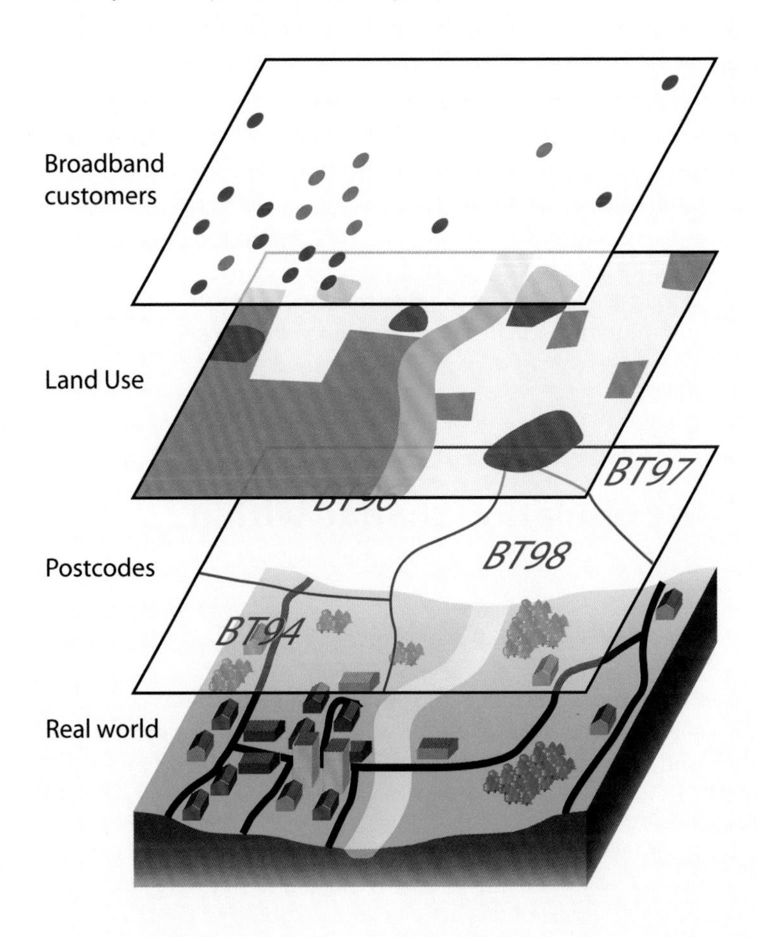

You can get a basic view of this when you use Google Maps or Google Earth. The software allows you to add different types of information onto a map by selecting a toggle box. If you want access to information you can choose whether to view details on traffic, weather, terrain, labels or access photos of the area.

GIS maps therefore allow geographers to interrogate key pieces of information associated with places. It allows you to look for patterns and solutions to issues using a map.

Figure 7

Different layers of information are plotted on top of each other for a particular location, building up a digital GIS map

Test yourself

What are the main types of GIS maps that can be generated? Use the Internet to research some examples. (3)

Migration

Migration is the movement of people from one place to another. Migration can take place over short or long distances and it can be a permanent one-way movement or it may be temporary. Some people will choose to migrate (voluntary migration), while others may be forced to do so (forced migration). Forced migrants sometimes become refugees or asylum seekers. Migrants usually need to cross some form of administrative boundary (either at a local government or national government level) for the migration to be counted in official figures.

Internal Migration This is when people move from one place to another within the boundaries of a national country. For example, someone who moved from Northern Ireland to England would be classified as an 'internal migrant'.	**International Migration** This is when people move from one country to another country internationally. For example, someone who moved from Northern Ireland to Australia would be classified as an 'international migrant'. If the migrant has the right documentation and visas this is a legal migration. If the migrant does not have permission to enter the country and tries to slip in past the border guards, this is an illegal migration.
Immigration This is when people move **I**nto a country.	**Emigration** This is when people are leaving or **E**xiting a country.

Migratory moves are rarely unplanned and often people from the same place will migrate along the same routes and towards the same places as people before them. This creates a migratory stream or flow from one place to another. The place that a migrant leaves is known as the **origin** and the place where they are intending to move to is called the **destination**. Sometimes there will be a large flow of people (the scale of migration will be large) from one place to another and sometimes the stream will only be a trickle (the scale of migration will be small).

Using GIS to investigate population change within Northern Ireland

Test yourself

Much of the population data that is collected within Northern Ireland can be accessed on the Northern Ireland Statistics and Research Agency (NISRA) website (http://www.nisra.gov.uk). There are many figures available based on past census information and mid-year estimates.

Recently, some of this information has been collated onto GIS maps as part of the NINIS area, which provides access to visual representations of some of the statistics they have. NINIS can be accessed at: www.ninis2.nisra.gov.uk. Clicking on the 'Interactive Content' link will show a menu of the various interactive maps available. For this activity you are going to need to click on the 'Population' section. A table of maps will appear and you need to click on the map labelled subset: 'Population Change', title: 'Components of Population Change'. Resource A is a screenshot of this map.

Use the map to find out the following:

1. Which Council area of Northern Ireland had the highest rate of population growth? (1)
2. Which Council area of Northern Ireland has experienced the highest amount of population decline? (1)
3. Click on Belfast and study the population change line graph below the map of Northern Ireland. Describe the trend of population change in Belfast over the last 10 years. (Click on Belfast and look at the graph of population change below the map.) (4)

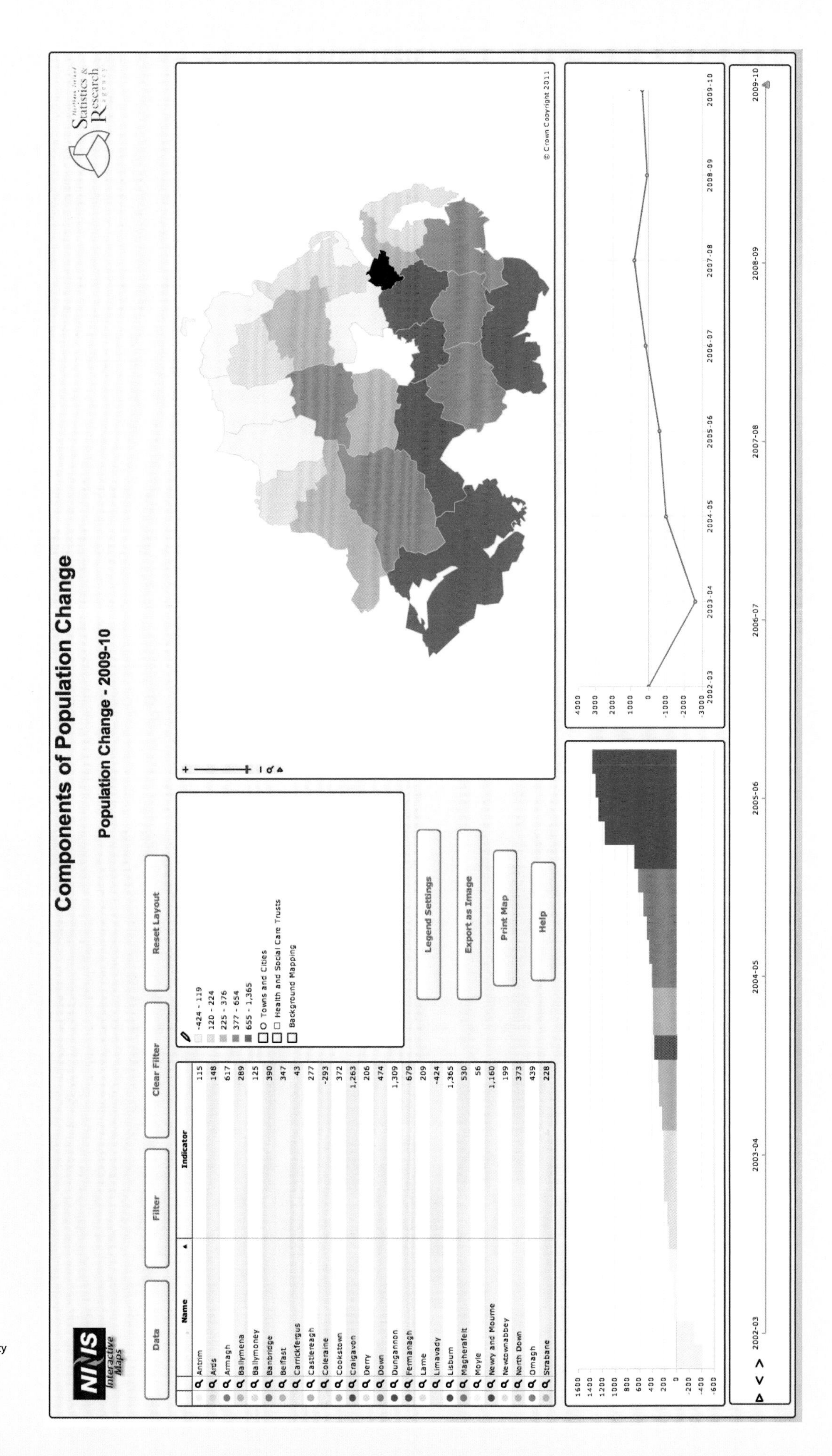

Resource A

Source: Neighbourhood Statistics (NISRA) Website: www.nisra.gov.uk/ninis

Using GIS to investigate migration flows within Northern Ireland

Test yourself

NINIS can be accessed at: www.ninis2.nisra.gov.uk. Clicking on the 'Interactive Content' link will show a menu of the various interactive maps available. For this activity you are going to need to click on the 'Population' section. A table of maps will appear and you need to click on the map labelled subset: 'Migration', title: 'Central and Eastern European Migrants'. Resource B is a screenshot of this map.

The map is titled 'Estimated size of A8 population by LGD in 2009'. The term 'A8' refers to the group of central and eastern European countries that joined the EU in May 2004: the Czech Republic, Estonia, Hungary, Latvia, Lithuania, Poland, Slovakia and Slovenia. A 'LGD' refers to a local district council area. A 'LGD' refers to a local district council area.

1. What do you think is meant by the term 'scale of migration'? (2)
2. What so you think is the 'origin of in-migration'? (2)
3. Use the map to answer the following:
   a. What areas of Northern Ireland experienced the least amount of migrants from the A8 countries? (2)
   b. What areas of Northern Ireland experienced the highest amount of migrants from the A8 countries? (2)
   c. Where is the greatest concentration of A8 migrants in Northern Ireland?
      i. How many live there? (2)
      ii. Why do you think the greatest concentration is found there? (2)
4. List three advantages of getting information from a GIS such as this. (3)
5. List three disadvantages of getting information from a GIS such as this. (3)

Further extension

1. Resource B on page 151 shows the estimated size of the A8 population in 2009. See if you can find more up to date information from the NINIS website about how the breakdown of this population might have changed.
   a. Describe any pattern that you notice between 2009 and any new figures that you find. (4)
   b. Explain reasons for any change that you find. What are the potential reasons why more people have either moved into or out of Northern Ireland? (4)
2. Use the following websites to find out more information about how GIS can be used in geography:

   National Geographic Education:
   http://www.mywonderfulworld.org/toolsforadventure/games/gis.html

   Ordnance Survey:
   http://www.ordnancesurvey.co.uk/oswebsite/gisfiles/
   and
   http://mapzone.ordnancesurvey.co.uk/mapzone/giszone.html

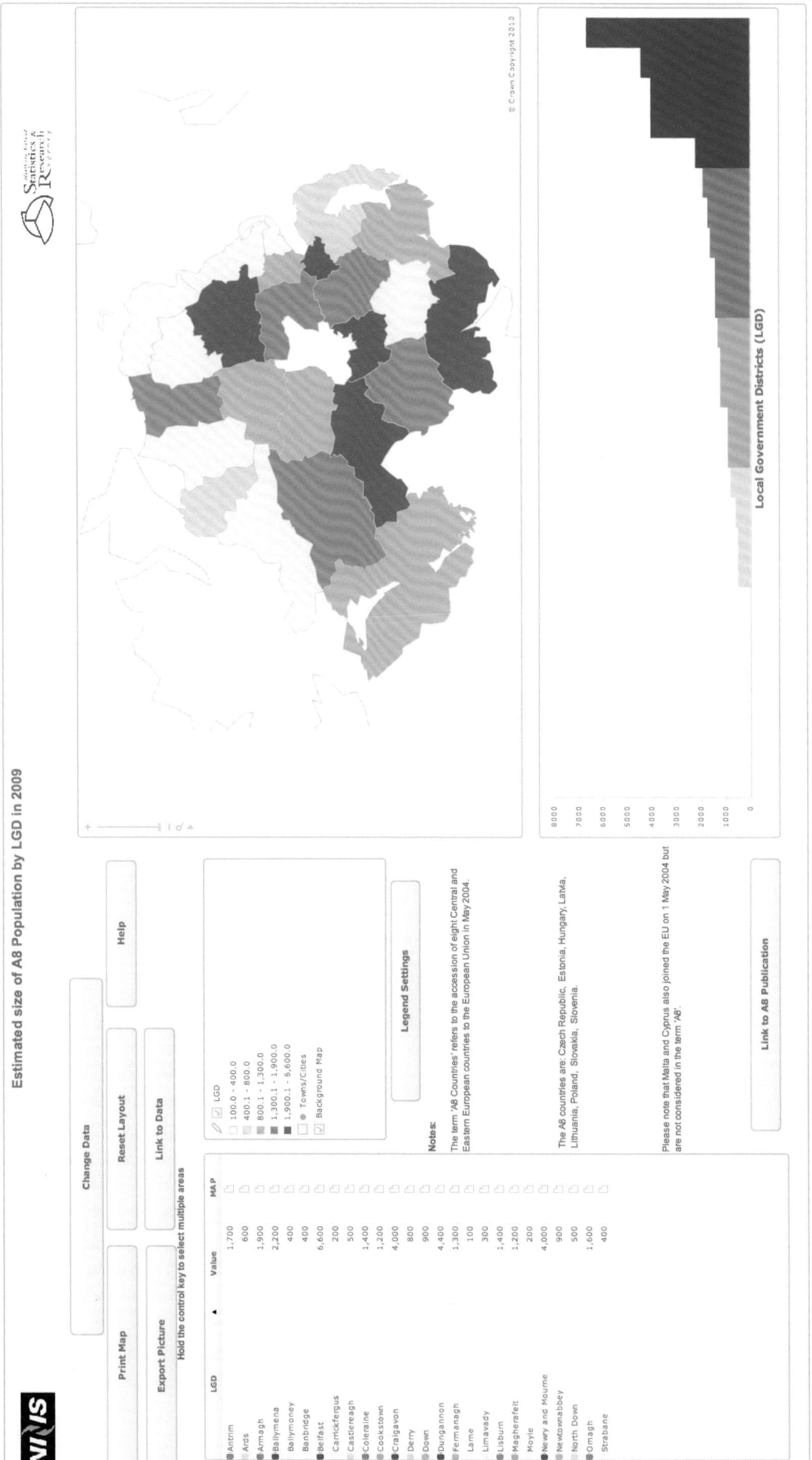

Resource B

Source: Neighbourhood Statistics (NISRA) Website: www.nisra.gov.uk/ninis

NINIS

Estimated size of A8 Population by LGD in 2009

Northern Ireland Statistics & Research Agency

Change Data

Print Map | Reset Layout | Help

Export Picture | Link to Data

Hold the control key to select multiple areas

LGD	Value	MAP
Antrim	1,700	
Ards	600	
Armagh	1,900	
Ballymena	2,200	
Ballymoney	400	
Banbridge	400	
Belfast	6,600	
Carrickfergus	200	
Castlereagh	500	
Coleraine	1,400	
Cookstown	1,200	
Craigavon	4,000	
Derry	800	
Down	900	
Dungannon	4,400	
Fermanagh	1,300	
Larne	100	
Limavady	300	
Lisburn	1,400	
Magherafelt	1,200	
Moyle	200	
Newry and Mourne	4,000	
Newtownabbey	900	
North Down	500	
Omagh	1,600	
Strabane	400	

LGD
100.0 - 400.0
400.1 - 800.0
800.1 - 1,300.0
1,300.1 - 1,900.0
1,900.1 - 6,600.0
Towns/Cities
Background Map

Legend Settings

Notes:

The term 'A8 Countries' refers to the accession of eight Central and Eastern European countries to the European Union in May 2004.

The A8 countries are: Czech Republic, Estonia, Hungary, Latvia, Lithuania, Poland, Slovakia, Slovenia.

Please note that Malta and Cyprus also joined the EU on 1 May 2004 but are not considered in the term 'A8'.

Link to A8 Publication

8000
7000
6000
5000
4000
3000
2000
1000
0
Local Government Districts (LGD)

The positive and negative impacts of international migration

International migration is the movement of people from one country to another. Over the last 300 years there have been many different streams of migration as people moved from one country to another.

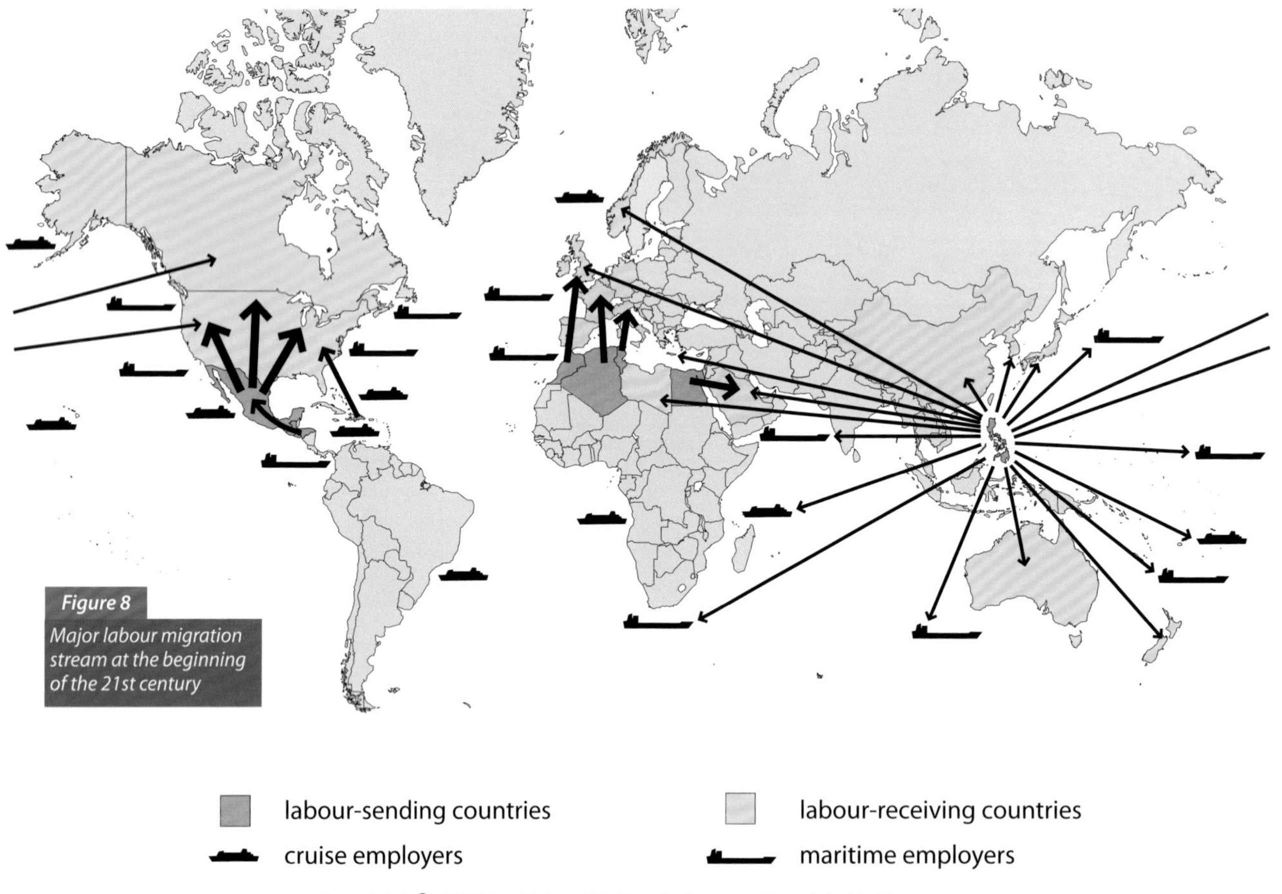

Figure 8
Major labour migration stream at the beginning of the 21st century

The decision to migrate from one place to another is rarely taken lightly and there are many different factors that will persuade someone to leave or stay.

Voluntary migration is when someone makes a 'free choice' to migrate. They are usually looking for a better life, a better standard of living and more personal freedom. They might be looking for the chance of a better job, to earn a better salary or even to pay less tax. They might be moving to a better climate or to a location with easier access to hospitals, schools or sources of entertainment. For many migrants these are **PULL factors**, which attract them to particular place.

Forced migration is when the migrant has no personal choice in the decision to move. This is caused by **PUSH factors**, where the person is pushed from an area. Push factors include natural disasters, economic problems, war and racial discrimination. Overpopulation can also push resources in an area to their limit creating famine, starvation and desertification.

International migration is not a new phenomenon; it has been responsible for the spread of people across the world since time began. No country and no part of the world has been unaffected by migration but some countries, such as the USA, Canada, Australia and New Zealand, have been more heavily influenced by migratory movements than others.

Migration can bring both positive and negative impacts both to the people in the origin country (out migration) and those in the destination country (in migration).

Case Study: International migration from Turkey to Germany

Immigration into Germany reached its peak in the 1990s and caused the German government to look carefully at their migration strategy. In January 2000 a new Nationality Act was passed and an additional Immigration Act came into force from 2005. 2010 population figures show that Germany currently has the highest number of foreign-born citizens with in the EU (close to 10 million people and making up 12% of their population).

One of the main reasons for this growth was due to the loss of many men of working age following the Second World War. The loss led to a serious skill shortage and too few people to fulfil the jobs that needed completed. This undermined Germany's economic prospects and it began to look for labour from outside the country. The German government started by making labour recruitment agreements ('Anwerbeabkommen') with Italy in 1955, Spain in 1960, Turkey in 1961 and Yugoslavia in 1968.

Many Turkish workers made the decision to travel to Germany and become a guest worker ('Gasterbeiter'). In 2006, there were 6.7 million foreigners living in Germany, which is about 8.2% of the total population. Of this, 1.7 million people held Turkish citizenship. However, by 2010 the German Embassy estimated that there were about 4 million Turks in Germany. In recent years there has been increasing integration of Turkish people into German society, for example footballers such as Mehmet Scholl and Mesut Ozil are from Turkish-German extraction. Ozil, who is third generation Turkish-German explains, "My technique and feeling for the ball is the Turkish side to my game. The discipline, attitude and always-give-your-all is the German part."

Recent statistical data from the German Federal Office for Migration and Refugees shows that in 2011 31,021 Turks entered Germany. The total number of people with immigrant backgrounds living in Germany increased by 216,000 between 2010 and 2011.

The majority of Turkish people in Germany have migrated into cities such as Dusseldorf, Cologne, Dortmund and Munich. Around 8% of the Turkish ethnic population have settled in Berlin (around 300,000 people) especially in a neighbourhood called Kreuzberg which is sometimes known as 'Little Istanbul'.

The table below helps to identify some of the main impacts that migration movements can have on (a) the economy and (b) services. These impacts can have both positive and negative effects on the area of origin (Turkey) and the area of destination (Germany).

	POSITIVE IMPACT ON ORIGIN (TURKEY)	POSITIVE IMPACT ON DESTINATION (GERMANY)	NEGATIVE IMPACT ON ORIGIN (TURKEY)	NEGATIVE IMPACT ON DESTINATION (GERMANY)
Impact on the Economy	• As more people are leaving Turkey this is taking pressure off Turkey's unemployment rate. • There is more money coming back into the economy as family members send home remittances. • The majority of guest workers only spend a few years working in Germany and when they return they bring more money and more skills with them, making them more valuable to the Turkish economy.	• After the war there was a serious labour shortage. In 1961, with the building of the Berlin Wall many workers were needed. Turkey provided a cheap, skilled, constant supply of eager workers willing to fill the gap. • Rises in school leaving age and earlier retirement ages meant that there was a smaller workforce. Turkish immigrants provided the additional workforce needed to keep production at maximum levels. • Working conditions and wages were the best in Europe. • The increased numbers of cheap labour allowed rapid expansion of German manufacturing.	• The people who migrate are usually the more educated and skilled workers so the country experiences a 'brain drain'. • The people who are left will be the less skilled and therefore productivity and innovation will decline.	• Even in 1961 there were still Germans who remained unemployed and many felt that the call for migrant workers was due to pressure from the USA to help stabilise Turkey. • In 1967 and 1990 recession led to unemployment. Many German people lost their jobs but Turkish people remained employed in their low paid jobs. This fuelled social and ethnic conflict in some cities. • Much money earned by the migrant workers (remittances) was sent back to Turkey and out of the German economy.
Impact on Services	• As more people leave Turkey, this can reduce the pressure on the country's services, allowing the government to concentrate on essential services. • As more money (remittances) is sent back home, families can afford to use more of the services available in the country. • In recent years nearly as many people are leaving Germany and going back to Turkey as are leaving Turkey to go to Germany. As people return home, they expect the same standard of living as they had in Germany and there is greater uptake of services.	• Turkish people usually took jobs that Germans did not want to do and many of these were providing low cost personal services for the German population. For example, driving trams and buses and cleaning.	• As more of the economically active people are leaving, this might cause a reduction in services. There will be fewer people to pay for and fewer people to use the services. For example, bus routes will stop and rural schools will close.	• More translation services are needed, which costs money. For example, more leaflets are needed in Turkish and there are more students in school for whom German is not the first language.

	POSITIVE IMPACT ON ORIGIN (TURKEY)	POSITIVE IMPACT ON DESTINATION (GERMANY)	NEGATIVE IMPACT ON ORIGIN (TURKEY)	NEGATIVE IMPACT ON DESTINATION (GERMANY)
Other Impacts	• The more migrants who experience life in other countries and return to Turkey, the more enriched the skill and knowledge base of the Turkish population becomes, allowing for more opportunities.	• Increase in people from other countries increases the cultural diversity in the country and brings new attitudes to fashion, religion, food and leisure activities. For example, Turkish kebabs, shaves and baths. • Germany currently has the lowest fertility rate in the EU, with only 1.36 children per woman. The migrants are young and within the reproductive years. Many of them will have children, which will help stabilise birth rates within the country.	• Many traditional Turkish people are concerned that the increased global attitude is diluting their Turkish religious beliefs and cultural traditions. They feel that these might need to be protected in the future. • Most of the Turkish migrants are young. • This means that the birth rate in Turkey will be reduced as people are giving birth overseas rather than at home.	• In the early 1990s a number of racial/ethnic attacks on Turkish people fuelled concerns of social problems within Germany. For example, an arson attack in Solingen killed five Turkish people who were all part of a family that had been living in Germany for 23 years. • Many migrant workers who had lived in Germany wanted German citizenship but had to wait a very long time as the German authorities refused to allow migrants to become naturalised citizens. • In 2010, the German Chancellor Angela Merkel announced that 'multiculturalism' (where people would live happily side by side) had failed and that immigrants would need to do more to integrate into society, including learning the German language. This made it more difficult for new immigrants to 'fit' into German society.

After the Second World War, Germany was divided into East and West. In the 1960s and 1970s much of the immigration into West Germany came from the Southern European countries such as Turkey, Yugoslavia, Italy, Greece and Spain. In 1990 East and West Germany were reunified to form Germany. Recently, changes to migration rules within the EU mean that countries from the East of Europe account for much higher percentages of immigrants into Germany, such as Poland, Romania, Bulgaria, Hungary and still from Turkey.

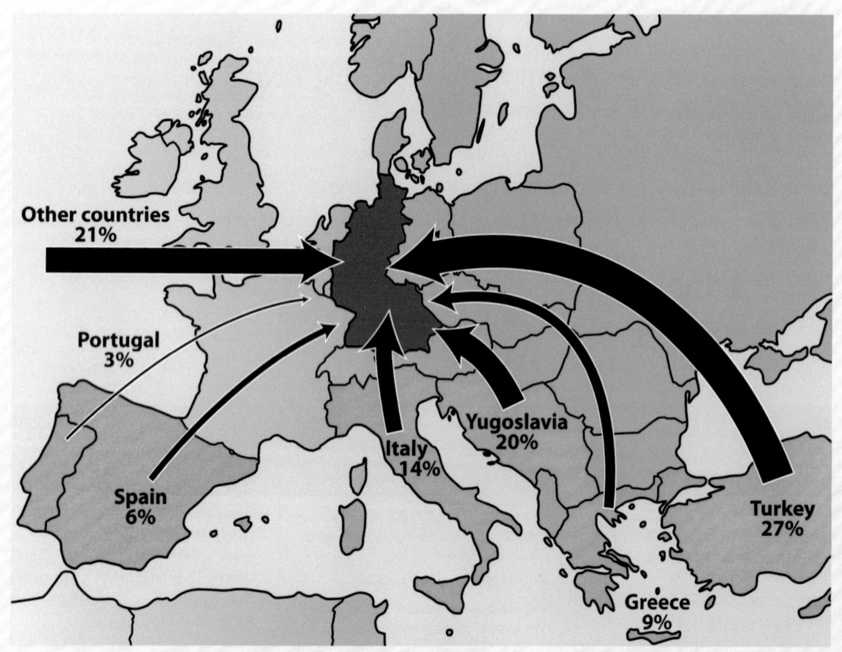

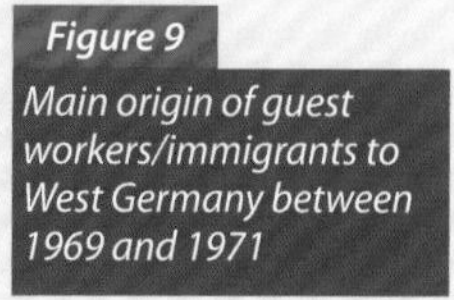

Figure 9

Main origin of guest workers/immigrants to West Germany between 1969 and 1971

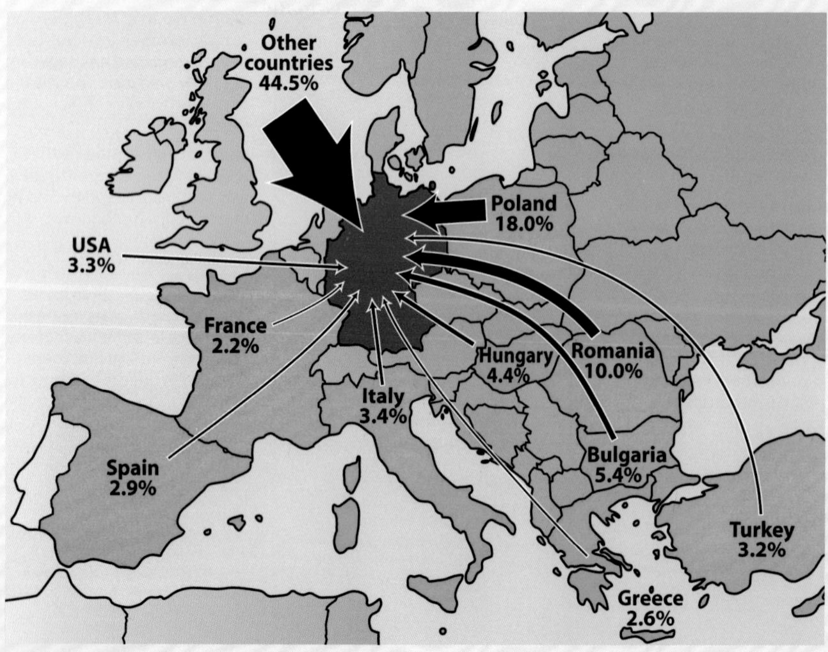

Figure 10

Origin of migrants moving into Germany in 2011

Source: Figures from the German Federal Statistics Office

Test yourself

1. Resource A shows the ten most common foreign citizenships in Germany at the end of 2006.
   a. Use the information from the table to draw a bar chart representing the top ten foreign citizenships in Germany. (10)
   b. Use the information from the table to draw a pie chart showing the total foreign population for Germany in 2006 (out of 6.7 million foreign people). (7)
   c. What pattern is noticeable about the location of the immigrants? (4)
   d. What information is not included in this table which might be useful for determining how many of the people in Germany are from Turkish ethnicity? (3)

Resource A

CITIZENSHIP	TOTAL
Turkish	1,738,831
Italian	534,657
Polish	361,696
Serbian-Montenegrin	316,823
Greek	303,761
Croatian	227,510
Russian	187,514
Austrian	175,653
Bosnian-Herzegovinian	157,094
Ukrainian	128,950

Source: Figures from the German Federal Statistics Office

2. What do you think the main push and pull factors are for Turkish people when deciding whether to migrate to Germany or not? (Push: why did people want to leave Turkey? Pull: why did they want to go to Germany?) (3 + 3 = 6 marks)
3. Why did people originally migrate from Turkey to Germany? (3)
4. On balance, do you think that this migration movement has been beneficial for Turkey? Explain your answer in detail. (5)
5. Do you think that this migration movement was necessary and beneficial for Germany? (5)
6. Some geographers argue that the biggest issue related to migration is the social consequences of how migration splits up families. To what extent would you agree or disagree with this statement? (8)

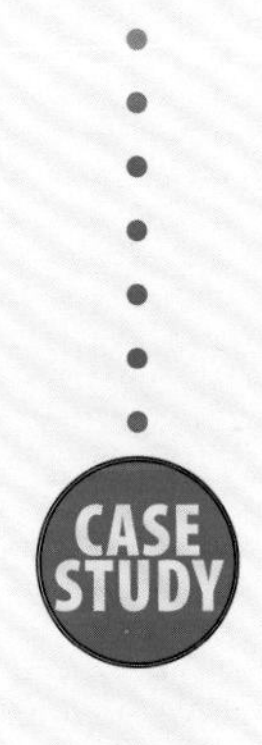

Population structure

A **population pyramid** is a graph that can be used to show the specific age breakdown of a population. It allows us to assess the different aspects of population structure and helps us to understand the birth rate, death rate and life expectancy of people in a particular country. It also allows us to compare the demographic development of one country to another. Really it is just two bar graphs drawn back to back.

Usually population pyramids are drawn with the male population on the left and female population on the right. Pyramids are typically divided into five year age groups (called cohorts). However, more modern, computer-generated graphs show the population for every year group. The majority of population pyramids will show each cohort as a percentage of the total population but some will show the raw numbers of people within each age category.

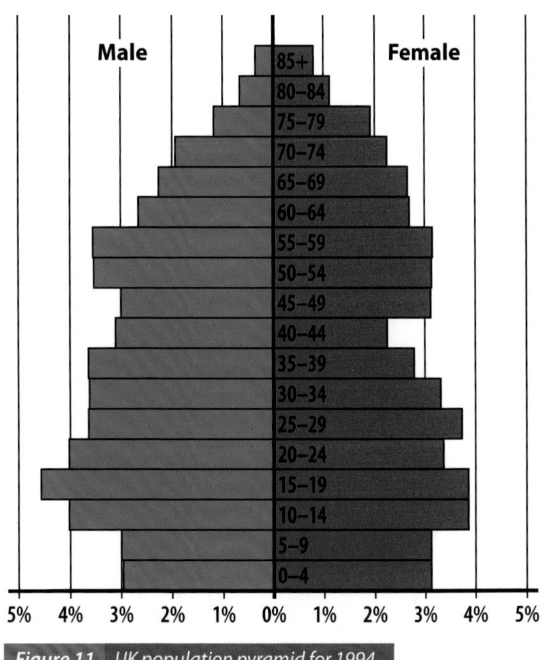

Figure 11 UK population pyramid for 1994

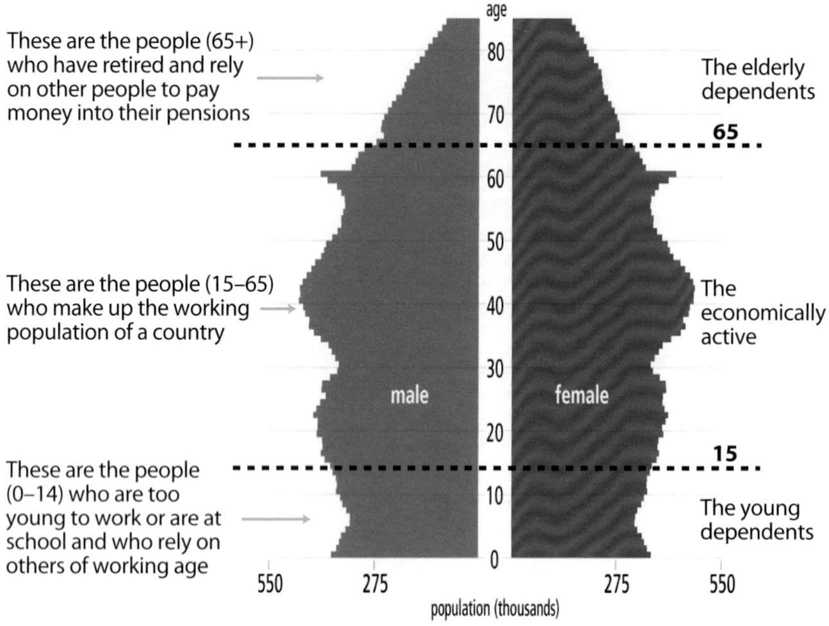

These are the people (65+) who have retired and rely on other people to pay money into their pensions

These are the people (15–65) who make up the working population of a country

These are the people (0–14) who are too young to work or are at school and who rely on others of working age

The elderly dependents

65

The economically active

15

The young dependents

age

male

female

population (thousands)

Source: Office for National Statistics

Figure 12

Age Structure of the United Kingdom 2007

Reading a population pyramid

Usually you will be using a population pyramid to decide whether the population is representative of a More Economically Developed Country (MEDC) or a Less Economically Developed Country (LEDC).

A MEDC population pyramid will have a much higher number of older people (at the top of the pyramid), even sides and a narrow base, whereas a LEDC pyramid will have a higher number of younger people (wide base of the pyramid), decreasing sides and a narrow top (with few people reaching old age).

Structured notes

Draw your own population pyramid

It is unlikely that you will be asked to draw out a full population structure during an exam, as it would take up too much time. However, a common task is for you to complete a pyramid. The best way to understand how a population pyramid works is to practice by drawing your own. You need some graph paper to complete this task. Follow the instructions below carefully.

You are going to draw a population pyramid that represents the population of Ballymena Borough Council as recorded in the 2001 census. The total population of this area was 58,610 people: 28,571 males and 30,039 females.

Resource A

The population of Ballymena Local Government District (from 2001 Census)

Source: Neighbourhood Statistics (NISRA) Website: www.nisra.gov.uk/ninis

AGE	MALES	% OF TOTAL	FEMALES	% OF TOTAL
0–4	1877	3.2	1853	3.2
5–9	2061	3.5	1908	3.3
10–14	2210	3.8	2100	3.7
15–19	2109	3.6	2003	3.4
20–24	1707	2.9	1661	2.8
25–29	1876	3.2	1889	3.2
30–34	2296	3.9	2336	3.9
35–39	2093	3.6	2239	3.8
40–44	1971	3.4	1981	3.4
45–49	1938	3.3	1941	3.3
50–54	1856	3.2	1901	3.3
55–59	1682	2.8	1794	3.1
60–64	1384	2.4	1482	2.5
65–69	1191	2.0	1311	2.2
70–74	940	1.6	1242	2.1
75–79	713	1.2	1026	1.8
80–84	426	0.7	717	1.2
85–89	180	0.3	436	0.7
90+	61	0.1	219	0.4

1. On your graph paper draw out the frame for a population pyramid. Draw a horizontal line along the bottom labelled '% of the total'. The maximum you need is 4%, so work out a suitable scale.
2. Next draw a vertical line up the middle of the page. Males will be on the left and females on the right.
3. Insert 19 separate bars to represent each of the age cohorts from 0–4 up to 90+. Label these clearly along the centre axis.
4. Use the figures in Resource A to draw the bars to represent the percentage total for the male and female populations for Ballymena.
5. Write a suitable title on your graph.

Test yourself

Using your population pyramid for Ballymena's **2001** census and the figures in **Resource A**, answer the following questions:

1. What pattern do you notice about the difference between males and females in Ballymena at ages 0–24? Use figures to support your answer. (3)
2. Describe the pattern showing males and females in Ballymena between ages 55–90. Use figures to support your answer. (3)
3. Give one reason why you think there are more females than males in the Ballymena population. (3)
4. Which age cohort is the biggest in Ballymena for both males and females? Can you suggest a reason why this might be the case? (4)
5. What evidence from the shape of the population pyramid might lead you to believe that Ballymena was part of a MEDC? (4)

Extension activity

AGE	MALES	FEMALES
0–4	2112	1987
5–9	1995	1962
10–14	1996	1994
15–19	2136	1904
20–24	2054	1878
25–29	1948	2032
30–34	1946	2021
35–39	2127	2240
40–44	2469	2514
45–49	2202	2284
50–54	2059	2012
55–59	1920	1916
60–64	1815	1867
65–69	1563	1723
70–74	1157	1363
75–79	909	1155
80–84	603	937
85–89	298	545
90+	89	312
Total 64,044	31398	32646

The population breakdown for Ballymena from the 2011 census is shown in Resource B.

Draw another population pyramid for the 2011 census, using the same scale so that you can compare how the Ballymena population has changed over the last 10 years.

Resource B

The population of Ballymena Local Government District (from 2011 Census)

Source: Neighbourhood Statistics (NISRA)
Website: www.nisra.gov.uk/ninis

Compare and Contrast the structures of LEDCs and MEDCs

There is a strong link between the shape of a population pyramid and the stage in the Demographic Transition Model that a country might find itself in.

DTM STAGE	POPULATION SHAPE	SHAPE DESCRIPTION	KEY INFORMATION
1 (LEDC)	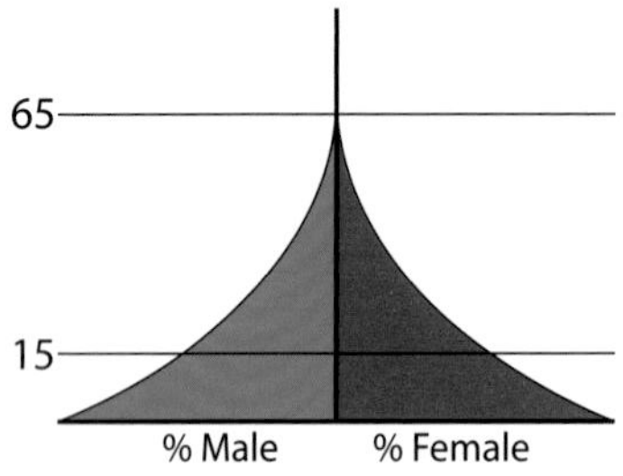 65 15 % Male % Female	**Concave**	High Birth Rate (<15). High Death Rate (15+) with decreasing number of people in each age cohort. Very short life expectancy (around 30 years).
2 (LEDC)	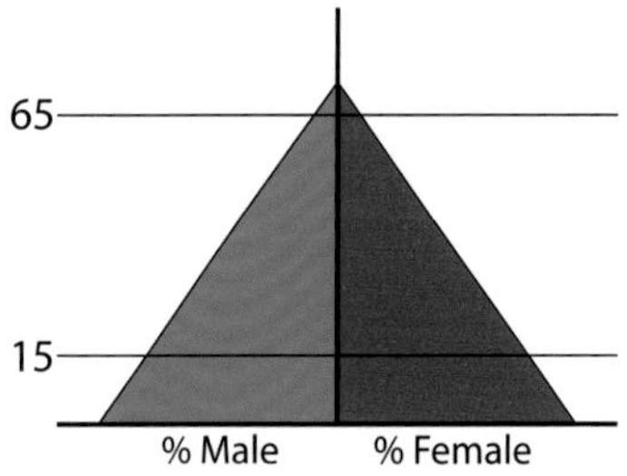 65 15 % Male % Female	**Triangle**	High Birth Rate (<15). Fall in Death Rate (15+) and slight increase in number of people surviving in each age cohort. Still a short life expectancy (around 40 years).
3 (LEDC/MEDC)	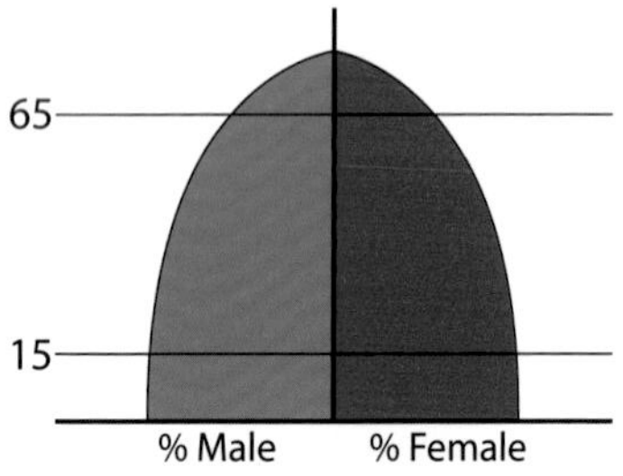 65 15 % Male % Female	**Tongue**	Falling Birth Rate (<15). Falling Death Rate (15–65). Life expectancy increases with more people living beyond 65.
4 (MEDC)	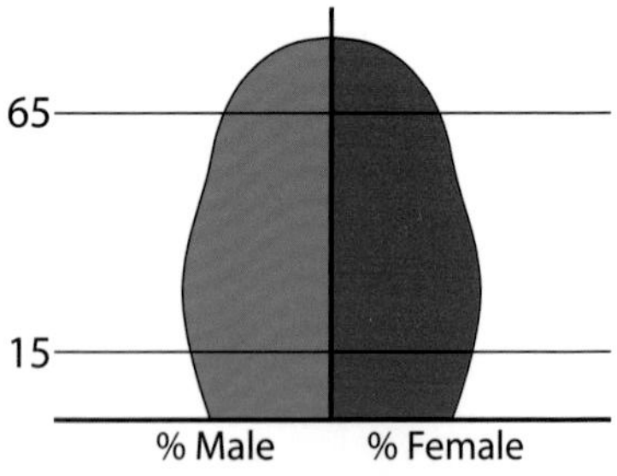 65 15 % Male % Female	**Leaf**	Low Birth Rate (<15). Low Death Rate (15–65). Life expectancy continues to increase with a high number of people living beyond 65. Average life expectancy reaches 75.
5 (MEDC)	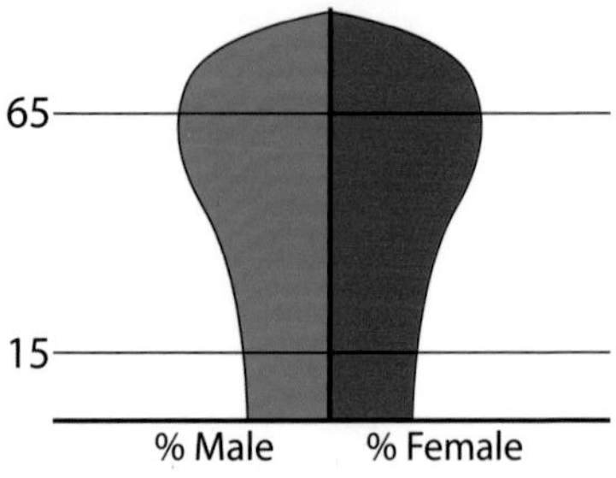 65 15 % Male % Female	**Hot air balloon**	Very Low Birth Rate (<15). Low Death Rate (15–65). Life expectancy continues to increase and average life expectancy pushes up to 85 and beyond. Population structure is very 'top heavy'.

Test yourself

Look at the population pyramids below and answer the questions which follow.

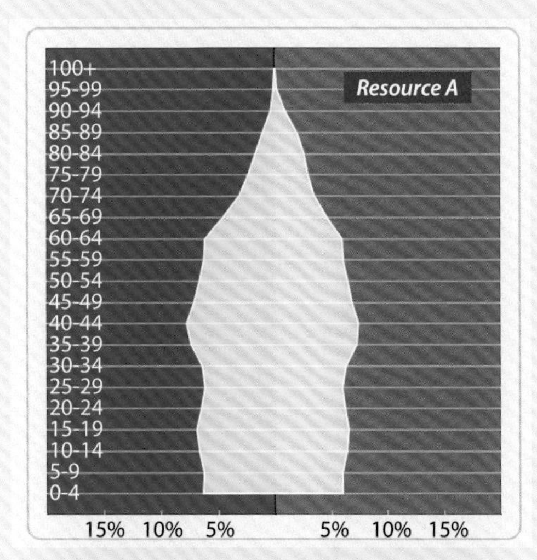

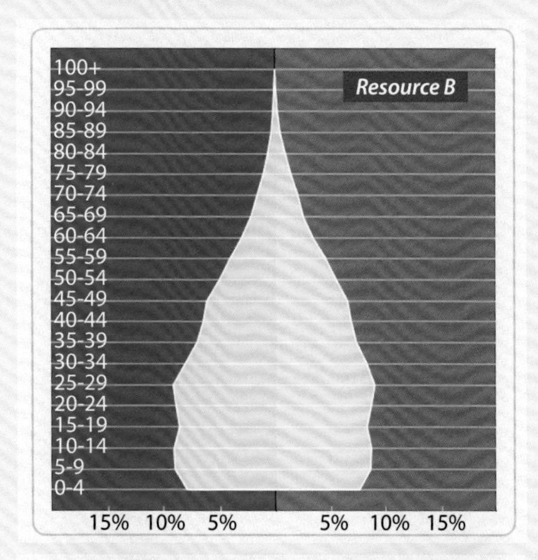

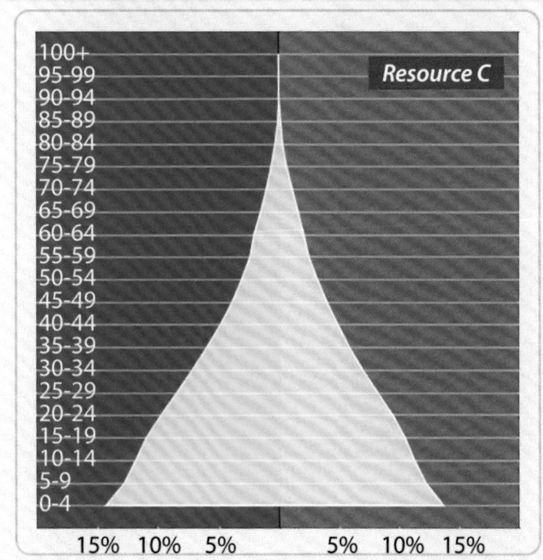

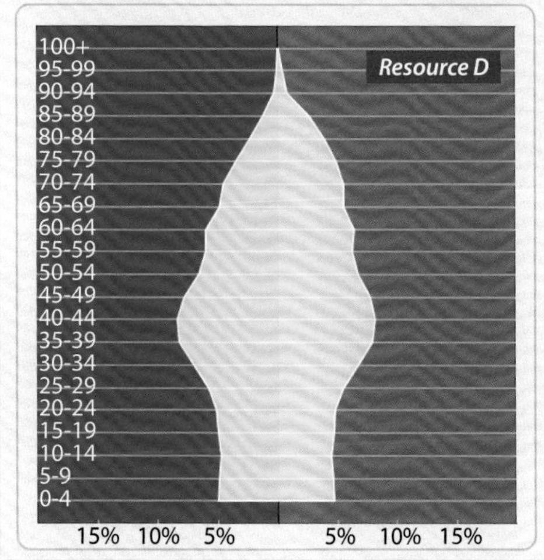

Resources A–D: Visualisation by Martin De Wulf, www.populationpyramid.net, Data from Population Division of the Department of Economic and Social Affairs of the United Nations Secretariat, World Population Prospects: The 2008 Revision

1. Resources A–D show the population structure for four countries that are at different places within their demographic development. See if you can match up the graphs with each of the following countries:

   Norway
   Brazil
   Ghana
   Italy (4)

2. Which stage do you think each of the four graphs represent? Give at least two pieces of evidence from each population pyramid to back up your answer.

   Your answers could be structured as follows:

   Graph A represents Stage 3. (1 mark)
   Evidence 1: Life expectancy is high because ... (2 marks)
   Evidence 2: Birth Rate is lower (at 6%) because ... (2 marks)

   (4 graphs × 5 marks = 20 marks)

3. The population structures of Graph C and Graph D are very different from each other. Suggest some reasons why the population behaves so differently in these two countries. (4)

4. Which two graphs do you think show the population structure for LEDCs and which two graphs show MEDCs? (2)

Test yourself

Resource A Estimated population pyramid for Scotland, June 2005

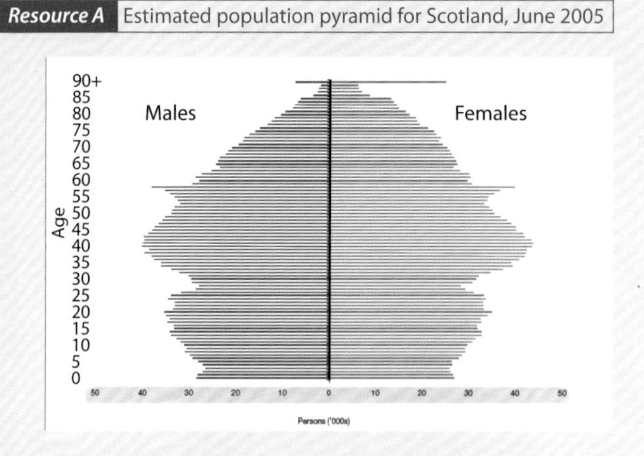

The population pyramid in Resource A shows two baby booms in 1947 and in the 1960s as bulges in the population at age 58 and between 35 and 45.

1. Is there any evidence in this population pyramid of out-migration? (3)
2. Is there a noticeable difference in the number of males and females aged 24–32 in Scotland at this time? Explain what could have caused this. (4)
3. The birth rate for Scotland has fallen rapidly in recent years. Using only evidence from the population pyramid, explain why this might be the case. (3)

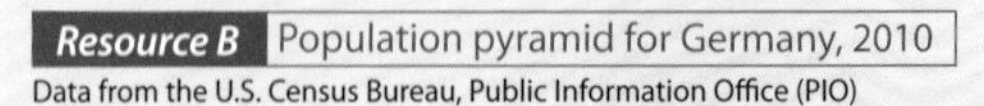

Data from the U.S. Census Bureau, Public Information Office (PIO)

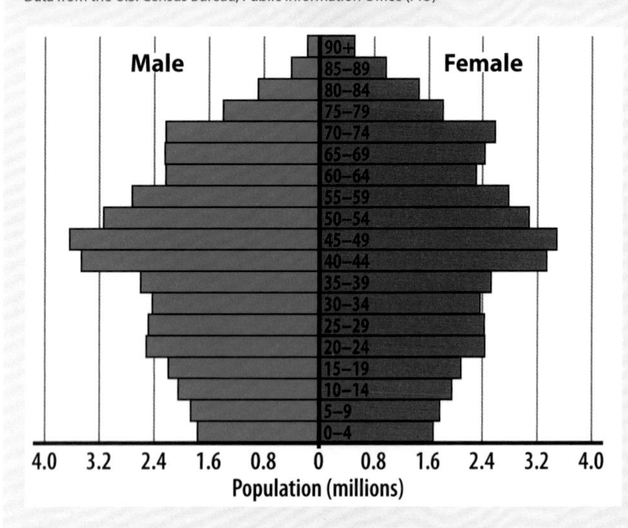

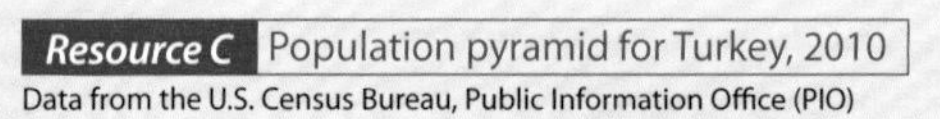

Data from the U.S. Census Bureau, Public Information Office (PIO)

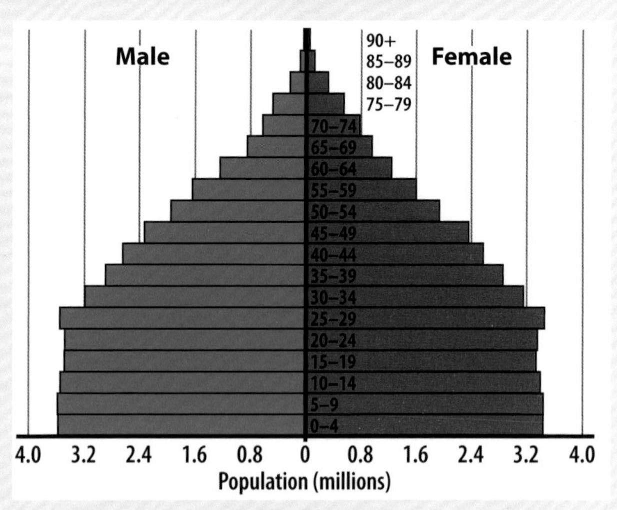

4. Take a look at the population pyramids for Germany and Turkey in 2010 (Resources B and C) and answer the questions which follow:
   a. In which country do you think the number of births is higher? Give evidence to back up your answer. (4)
   b. In which country are there more people aged 80 and above? Give evidence for your answer. (4)
   c. Which stage of the DTM do you think Germany would be found at? Give some reasons for this. (3)
   d. Which stage of the DTM do you think Turkey would be found at? Give some reasons for this. (4)
   e. What are the main differences between the population structures for each country? (4)
   f. Using only the population pyramids to help you, explain which country is more likely to have experienced in-migration. (3)

The social and economic implications of aged and youth dependency

A MEDC pyramid for an aged dependent population (the UK)

Figure 13 shows a population pyramid for the UK in 2010. It is around Stage 4 of the Demographic Transition Model. The pyramid has steep sides, as very few people are dying (death rate = 8/1000) and the majority of children that are born (birth rate = 12/1000) are surviving until they are 65 years old, if not older. This extended life expectancy is due to good health services.

The UK could therefore be described as an 'ageing' population, with a sizeable proportion of its population retired and older than 65.

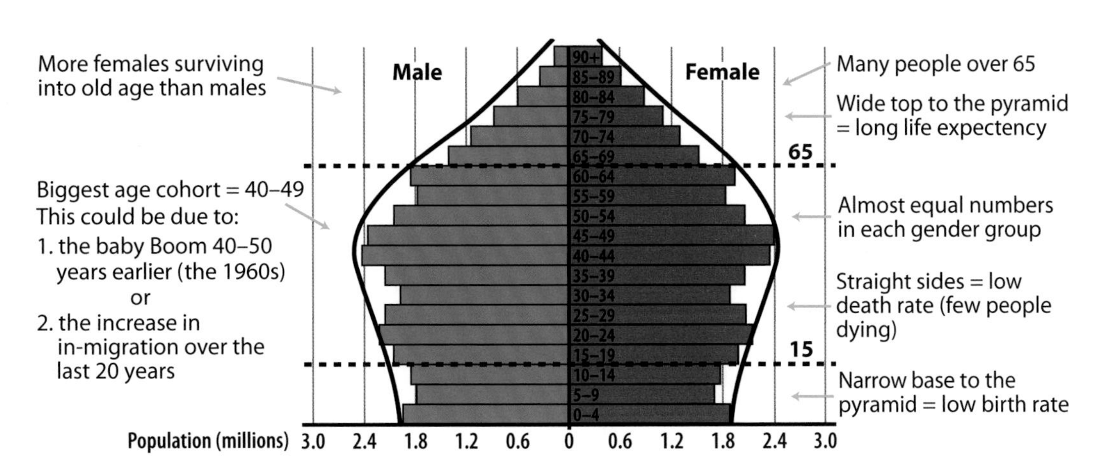

Figure 13
Population pyramid for the UK (MEDC) in 2010

Data from the U.S. Census Bureau, Public Information Office (PIO)

The UK is often described as having a high dependency ratio. Dependency helps to indicate the balance of people who are in work (the working population) against those who are not in work (the young and the elderly). In other words, the people who pay taxes for services and those who do not. Those who do not are also mainly dependent on these taxes for schools, pensions and healthcare. The dependency ratio for a country can be calculated by dividing the number of people that make up the dependent population by the working population and multiplying by 100.

$$\text{Dependency ratio} = \frac{\text{youth dependent population (0–14)} + \text{aged dependent population (65+)}}{\text{working population (15–64)}} \times 100$$

MEDCS usually have a dependency ratio between 50 and 75. This means that every 100 people in work need to earn enough money to pay for about 50 to 75 dependents. LEDCs usually have a dependency ratio above 100.

The dependency ratio for Northern Ireland (2009)

$$\text{Dependency ratio} = \frac{\text{youth dependent population (0–14)} + \text{aged dependent population (65+)}}{\text{working population (15–64)}} \times 100$$

$$\text{Dependency ratio} = \frac{382{,}127 + 301{,}867}{1{,}104{,}902} \times 100 = 61.9$$

Figures from: Neighbourhood Statistics (NISRA) Website: www.nisra.gov.uk/ninis

The social implications of an aged dependency

There are a number of reasons why a country might be experiencing an aged dependency. In MEDCs death rates have fallen due to improvements in:

- hygiene standards, sanitation, water and sewage treatment, and health education.
- medicine, resulting in cures, immunisations and vaccines.
- access to hospitals and doctors.
- diet and access to food supply.

As the population continues to get older, MEDCs often face a range of social issues that will need to be addressed by the government.

1. **Care for the elderly:** As people get older they might not be able to look after themselves as well as they once could. Some might need a little support from a care worker who visits daily or have their meals provided by a meals at home service. Others may decide to move into sheltered accommodation, a residential home if they need further support, or a nursing home if they need more complex care, such as increased medical assistance. All of this care can be quite expensive.
2. **Impact on family life:** As life expectancies increase, older family members are going to live longer and this can put pressure on family life. For example, 65 year old children may be responsible for looking after 90 year old parents. Families might have to take hard decisions about how to best care for their elderly relatives; is it best to care for them at the parent's own home, the relative's home or will they receive better care in a residential home?
3. **Medical issues:** As medicine advances it provides mores cures and improved methods of treatment for illnesses that previously claimed people earlier in their lives. People are now living long enough to suffer from 'degenerative' or long term illnesses (such as Alzheimer's or Parkinson's disease).
4. **Loneliness:** Many elderly people end up living alone when their partner dies. Women experience this more as their life expectancy is nearly 5 years older than for men.
5. **Ageism:** Some elderly people who reach the age of 65 or 66 would prefer to continue to work but due to corporate or government policy they may be forced to retire. Some older workers also find it difficult to get another job and can face bias due to their age.
6. **Security:** Some older people feel more vulnerable to crime and vandalism.

The economic implications of an aged dependency

There can also be some economic and financial implications involved when a country has an increase in its older population.

1. **Residential care:** The government needs to set aside an increased amount of money to create and maintain residential accommodation for elderly people. As people live for longer their needs will change and they will require greater access to ground floor accommodation, lifts, single storey houses or sheltered accommodation.
2. **Health care:** Quality healthcare for elderly people can be expensive. The government needs to set aside an increased amount of money to cover prescriptions, dental treatment, home visits and home help.
3. **Benefits:** Each elderly person receives a state pension. Thirty years ago, when life expectancies were lower, it was common for a man to retire at 65, claim his pension for 7 years and die around age 72. However, today a man who retires at 65 might live on to 82 years and the government has to continue to pay out the pension for an additional 10 years. The problem for women is further compounded, as until very recently the retirement age for women was 60, so a typical woman might be claiming a pension for up to 30 years. Other benefits such as free public transport, TV licences and winter fuel allowances also add to the financial burden.

A LEDC pyramid for a youth dependent population (Kenya)

Figure 14 shows a population pyramid for Kenya in 2010. Over the last 20 years Kenya has remained around Stage 2 of the Demographic Transition Model. The pyramid has a pronounced concave pyramid shape, where the number of people in each age cohort continues to get smaller as the population gets older. Birth rates remain high (birth rate = 35/1000) but death rates have plummeted in recent years (death rate = 8/1000). However, it would be fair to say that the effect of this improvement in life expectancy has still to be felt fully throughout the Kenyan population.

The wide base of the pyramid indicates the high birth rate (and a total fertility of 4.4). However, the infant mortality rate in Kenya remains relatively high (about 47 for every 1,000 live births).

One of the largest changes in the study of population over the last 15 years has been the rapid decline of death rates in LEDCs. The birth rate for Kenya has maintained a steady but high rate since 1995 (from 36/1000 to 35/1000 in 2010).

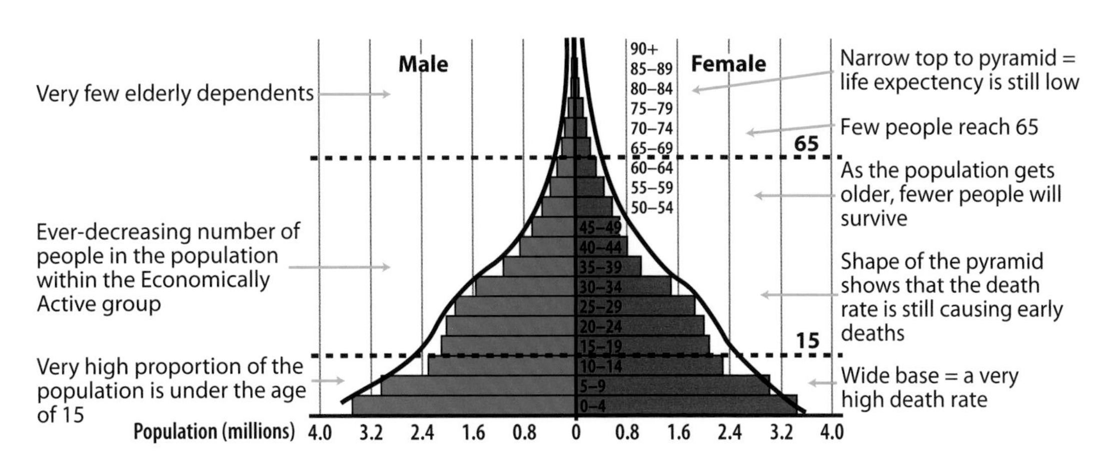

Figure 14

Population pyramid for Kenya (LEDC) in 2010

Data from the U.S. Census Bureau, Public Information Office (PIO)

Many of the young people who have been born in the last 15 years have survived. Improved investment and facilities in healthcare, developments in education and a better understanding of how diseases such as AIDS/HIV, malaria and cholera spread has led to a dramatic drop in Kenya's death rate (from 10/1000 in 1995 to 8/1000 in 2010). The difference between the death and birth rate means that the population is increasing faster than ever, causing Kenya's population to grow at 2.7% per year.

Kenya can therefore be described as a youthful population, with a sizeable proportion of the population under the age of 15. As the population continues to grow and expand due to more and more babies being born and surviving, the country will face a range of social and economic issues that will need to be addressed by the government.

The social implications of a youthful dependency

There are a number of reasons why a country might be experiencing a youthful population. In LEDCs birth rates are high due to the following:

- The infant mortality and child mortality rates in most LEDCs remains high. As parents are not always guaranteed that their children will reach adulthood, they tend to have larger families, to ensure some survive.
- Many people in LEDCs are subsistence farmers and they need children to help provide a good supply of labour. Parents also try to ensure that someone is available to care for them when they grow old.
- In many LEDCs people do not have the same access to education as in MEDCs. This has a big impact on family planning, as many do not know how to use family planning measures effectively. In addition, they do not have much disposable income available to spend on contraceptives/birth control.
- In many LEDCs having a large family is a cultural strength which can increase the respect and tribal role for a man.

As more people are born and populations increase, LEDCs often face a range of social issues due to the youthful nature of their population:

1. **Lifestyle:** Many of these children are born into a life cycle of poverty, which is nearly impossible to break. With larger families, any property and possessions has to be shared amongst more people and each child receives less.
2. **Care for young people:** Many young people have been made orphans by the death of their parents due to HIV/ AIDS, Malaria or Tuberculosis. Orphanages will need to be set up to help look after these children.
3. **Overcrowding***:* In many cases there are too many children living in small and cramped conditions, which can allow illness and disease to spread.
4. **Medical issues:** There are few doctors or hospitals in most LEDCs and as a result people often die from relatively straightforward illnesses. Children cannot afford even the most basic medicines. There is no 'free' healthcare in the majority of countries and many children suffer as there is no one to look after them properly. Access to hospitals is extremely limited and many people have to travel long distances to reach them.

5. **Opportunities/education:** Few children in LEDCs have the opportunity of an education and cannot read or write. They find it very difficult to get 'formal' jobs, preventing them from improving their living conditions. They remain poor through their lives, are often unemployed, forced to remain in squalid conditions in slums and shanty towns, and some turn to crime and vandalism.

The economic implications of a youthful dependency

There can also be some economic and financial implications when a country has an increased number of young people.

1. **Education:** Many LEDCs were struggling to educate their population even before the population expansion. Now, the few resources that were available for schools and universities have to be stretched much further.
2. **Healthcare:** Many people cannot afford even the most basic healthcare and they rely on patchy government support or aid agencies to help them. Medicines are basic and can be expensive.
3. **Opportunities:** There is a severe lack of jobs and opportunities for people in LEDCs. With a massive increase in the population, the children will all grow up together and be looking for the same jobs. This means that many will go into the informal sector or be forced into a life of crime.

Check your learning

Now that you have studied Part 1: Population Growth, Change and Structure, return to page 138 and answer the Key Questions for this section.

Test yourself

1. Draw the shape of a population pyramid that you would expect to see in an MEDC. (3)
2. Why might you expect the structure of the population to show an ageing population? (3)
3. What are the social and economic consequences of an 'aged' population? (3 marks for social consequences and 3 marks for economic = 6)
4. Draw the shape of a population pyramid that you would expect to see in an LEDC. (3)
5. Why might you expect the structure of the population to show a 'youthful' population? (3)
6. What are the social and economic consequences of a youthful population? (3 marks for social consequences and 3 marks for economic = 6)
7. In your opinion which is better: an aged or a youthful population, or something in between? Explain your answer. (6)

EXAM PRACTICE QUESTIONS

Some of these questions are from previous CCEA GCSE examination papers and others have been written in the same format to give you practice at answering 'exam style' questions.

Try to answer the questions with as much detail as possible. Also consider the number of marks that each question receives, as this will give you a good indication of the amount of depth that your answer needs.

Resource A

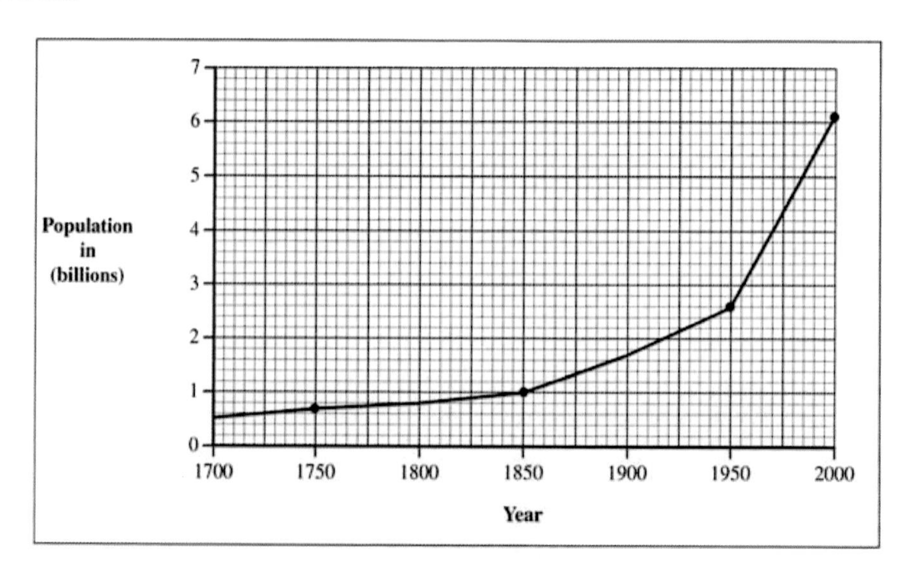

Resource B

Source: Principal Examiner

Resource C

LOCAL GOVERNMENT DISTRICT (Council Area)	NUMBER OF MIGRANTS INTO N. IRELAND FROM OUTSIDE THE UK (mid 2005-mid 2006)
Antrim	955
Armagh	956
Belfast	5880
Derry	1571
Dungannon	1920
Moyle	229
Newry and Mourne	2053
Strabane	483

Source: Neighbourhood Statistics (NISRA) Website: www.nisra.gov.uk/ninis

Resource D

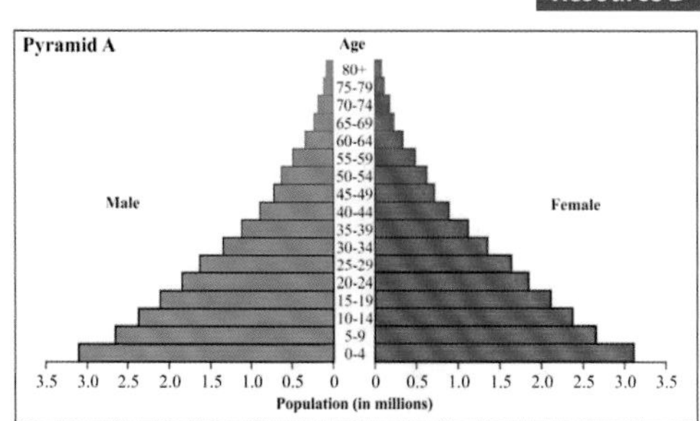

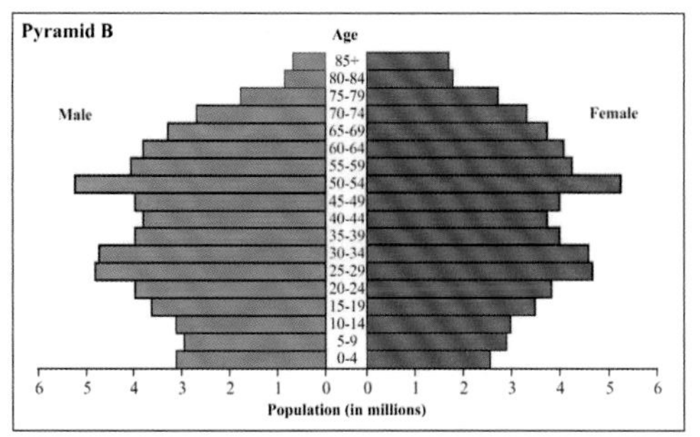

Resource E

A1 — A8 Population by Local Government District 2007

A8_Migrant_Population_2007(2) [Compatibility Mode]

A8 Population by Local Government District 2007

LGD Name	A8 Population (2007) A8Pop_d07	2007 Mid Year Estimate Mid_Year_Est_d07	% of Population Perc_Pop_d07
Antrim	1500	52600	2.9
Ards	500	77100	0.6
Armagh	1200	57700	2.1
Ballymena	1600	62100	2.6
Ballymoney	300	29700	1.0
Banbridge	300	46400	0.6
Belfast	5000	267500	1.9
Carrickfergus	100	40000	0.2
Castlereagh	300	65600	0.5
Coleraine	1200	56800	2.1
Cookstown	900	35400	2.5
Craigavon	3100	88800	3.5
Derry	600	108500	0.6
Down	600	69200	0.9
Dungannon	3700	54300	6.8
Fermanagh	1000	61300	1.6
Larne	100	31300	0.3
Limavady	300	34400	0.9
Lisburn	1000	113500	0.9
Magherafelt	900	43100	2.1
Moyle	100	16700	0.6
Newry and Mourne	3000	95500	3.1
Newtownabbey	700	81700	0.9
North Down	300	78700	0.4
Omagh	1200	51500	2.3
Strabane	300	39400	0.8
Northern Ireland	**30000**	**1759100**	**1.7**

DC | Metadata

Source: Neighbourhood Statistics (NISRA) Website: www.nisra.gov.uk/ninis

Foundation Tier

1. Study Resource A which shows the world population growth from 1700 to 2000.

   Complete the statements taking your answers from the list below.

   fastest slowly 1850 1920 more less

   The world's population grew _____________ from 1700 to 1850. By _________ the world's population reached 1 billion. The _______________ increase was from 1950 to 2000. The population of the world is now _________ than 6 billion. (4)

2. State the meaning of the term migration. (2)
3. International Migration is when people move from one country to live in another. Explain one positive impact international migration can have for a country that gains lots of migrants. (3)
4. Study Resource B which gives information about migration to Australia in 2004–2009.

   Decide if the following sentences are True or False:

   i. Over 54,000 migrants moved from Asia to Australia.
   ii. No migrants came from Antarctica.
   iii. Africa supplied most immigrants to Australia.
   iv. More migrants came from South America than the UK. (4)
5. Many countries have a higher percentage of their population aged under 15 years. Explain one impact that a large young population may have on a country. (3)
6. Study Resource C which shows information obtained from NINIS, an online Geographic Information System, showing migrants into districts in Northern Ireland.
   i. State the number of migrants from outside the UK who are living in the Moyle Local Government District. (1)
   ii. Name a computer program which could be used to present this table of data as a graph. (1)
   iii. Using the table, describe the pattern of the number of migrants to Northern Ireland from outside the UK. (3)
7. For one named country within the European Union you have studied describe one positive and one negative impact of migration on your country.
   i. Country (1)
   ii. Positive impact (3)
   iii. Negative impact (3)

Questions and resources from CCEA GCSE Geography Papers
June 2010 (Question 2), June 2011 (Questions 4, 7; Resource B) and Specimen 2010 (Questions 1, 3, 5, 6; Resources A, C, © CCEA

Higher Tier

1. Study the Resource A which shows the world population growth from 1700 to 2000.
   i. State the world's population in 1850. (1)
   ii. State the year when the population of the world reached 6.1 billion people. (1)
   iii. Explain fully why the world's population increased rapidly from 1900. You should give two reasons in your answer. (6)
2. Study the Resource D which shows two population pyramids.
   i. State the two aspects of population structure that are shown on a population pyramid. (2)
   ii. State whether pyramid A or B is more typical of a LEDC. Give a reason for your choice. (3)
   iii. Explain one impact that a large youth dependent population may have on a country. (3)

3. Study Resource B which gives information about migration to Australia in 2004–2005.
   i. State the meaning of the term immigration. (2)
   ii. Describe what Resource F shows about the origins of Australian immigrants in 2004–2005. (4)
   iii. Name an appropriate mapping method for this data. (1)
   iv. State three advantages of using GIS to analyse migration. (3)
4. Study Resource E which contains data obtained from an online GIS – Northern Ireland Neighbourhood Information Service (NINIS).

   The A8 Countries are eight central and Eastern European countries which joined the European Union in May 2004. These A8 countries are the Czech Republic, Estonia, Hungary, Latvia, Lithuania, Poland, Slovakia and Slovenia.
   i. Using Column B, describe briefly the distribution of migrants from the A8 countries living in different districts in Northern Ireland. (3)
   ii. State one other way of displaying information in an online GIS (other than tables of data). (1)
   iii. State one advantage of using a GIS to investigate the number and percentage of migrants into Northern Ireland. (2)
5. Explain one impact that a large aged dependent population may have on a country. (3)
6. Discuss the positive and negative impacts of migration for one country within the European Union which you have studied. (8)

PART 2: SETTLEMENT SITE, FUNCTION AND HIERARCHY

(a) The site and location of a settlement
Site: defensive site, wet point site and bridging point
The location of a settlement

(b) The four measures of settlement hierarchy
Population size
Function, high and low order
Range
Threshold

(c) The characteristics and location of land-use zones in MEDC cities
Central business district (CBD)
Inner city
Suburban residential
Industrial zones
The rural-urban fringe

(d) Skills: Using maps and aerial photos to identify settlement characteristics
Site characteristics
General functions
Position in hierarchy
Land-use zones for a range of settlements

Key words

Settlement
Settlement Hierarchy
Sphere of influence
High and Low order functions
Range and Threshold
Site (wet point, bridging point and defensive)
Function
Central Business District
Rural-urban fringe
Inner City
Suburbs

Key questions

By the end of this section you will be able to answer the following questions:

1. State the meaning of the term site.
2. What is the meaning of the term sphere of influence?
3. State fully one reason why Belfast has a larger sphere of influence than Ballymena.
4. State two pieces of evidence which could be found on a map to indicate that an area was part of the CBD of a city.
5. What is the meaning of the term urban-fringe?
6. Explain why larger settlements have more services than smaller settlements.
7. Explain one advantage and one disadvantage of living in suburban areas.

The site and location of a settlement

A **settlement** is a place where people live and work. Settlements come in all shapes and sizes – they can be as large as megacities such as London (England) or Sao Paulo (Brazil), or can be as small as individual houses dispersed in a farming area such as the Lake District (England). The story behind the development and growth of each settlement is different and depends on the reasons why the particular site was chosen in the first place.

The **site** of a settlement describes the exact place where a city, town or village might be located. Historically, this location was often chosen for its physical advantages, which were important when the site was first established. Some of these physical advantages are explained below:

Wet point site	Water is a necessity for all life. Settlers (the people choosing a site for settlement) need a clean drinking supply that is easy to reach. Remember piped water is a relatively modern invention and previously it had to be fetched by hand.
Dry point site	A site should be close to water, but not so close that it will flood on a regular basis. Raised areas beyond the flood plain of a river are popular sites.
Building materials	Early settlements were often located near to woodlands, which provided access to wood for building and fuel. Later, stone was needed to make settlements permanent and secure from the elements and invaders.
Defence	Sometimes settlers would need protection against other tribes or families. A place on a hill or rock would give good views all around (for example, Edinburgh or Carrickfergus) and a place built inside a river meander also meant that attack could only come from one easily defended access point.
Fuel supply	A fuel supply was important for cooking and to ensure that people could survive cold winters. In Ireland people mostly used wood or peat.
Food supply	Land is needed for both arable (crop) and pastoral (animal) farming to provide food. Sometimes trees need to be cut down to provide this land.
Nodal Point	A nodal point is a place where several natural routes join to create a meeting point. For example, Ballymena means 'Middle Town' in Irish as it was a natural middle point for people travelling west from Larne, north from Belfast or Carrickfergus, or south from Coleraine.
Bridging Points	400 years ago, physical barriers such as deep or wide rivers provided a real challenge for settlers. Over time different methods were used to improve access and eventually bridges made rivers easy to cross.
Shelter and aspect	A south facing aspect allows a place to receive more sunshine, heat and light compared to a north facing place. In Ireland it has always been an advantage to be sheltered from the prevailing south-westerly winds.

The **location** or **situation** of a settlement is where the settlement is located with reference to what surrounds it. These surroundings include its physical features, its communication links and its neighbouring settlement. Northern Ireland contains many good examples of different types of settlement site. These include the following:

WET POINT SITE	DEFENSIVE SITE	BRIDGING POINT
Example: Ballymena	*Example: Carrickfergus*	*Example: Belfast*

The town of Ballymena was built up around the river Braid. Local settlers in the area chose to live in areas of good farmland that provided easy access to a constant water supply. In later years (from 1800 onwards) this location was to become even more important as the access to rivers provided a power supply for the new industrial textile machinery in the linen industry.	When a settlement is being pioneered for the first time, often they have to make sure that they are in a position to protect themselves from invasion and attack. Carrickfergus is one of the oldest towns in Northern Ireland. John de Courcy started building the castle on the rock with a view up and down Belfast Lough in 1177.	Many major towns and cities around the world today are located on rivers and were first sited as bridging points. The name for Belfast comes from the Irish *Beal Feirsde*, which means 'the mouth of the river on the sandbar'. Many believe that this is a reference to where the River Farset flows into the Lagan, where a sandbar was once located. The first bridge was built across the Lagan in 1688.

It is the location that will encourage a place to continue to grow from a small settlement into a large settlement. For example, both Ballymena and Belfast are built alongside rivers in County Antrim but some of Belfast's other locational advantages meant that it grew to be over ten times bigger than Ballymena. In 2010, the population of Belfast was 480,000 and the population of Ballymena was 32,000.

It is important to remember that most cities, towns and villages actually grew up a very long time ago. As these places grew from villages into towns and then eventually cities, they changed and many 'forgot' the actual reasons why they were built on their site in the first place. Through the nineteenth, twentieth and twenty-first centuries, many of the physical locational advantages were replaced by human factors.

Structured notes

1. Write a definition for each of the following words:
    - Settlement
    - Site
    - Location (or situation)
2. Make a copy of the table below and complete it by following the instructions below.

LOCATION ADVANTAGES	PHYSICAL OR HUMAN ADVANTAGE	SHORT DESCRIPTION	SKETCH TO ILLUSTRATE	NORTHERN IRELAND EXAMPLE
Wet point site				
Dry point site				
Building materials				
Defence				
Fuel Supply				
Food Supply				
Nodal Point				
Bridging Points				
Shelter and aspect				

    i. For each of the location advantages in Column 1, decide whether you think this is a physical or human advantage.
    ii. Use three or four words to write a short description of each advantage.
    iii. Draw a simple sketch to illustrate each advantage.
    iv. Identify a town or city in Northern Ireland that you think is a good example of each locational advantage (try to use each place just once).

3. Write a list of things that might be important if building a new settlement in Northern Ireland today. What differences might there be between a built settlement today and one built 200 years ago?
4. Some geographers say that political and economic factors are now the most important 'new' locational advantages for a place. Think about how these might work and add them to the bottom of your table.

Test yourself

Take a look at the information below, which shows details about the original site of Belfast. Explain why this site was an attractive place to build a settlement for early settlers. (5)

Resource A

"Belfast owes it existence to the sea and the river Lagan – Growing up around the lowest crossing point of the river, the town became the connection between a vibrant hinterland and wider worlds across the seas".

"Belfast was born at the meeting of three rivers: the Lagan, Farset and Blackstaff. The town's name comes from the Irish – *Beal Ferste* – meaning the mouth of the Farset. People have lived here since earliest times. The first settlement grew up around the crossing point of the Lagan nearest the sea. As a town developed, its potential as a port and place of trade was recognised and efforts were made to improve access for ships. The small vessels of the seventeenth century could moor at the foot of High Street, but as they got bigger they could not navigate the broad, shallow Lagan with its expansive mud banks".

Source: Northern Ireland Community Archive (www.niarchive.org)

Resource B

Drawing of Chichester Quay (now High street) in Belfast

Source: Ireland Illustrated, GN Wright, London, 1831. Drawn by TM Baynes and engraved by J Davies.

Resource C

Location map of Belfast

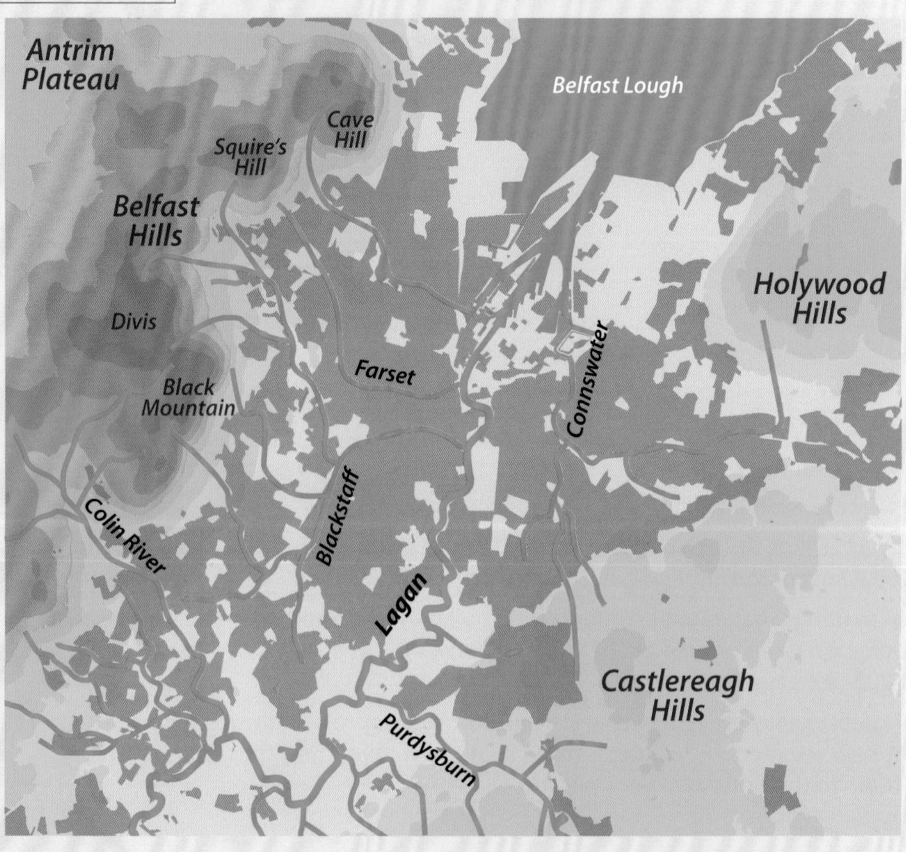

The four measures of settlement hierarchy

A settlement hierarchy is used to rank the importance of places within a particular area. It is represented as a pyramid, with the most important settlement at the tip and the least important at the base. This means that cities and conurbations are at the top, with isolated dwellings and small settlements at the bottom.

What makes one place more important than another place? Is it the size of the settlement, the number of residents, the number of people who work there or is it the distance that people are prepared to travel to reach it? This is the big question and geographers often disagree about the answer.

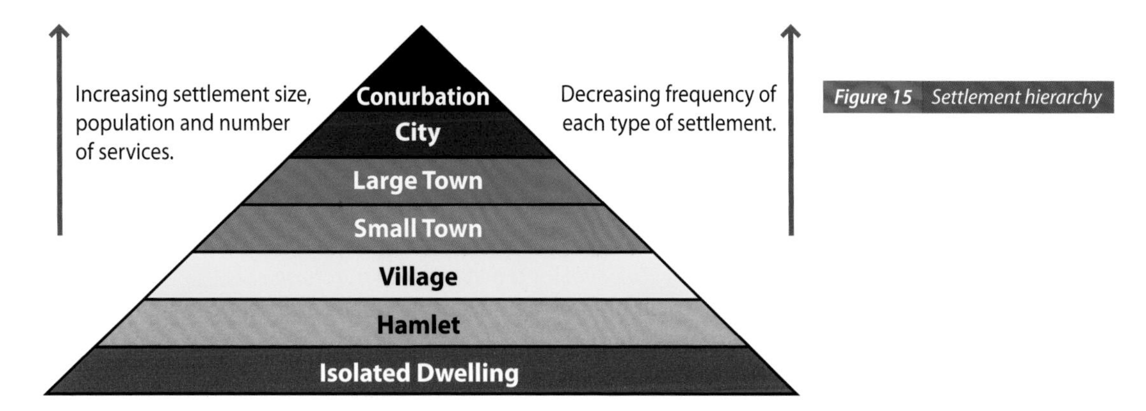

Figure 15 *Settlement hierarchy*

Population size

The main way to decide the importance of a settlement is to count the number of people who live there. The population of a place indicates which category it belongs to.

For example, in Northern Ireland the following categories are generally accepted:

CATEGORY	POPULATION
Conurbation	More than 1 million people
City	100,000 to 1 million people
Large Town	Less than 100,000 people
Small Town	Less than 20,000 people
Village	Less than 2,000 people
Hamlet	Less than 100 people
Isolated Dwelling	2–4 people

However, the problem is that these indicators are not agreed throughout the world. In some bigger countries, such as the USA and China, a town will be much bigger than 100,000 people. Additionally, some countries, such as Germany, make no distinction in their language between 'town' and 'city'. The main distinction is only between a more rural area (hamlet and village) and a more urban area (town and city). This means that using the population of a settlement alone can become quite complicated.

Function, high and low order

The **function** of a settlement relates to its economic and social development, and refers to its main activities. Today, most larger settlements tend to be multifunctional (they have several functions), although one or two functions might be more important than others. In some cases, the original function may no longer apply. For example, the town of Carrickfergus in County Antrim no longer needs a defensive function and the town of Annalong in County Down used to concentrate on fishing but is a tourist destination today.

The following are some of the main functions of a settlement:

- **Market town:** Where farmers buy and sell goods and services.
- **Mining town:** Where fuel and minerals are extracted.
- **Industrial town:** Where raw materials are processed into manufactured goods.
- **Port:** A transport hub for ships on the coast, a river or a lake.
- **Route centre:** Where settlement is located on junctions of several roads.
- **Service centre:** Where the area's specific needs and services are provided for. For example, farming equipment in an agricultural area.
- **Cultural/religious settlement:** Where people from other parts of the world come to live and study.
- **Administrative centre:** Where government offices and general office buildings are located.
- **Residential town:** Where people live but generally work elsewhere.
- **Tourist resort:** Where people visit to enjoy themselves and their recreational needs are catered for.

When people aim to access goods or services in a particular settlement these can be divided into **High** and **Low** order goods.

Generally people are happy to travel long distances to purchase or use high order goods and services, such as a car or to go on holiday. Low order goods and services are more local services, such as milk and newspapers, and people want access to these close to where they live.

Range

The range is the maximum distance that people will be prepared to travel to obtain a service. As discussed above, people will travel further for high order goods or services than for low order ones. Therefore, the higher the number and complexity of the goods and services that a place provides determines its appeal to people and the distance they will travel to reach it.

Figure 16
Services usually found in a settlement hierarchy

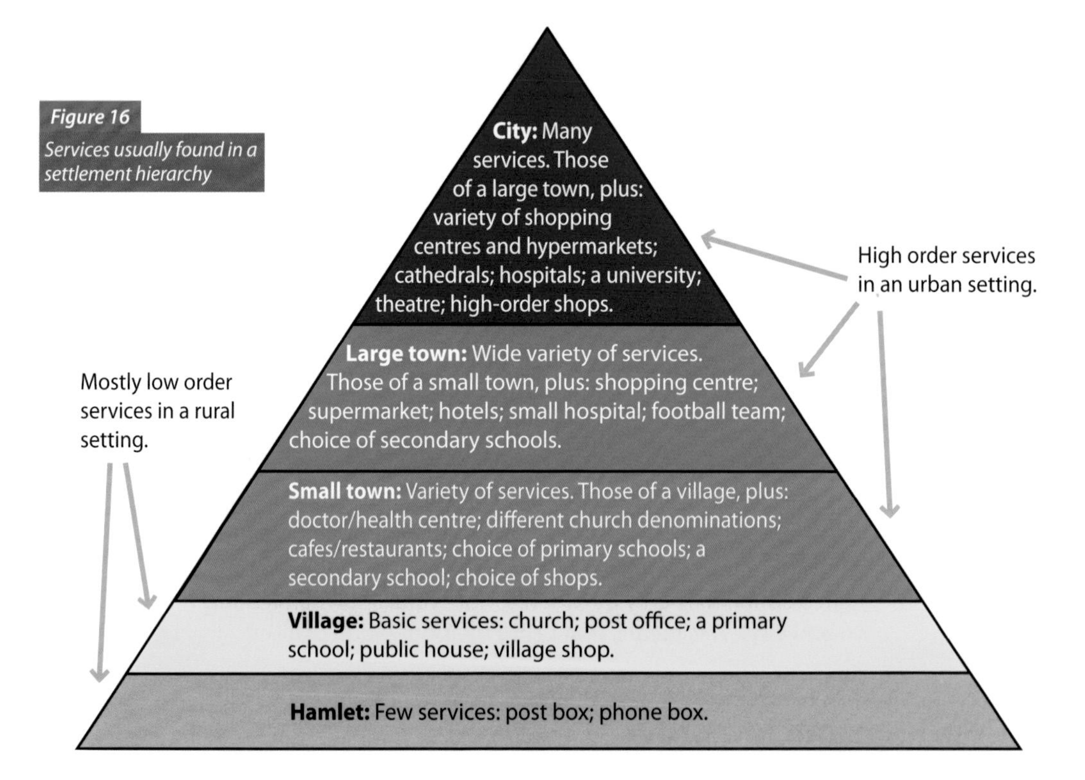

The sphere of influence (or market area) is the area that is served by a particular place. This area depends on the size and services of a town and the level of competition from rival settlements. For example, in terms of shopping, the sphere of influence of Ballymena in County Antrim is much larger than the similar sized towns of Antrim or Larne. More people would be prepared to travel a longer distance (the range is bigger) to Ballymena to shop than these other towns because they believe the shops cater for more high order goods and have a greater variety.

Threshold

The threshold population is the number of people that are needed to ensure that there is enough demand for a particular service. Increasingly, high street shops will do a population threshold analysis before deciding to set up a shop in a particular place. For example, a threshold population of approximately 300 people would be sufficient for a small village shop but around 70,000 people might be needed before a big retailer such as Marks & Spencer or Sainsbury's would think about setting up in a town.

Test yourself

1. Using the information from this section, write your own definition for each of the following:
   - City
   - Large Town
   - Small Town
   - Village
   - Hamlet

   (2 marks × 5 = 10)
2. Draw your own settlement hierarchy pyramid and try to identify some places from Northern Ireland that you think are good examples of each settlement type. (5)
3. Which of the four measures of settlement hierarchy do you think is the most important in determining whether a place will grow into an important settlement? Explain one reason why you think this. (4)
4. Which of the four measures of settlement hierarchy do you think is the least important in determining whether a place will grow into an important settlement? Explain one reason why you think this? (4)
5. "The concept of settlement hierarchy is made up through a combination of a number of factors". Discuss to what extent you agree or disagree with this statement. (6)

The characteristics and location of land-use zones in MEDC cities

Cities are fairly complex places. They develop over long periods of time and are always trying to redevelop and reinvent themselves. This section will identify the different features within the city and group them into broad categories to help explain how and why they have built up in this way.

The story of how a city develops is linked very closely with the history of the place. The patterns of urban land-use result from many different choices made over time by individuals, planners, architects and local government. These decisions are mostly influenced by economic factors and the land-use for a site is usually chosen for its ability to provide the greatest profit. Urban land-use is closely related to land values and these land values usually reach a peak at the city centre and decline as you move away from it. Not every city has developed in the same way and sometimes cities in MEDCs show different patterns and shapes.

Now, let us take a closer look at how land-use might change in different places across a city from the central business district (CBD), through the inner city, industrial zones, suburban residential area and to the rural-urban fringe.

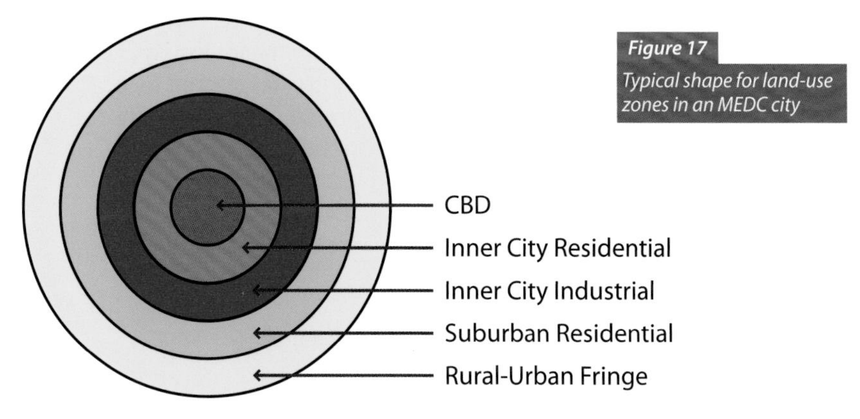

Figure 17
Typical shape for land-use zones in an MEDC city

Figure 18
Concentration of high rise buildings emphasises the importance of Toronto CBD

Central business district (CBD)

The central downtown area contains the main commercial streets and public buildings. This is the core of the city's business and civic life and is the place where business and retail meet. In the past, the CBD was an area of varied and competing land-use, but over the years the rising value of CBD land and property has forced out most of the residential and industrial users. Peak land values occur at the centre, where there are department stores, supermarkets, and headquarters and offices for companies with large turnovers and high profits. The smaller trader is forced out to the edge of the CBD. The CBD is an area where vehicle traffic and pedestrian traffic is likely to be most concentrated.

Figure 19
The cranes of Harland and Wolff show that some industry remains in Titanic Quarter (Belfast)

Inner city industrial zone

Most old cities in the UK grew up because of heavy industries near the CBD. The inner city was the main source of income for the city. In the nineteenth and twentieth centuries many people migrated into cities to work in new factories in the inner city areas that generally surrounded the CBD. Many of the streets in inner cities were built up with terraced houses that were close to the main source of employment or factory.

Figure 20
Old residential accommodation in Battenburg Street, West Belfast

Inner city residential

As factories grew, so too did the number of houses that were needed for their workers. People could not afford transport costs so they needed to live within walking distance of their place of work. In many cases, it was actually the factory owners who built and owned the buildings that their workers would live in. This meant that losing your job meant losing your house as well. In recent times, industry has started to locate to 'brown-field' sites in redeveloped parts of the city or to 'green-field' sites that are in the rural-urban fringe, where industry has space to grow and develop.

Suburban residential

Residential areas account for the largest single use of land area in any city (about 45% of the total area in towns over 10,000 people in the UK). Large-scale suburban residential development took place in most cities throughout the twentieth century. Some of this started in cities as early as the 1920s and much of it was fuelled by the increase in car ownership and public transport. People no longer had to live within walking distance of where they worked and could move to houses with more space at the edge of the city. People wanted access to better, bigger homes, with more space to allow their families to grow.

Figure 21

Three-storied suburban houses built in the 1920s in part of North Belfast

The rural-urban fringe

From the 1960s, many people had increased disposable income and they wanted more from where they lived. They wanted more than the '2-up 2-down' houses that they had been brought up in, desiring more rooms and green space. Urban sprawl continued, with land in these areas being used for private developments or outer-city council housing estates. Many of these houses were of a high quality, with lots of space and no pollution. Transport costs were high but people who lived here were often able to afford a car, if not two, for personal transport. The city was continuing to expand into the countryside and the boundary between the urban and rural area was rapidly becoming blurred. In many cases, big multi-tenant shopping centres were also built in this area because the land was much cheaper than in the CBD and it had become accessible to the increasing numbers living in the fringe. In recent years, industry has started to relocate here as well, in order to avoid the high costs and congestion of the inner city.

Figure 22

Modern housing development in the rural-urban fringe of South Belfast

Test yourself

With reference to each of the different land-use zones in an MEDC city, describe and explain at least three different building types that you would expect to find. Make a copy of the table below to help format your answer.

(3 marks × 5 = 15 for Column 1) + (2 marks × 5 = 10 for Column 2) = 25

LAND-USE ZONE	THREE BUILDING TYPES	WHY ARE THESE BUILDINGS LOCATED HERE?
1. CBD		
2. Inner city (residential)		
3. Inner city (industrial)		
4. Suburban residential		
5. Rural-urban fringe		

Land-use in Belfast (an MEDC city)

Figure 23
Land-use zones in Belfast (simplified)

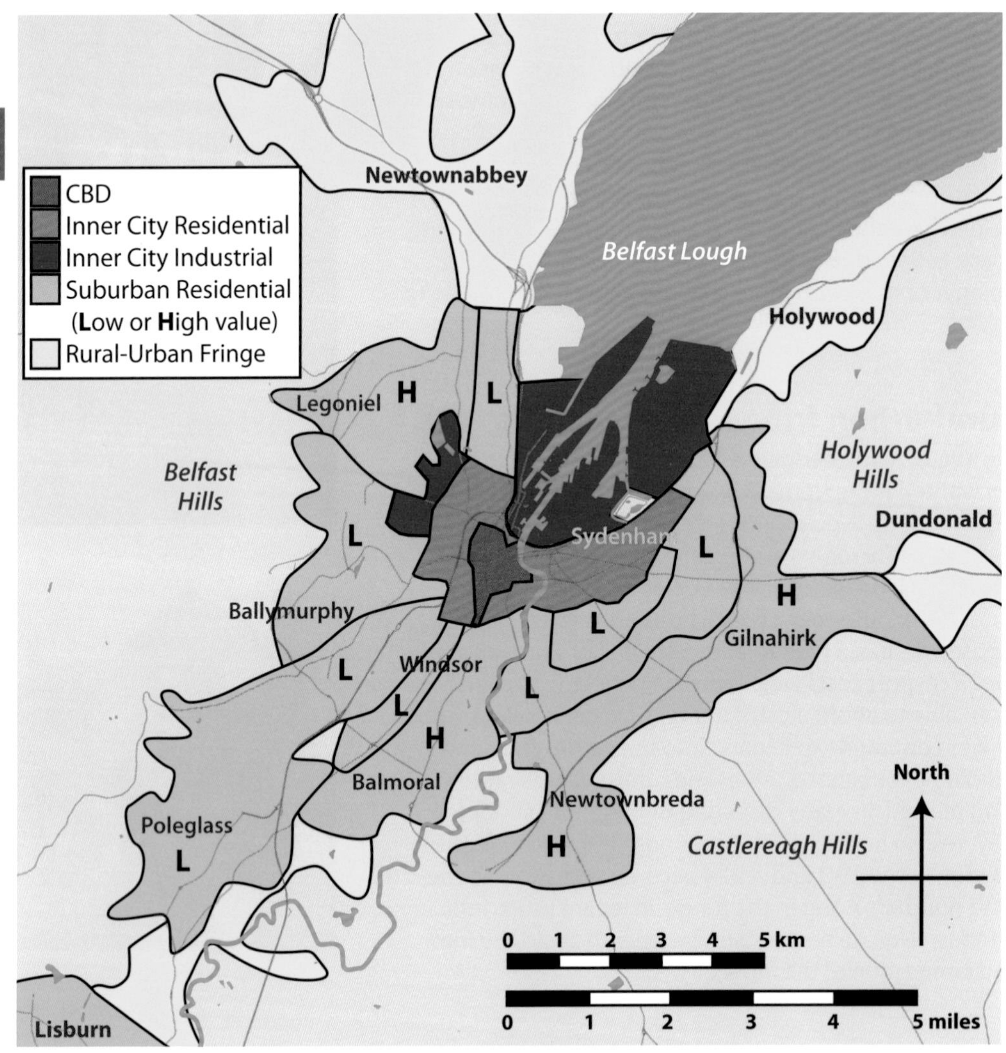

Belfast has grown from being a fortified crossing point of the river Lagan in the seventeenth century into the second biggest urban area in the island of Ireland. The population is estimated at 270,000 people. However the greater Belfast Metropolitan area encompasses around 640,000 people.

Figure 23 shows some of the major land-use types across the city. The river Lagan is a natural division through the city, however much of the pattern for growth across Belfast has actually been dictated by its position on a flood plain between the Belfast and Castlereagh Hills.

The central business district

Belfast's CBD originally grew on the western bank of the river Lagan but over time it moved further inland (especially after the building of the City Hall in 1900).

The inner city (residential and industrial)

In Belfast, the inner city residential and industrial areas are intertwined. The inner part of the city that surrounds the CBD was largely made up of old, established manufacturing industry. Belfast grew rapidly in the early 1800s due to the textile industry, with linen mills being built along rivers in North and West Belfast (on the Shankill, Crumlin and York Roads). This was followed by the engineering works that grew up along the Springfield and Newtownards Roads (such as Mackies and Sirocco) and the shipbuilding and its associated industries (such as rope-making) that developed around the mouth of the Lagan.

Figure 24

Aerial view of Belfast looking towards East Belfast (Short Strand and Albertbridge Road)

Courtesy of Belfast Harbour

Much of the housing in this area can be found very close to the factories. Houses were built in straight terraced house formations. In recent years, there has been much re-development and gentrification of these areas and government-led programmes such as Laganside and Titanic Quarter have changed the face of the city in these parts.

Suburban residential

The population of central Belfast grew very quickly from 1900–1950, as new house building to the west and east of the original inner city dwellings enticed more and more people from the countryside into the city. These houses were slightly larger than the terraces and allowed residents a little more space. Many were semi-detached and included a small garden and even a garage (for example, Oldpark, Antrim Road and Beersbridge Road). Some very high cost houses also began to be built in the more affluent south of the city (Malone Road).

Rural-urban fringe

In the late 1960s and early 1970s, partly as a response to 'The Troubles' and partly a response to the greater affluence of people across Belfast, more and more people were looking for a move beyond the city. Housing developments in Carrickfergus, Lisburn, Holywood, Bangor and the new town of Newtownabbey all enticed people to move towards the very periphery of the city. However, people still wanted to be connected to the city and with better public transport links and more personal transport (cars and motorways), they could now live on the fringes but still enjoy the benefits of city life. It was not long before new shopping centres (such as Abbey Centre and Sprucefield) were also set up on fringe locations in direct competition to the CBD.

Skills: Using maps and aerial photos to identify settlement characteristics

Resource A is a 1:50,000 scale map showing an extract of the Coleraine, Portrush and Portstewart area.

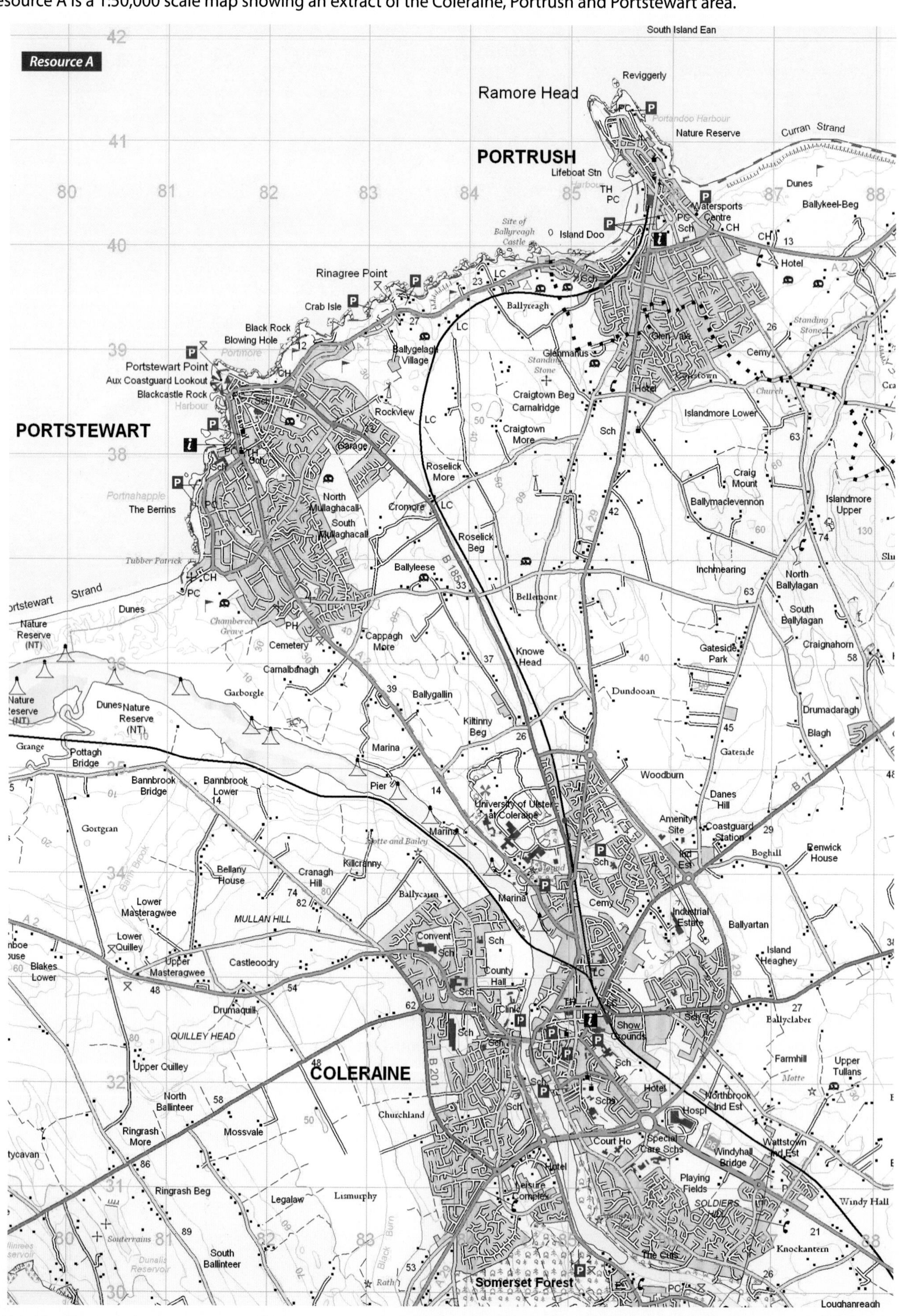

1. Site characteristics

Using Resource A look up the six figure grid references in Resource B. Think about the reasons why each site was built at its location. Make a copy of the table in Resource B and identify the most important reason in each case. The possible answers in Resource C might help you. (9)

Resource B

GRID REFERENCE	POSSIBLE LOCATION ADVANTAGE
845399	
864351	
805322	
811349	
846325	
824355	
850334	
870322	
852307	

Resource C

Wet point site	Dry Point site	Building materials
Food supply	Nodal Point	Bridging Point
Defence	Shelter/ Aspect	Fuel Supply

2. General Function

It can be quite difficult to identify the function of a town just by looking at a map. However, sometimes there are some clues that can help you to identify what the settlement's main activities might be. For this task we are going to look at the three major settlements in Resource A and try to identify some of the main functions, both past and present, for these places.

Make a copy of the table in Resource D. For the settlements listed in Column 1, look up each grid reference and identify a feature which might suggest a clue and a possible function for the place. The possible function answers in Resource E might help you. (23)

Resource D

SETTLEMENT	GRID REFERENCE	CLUE (WHAT DO YOU FIND HERE?)	POSSIBLE FUNCTION
Portrush	858405		
	857402		
	875408		
	875398		
	857407		
Portstewart	815388		
	816381		
	824382		
	815379		
Coleraine	860337		
	859332		
	858316		
	Area 8432		
	Area 8431		

Resource E

Industrial town	Port	Cultural/religious	Mining town
Service centre	Tourist resort	Route centre	Residential

3. Position in Hierarchy

It can be quite difficult to identify the particular position of a settlement within a hierarchy. The following exercise will give you an example of one way to do this for an area.

Take a look at the settlements that are found in the following grid squares:

Resource F

SETTLEMENT	GRID SQUARE
Bellemont	8436
Craig Mount	8637
Carnalridge	8538
Windyhall	8731
Portstewart	8138
Portrush	8540
Coleraine	8432

a. Make a copy of the table in Resource G. In the number of points column, give each settlement one point for every church, chapel, school, public house, post office, hotel, public telephone, car park, information centre, bus, train station and any other service shown on the map.

Resource G

SETTLEMENT	SERVICES IDENTIFIED	NUMBER OF POINTS	RANK ORDER	TYPE OF SETTLEMENT (for example, hamlet, village, small town, large town or city)
Bellemont				
Craig Mount				
Carnal Ridge				
Windy Hall				
Portstewart				
Portrush				
Coleraine				

b. Rank the settlements in order, putting the one with the most services (points) first and the one with the fewest services (points) last. (7)

c. Using your definition of settlement hierarchy, try to identify each settlement's rank in the hierarchy. (7)

d. What other services would you expect Coleraine to have if it was to be given status as a city? (2)

Resource H is a 1:50,000 scale map showing part of north Belfast.

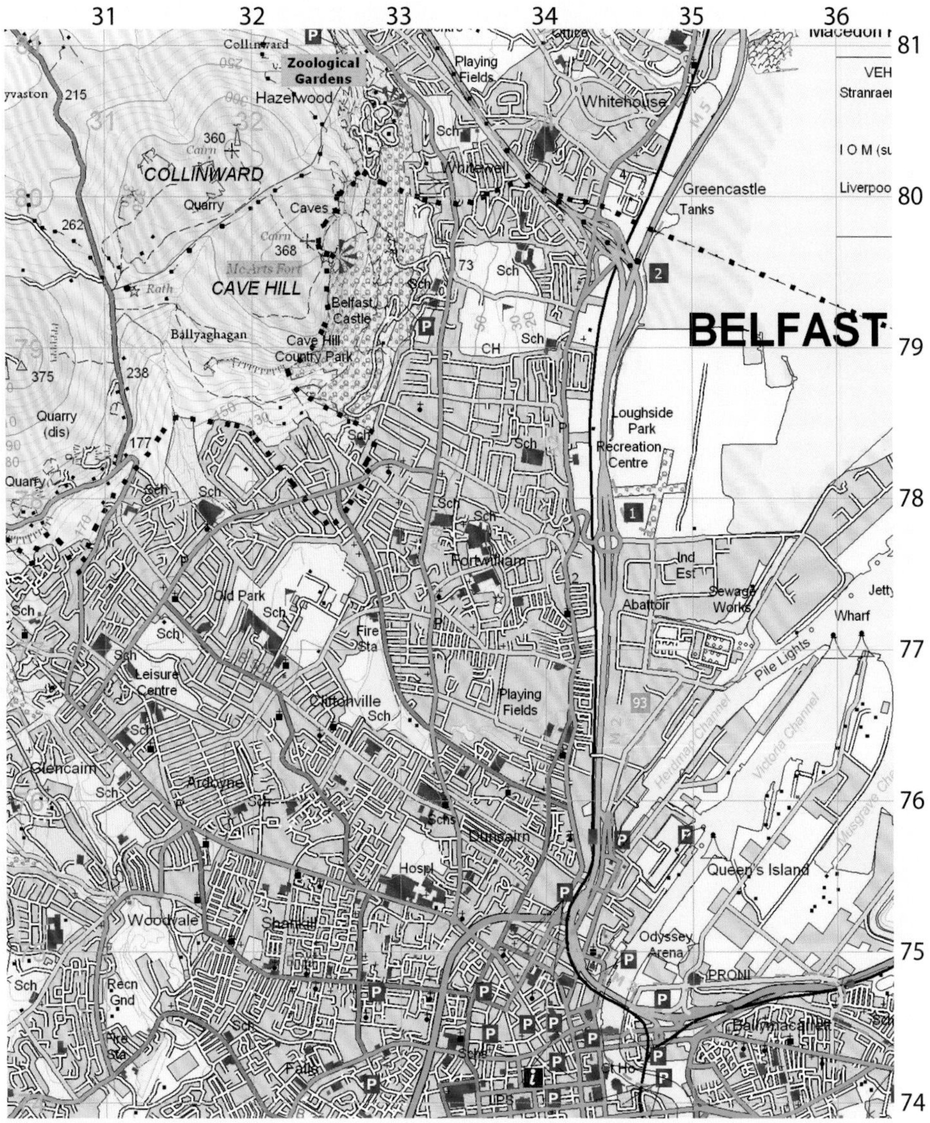

Resource H

4. Land-use zones

a. Make a copy of the table in Resource I and using Resource H, identify (by grid reference) areas of North Belfast that show the land-use zones listed in the first column. The second column gives you a list of characteristics to look out for on the map, which are common within this land-use zone. (10)

Resource I

LAND-USE ZONE	CHARACTERISTICS TO LOOK OUT FOR ON MAP	GRID REFERENCE
CBD	Tightly packed roads, town hall, bus and railway stations concentrated in one area	
Inner city (residential)	Straight roads built for terraced houses, close to factories and industrial areas	
Inner city (industrial)	Evidence of industrial areas and factories	
Suburban residential	Residential areas moving further away from the city centre towards the suburbs, possible crescent shaped roads and cul-de-sacs	
Rural-urban fringe	Developments and estates beyond the edge of the city, possibly on the other side of a belt of green land (green belt)	

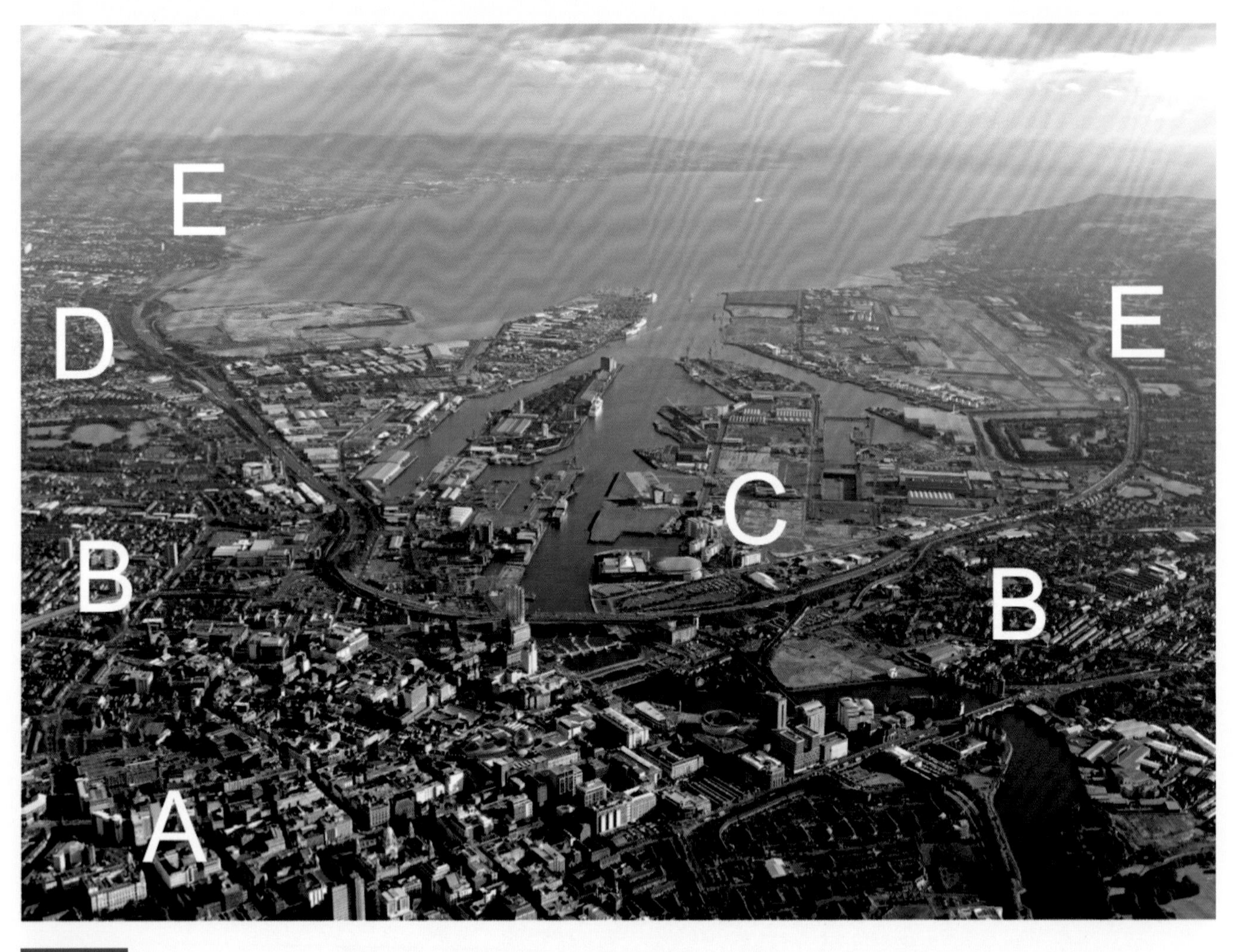

Resource J

Aerial view of Belfast

Courtesy of Belfast Harbour

b. Use Resource J to identify the potential land-use zones for each of the letters A–E indicated. (5)

Check your learning

Now that you have studied Part 2: Settlement Site, Function and Hierarchy, return to page 167 and answer the Key Questions for this section.

PART 3: URBANISATION IN LEDCS AND MEDCS

(a) **The causes of urbanisation**
Push and Pull factors
Natural Increase

(b) **The location, growth and characteristics of shanty towns**
Case study: Sao Paulo (an LEDC city)
Describe
Explain

(c) **Urban planning and regeneration of the inner city Case Study: inner city Belfast (an MEDC city)**
A scheme that aims to regenerate and improve the inner city zone in terms of:
Housing
Employment opportunities
The environment

(d) **Evaluate the sustainability of this urban planning scheme**
Assess the extent to which this project is sustainable

Key questions

By the end of this section you will be able to answer the following questions:

1. What is urbanisation?
2. Explain the causes of urbanisation. Give two reasons.
3. State one push and one pull factor which helps to explain why people leave the countryside and move to the city.
4. For an LEDC city you have studied, describe and explain the growth of its shanty town areas.
5. Describe two ways that inner city areas could be improved.
6. For a named city you have studied, describe one urban problem and state fully one way planners have tried to overcome this.
7. For an MEDC city you have studied, state fully how an inner city planning initiative has been sustainable.

Key words

Urban Sprawl
Urbanisation
Counter Urbanisation
Shanty Town

The causes of urbanisation

The process of urbanisation is when people move from rural (countryside) areas into urban (town or city) areas. This is a migration move which increases the proportion of people living in towns and cities. In 1800 only 3% of the world's population lived in an urban area. By 1950 this was 29% and by 2006 it had exceeded 50%. In other words, more people on the planet live in towns and cities than live in rural areas. Many of the MEDC countries have had a high urban proportion for many years. Recently, there has been much more urbanisation within LEDCs as many people give up rural life in the countryside and migrate into the city.

Test yourself

Make a copy of Resource A. Complete the table by writing whether each city is in an MEDC or LEDC. The population for each city is given in the brackets (in millions).

Resource A

The ten most populated cities in the world in 1950 and 2000

1950	MEDC/LEDC	2000	MEDC/LEDC
1 New York (12)		1 Tokyo (27)	
2 London (9)		2 Sao Paulo (16.6)	
3 Tokyo (7)		3 New York (16.4)	
4 Paris (5.5)		4 Mexico City (15.9)	
5 Moscow (5.4)		5 Mumbai (15.4)	
6 Shanghai (5.3)		6 Shanghai (15.3)	
7 Rhine-Ruhr (5)		7 Beijing (12.7)	
8 Buenos Aires (5)		8 Los Angeles (12.5)	
9 Chicago (4.9)		9 Kolkata (11.8)	
10 Calcutta (4.4)		10 Seoul (11.7)	

(½ mark each × 20 = 10)

Push and pull factors

Urbanisation and migration are very closely linked together. The main reasons why people make a decision to move from the rural countryside into the city are also the reasons why people migrate or move from one place to another.

Rural 'push' factors

Many people leave the countryside because they feel forced out of these areas because:

- there are few employment opportunities.
- there is pressure on the land (for example, the division of the land amongst children means that each area of land is too small to live on).
- many families do not own any land.
- there is overpopulation, resulting in high birth rates.
- people are starving, either due to too little output from the farm or crop failure and drought.
- food production is limited due to overgrazing, which has led to soil erosion.
- mechanisation means that there is less need for labour on the farms.
- farming can be hard work with little reward. In LEDCs this lack of money means a lack of machinery, pesticides and fertiliser.
- local communities (for example, Amazon Amerindians) are forced to move due to pressures of framing and logging.
- there are few services (schools and hospitals).
- there is little investment by the government in rural areas.

Urban 'pull' factors

Many people are attracted to cities because:

- they are looking for better paid jobs.
- they want better housing and a higher standard of living.
- they want access to better services (such as schools and hospitals).
- there are more reliable food supplies.
- religious and political activities can be carried out more safely in larger cities due to 'safety in numbers'.

Natural increase

The people who migrate into the city are generally young (and fertile), and once they have become settled, they often have children.

Although the process of urbanisation is still continuing in MEDCs, the rate of increase is much slower than in LEDCs. Many MEDC countries have reached their threshold in relation to urbanisation and some people have started to move back into the countryside from the urban areas, in a process known as **'counter urbanisation'.**

In many LEDCs, however, the rate of urbanisation has started to increase. More and more people have been persuaded that there are more benefits from living in urban rather than in rural areas, and they have been arriving in developing cities in huge numbers. As a result, a lot of pressure has been put onto the infrastructure and services of these cities.

The location, growth and characteristics of shanty towns

Case Study: Sao Paulo (an LEDC city)

For this case study we are going to take a look at Sao Paulo in Brazil. Sao Paulo is a good example of a city that has grown very quickly and where this growth has created some problems within the city. The most notable problem is the increase of shanty town areas within the city.

Sao Paulo location fact file

- South East Brazil
- Established 1554 (Portuguese colony)
- 30 miles from the Atlantic Ocean
- Situated on the Tiete river

The location of Sao Paulo

The site for Sao Paulo was first settled as part of the Portuguese colonial development of the 'new world' when a Jesuit mission station was set up here in 1554. The settlement quickly grew into a major city within Brazil.

Figure 25
The location of Sao Paulo

The growth of Sao Paulo

Sao Paulo was officially made a city in 1711. Through the nineteenth century it started to experience economic prosperity due to its role exporting agricultural products such as coffee. This prosperity attracted people to the city.

By the twentieth century the city had received much investment from European settlers and a building programme brought CBD development reminiscent of New York.

By 1940, the population of Sao Paulo was over 1.3 million and it continued to grow at a fast rate. Sao Paulo overtook Rio de Janeiro as Brazil's largest city, with a population of 3.7 million by 1960. From 1940 to the present day the population of Sao Paulo has been increasing at a very fast rate. Today it has an estimated population of 12 million people (or 20 million in the wider metropolitan area) and is widely regarded as the world's fifth largest city. It is the most important city of commerce and retailing in South America and, some would argue, the entire southern hemisphere.

Figure 26
Central area of Sao Paulo

Causes of urbanisation

Sao Paulo's rapid increase in population was caused by:

- massive amounts of rural to urban migration from the poor, agricultural north-east regions of Brazil to Sao Paulo.
- high amounts of international migration as many people from Portugal, Italy and Germany made their way to Sao Paulo.
- high population Natural Increase due to high birth rates and a lowering of the death rate (stage 3 of the Demographic Transition Model).

The following **pull factors** brought people to Sao Paulo:

- good employment opportunities, with 50% of all Brazilian industry clustered in and around Sao Paulo. These industries included construction (expanding the city), manufacturing (textiles and machinery), and mining and extraction (iron ore).
- good transport links (roads and railways) made it easier to migrate.

The following **push factors** moved people out of rural areas:

- farming was poorly paid and hard work. There was a perception that work in the city was easier and better.
- rural areas have poor services compared to the big city.
- famines and drought put a strain on life in rural areas.

Characteristics of the Sao Paulo shanty town

This rapid increase in the population of Sao Paulo caused problems. With an estimated 2,000 new migrants arriving into the city each week, huge pressure was put on resources and space within the city. This led to the creation of large 'shanty town' areas or 'favelas'.

Figure 27 *Contrasts between the rich and the poor in Sao Paulo*

Figure 28 *Paraisopolis, one of the biggest favelas in Sao Paulo*

A favela is the name given to a 'spontaneous' settlement in Brazil. Local councils in Sao Paulo define these as "an illegal occupation of terrain in a city where dwellers often have to live without basic infrastructure, such as water, sewage, electricity, garbage collection, mail, etc". These favelas have been growing at an even faster rate than the city as a whole. Between 1980 and 1990, the city of Sao Paulo increased in size by 22.9%, however, the percentage increase of population in the favelas was by 120.6%. Between 1991 and 2000, as the city as a whole increased by 15.7%, the favelas continued to increase by 57.9%.

One of the biggest favelas in Sao Paulo is Paraisopolis. It has an estimated 20,000 households and a population of over 100,000. When people first arrive in a new city such as Sao Paulo they need to find somewhere to sleep. Many of the new migrants do not have very much money and cannot afford to rent a room in a formal housing area. Instead, they gravitate towards the more informal shanty town areas and they create a makeshift home, constructed out of whatever materials that they can find, for example, bits of wood, plastic, corrugated iron and disused bricks. Many of these 'temporary' shelters are built illegally on wasteland near transport links, rivers, rubbish dumps, in marshy land or at the very edge of the city.

Figure 29 *Typical shape for a LEDC city*

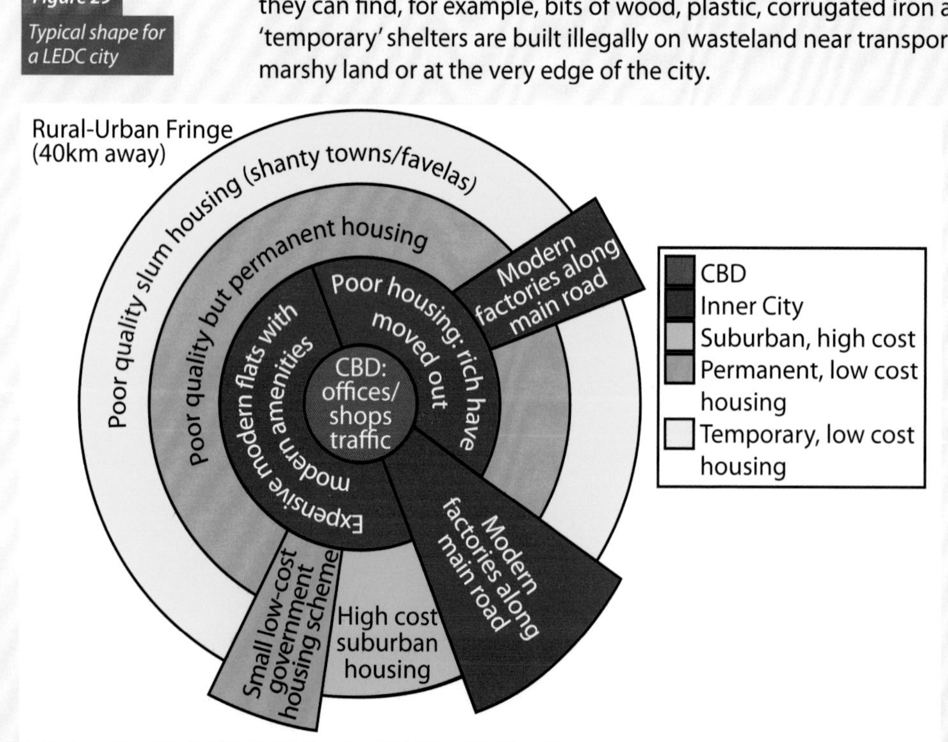

Unlike MEDC cities, the poorest housing tends to be found on the edge of the city and as you move into the centre, the quality of the housing will improve.

Transport and Traffic

Much of the transport infrastructure in Sao Paulo cannot keep up with the pace of urban growth across the city. There is a huge dependence on public transport in the city. However, many argue that the transport facilities are not sufficient for the needs of the population. Congestion and traffic jams are huge problems and cause much noise and air pollution.

Services

Basic services are limited in the favelas, with access to electricity, clean water, schools or doctors sporadic. Sewage often contaminates the water supply, leading to health issues and disease, such as typhoid and dysentery. Diseases can spread very quickly through the overcrowded conditions and few people can afford medical assistance.

Employment

There are not enough jobs and many of the migrants who arrive suffer from unemployment. They are forced into working in the informal sector, where they will provide basic services for a basic fee, for example roadside fruit selling, cutting hair, shoe shining, collecting rubbish, recycling and prostitution. In addition, with a youthful population structure, many of the young children are poorly educated and struggle to obtain formal employment when they do leave school.

Segregation

There is a very big divide between the wealthy and poor within the city. There is very little social interaction between the two social groups. The rich rarely go near the favelas and the poor are often discouraged from going into rich areas by security guards.

Housing

Housing highlights the segregation between rich and poor. Rich residents live in expensive housing, either in apartments close to the CBD or in a suburban block towards the edge of the city. The rich generally live affluent lifestyles in large, gated community areas, with luxury facilities such as swimming pools and tennis courts.

Poor residents live either in permanent but poor quality housing between the inner city and the suburban favelas or in the very poor quality conditions of the shanty towns. One third of the population of Sao Paulo is estimated to live in one roomed dwellings.

Crime

Many residents and visitors see crime as one of Sao Paulo's main issues. The favelas are perceived to be areas of organised crime, violence and drug-trafficking, though it has been argued that crime has decreased in recent years. Many affluent people move to other safer communities within Sao Paulo and away from the favelas. These are often gated communities.

Pollution

An industrial haze, intensified by traffic fumes, often hangs over the city. The city also produces large amounts of waste. In favelas, the rubbish is unlikely to be collected and its presence, combined with polluted water supplies and sewage in open drains, can cause further health hazards.

Test yourself

1. List any push factors that might influence people to move from the rural areas of Brazil and into Sao Paulo. (3)
2. Describe how the population of Sao Paulo has changed over the last 50 years. (4)
3. Explain some of the reasons why the population of Sao Paulo has changed over the last 50 years. (4)
4. What is a shanty town and why are they often found in LEDCs? (3)
5. Describe and explain the usual location of shanty towns within an LEDC city. (6)
6. Describe and explain three key characteristics that you would expect to see in a favela in Sao Paulo. (6)
7. What do you think are the main challenges facing those who live in a favela? (3)

Urban planning and regeneration of the inner city

Redevelopment is when an area is demolished and redesigned. For example, in the inner city a street of terraced houses might be knocked down and replaced with a block of flats or other cheaper housing.

Regeneration happens when an area is being upgraded. The aim is to improve the social and economic look of a place. It usually happens in an area where there is dereliction, pollution or out migration. It might allow some buildings to be restored, for example old buildings being repurposed for a different use.

Gentrification is when an area is demolished and redesigned. However, the aim is that the original residents will not move back into the area. Instead, it aims to attract different, richer people, by building new, more expensive accommodation.

Urban planning is a very important aspect and initiator of change in urban environments. Planning allows any changes to occur in an organised, methodical manner that will add value to the place or area.

Case Study: inner city Belfast (an MEDC city)

In many MEDC cities across the world an increasing number of programmes have been set up to deal with some of the issues prevalent within the old, poor, dilapidated inner city areas.

Image courtesy of Titanic Quarter Limited

Figure 30
Planning and regeneration of the Belfast Harbour/Titanic Quarter area

Belfast is a good example of city that has managed to shake off some of the dust of its troubled past through urban redevelopment and gentrification programmes over the last 15 years. Following the ceasefires in Northern Ireland in 1995, Belfast became a more stable city and more open to investment. This began with the Laganside Corporation, which was set up in 1989, with an aim to "contribute to the revitalisation of Belfast and Northern Ireland by transforming Laganside to be attractive, accessible and sustainable, recognised as a place of opportunity for all."

The Laganside development programme started with the Lagan Weir, built (at a cost of £14 million) to control the amount of water in the river upstream. Its purpose was to keep the mud flats covered at all times and to remove the tidal nature of the lower parts of the river. However, the Laganside programme did not deal with the river above the weir and only tackled some of the inner city areas downstream.

The next step was the redevelopment of the Belfast Harbour area, part of the inner city of Belfast. Much of Belfast's traditional manufacturing base had grown up in an area called Queen's Island. However, the decline of Harland and Wolff, from over 20,000 workers at its height to around 500 workers by 2002, meant that there was a large area in urban Belfast (around 185 acres) that was not being used to its full potential. By 2002, much of what had been a vibrant and busy industrial factory space was fast becoming a sad brownfield site close to Belfast's CBD. Land was lying abandoned, buildings were quickly becoming derelict and the whole site was offering little or no employment to people living in the local area. Planning legislation was quickly put into place and an opportunity was taken to redevelop this historic part of Belfast.

Figure 31
Model of what the completed Titanic Quarter might look like

Image courtesy of Titanic Quarter Limited

Development of a strategy for Belfast Harbour area

Building began in 2006, with the aim to reinvent the area as the Titanic Quarter: a fresh, modern space where people would come to live, work and play. It also planned new transport links to make this area easily accessible from all parts of the city. Titanic Quarter is currently being developed jointly by Belfast Harbour and Titanic Quarter Limited.

The Titanic Quarter Development Strategy clearly sets out a comprehensive framework of aims that have been built into the BMAP (Belfast Metropolitan Area Plan) 2015. These aims include the provision of:

- up to 5,000 dwellings (including apartments and townhouses).
- a high quality business quarter.
- commercial development.
- a major third level campus (Belfast Metropolitan College).
- a Titanic signature project.
- a restored former Harland and Wolff Headquarters.
- a 'gateway' hotel (Premier Inn) at the entrance to Titanic Quarter at Abercorn Basin.
- a cruise liner berth and ancillary facilities for tourism.
- leisure developments, such as restaurants, cafés, bars and health and fitness clubs.
- local services and business support, such as local retail, health care, crèches and day nurseries.

Housing

Courtesy of Chris Bennett

Figure 32
The first phase of high quality housing at Titanic Quarter (The Arc)

The 5,000 dwellings in the Titanic Quarter scheme consist mostly of apartments and townhouses, and should provide residential accommodation for about 20,000 people. The vision for the area is to create a "sustainable vibrant community with mixed tenure and mixed income groups." However, so far only luxury apartments have been built, which are clearly aimed at young professionals. Some social and more affordable housing is planned within the next development phase, as the economic climate within the UK improves further.

Employment

The £7 billion development is expected to create at least 25,000 jobs over the next 15 years. In the first instance the jobs were in construction but as construction is completed on Phase 1 and 2 of the project, jobs are being generated in IT and financial services (Microsoft, Google, NI Science Park and Citibank/Citigroup), education (Public Record Office for Northern Ireland and Belfast Metropolitan College), hotels (Premier Inn), museums and the Titanic Signature project, bars and restaurants, retail units, industrial space, car show rooms (Audi) and the film industry (in the old Harland & Wolff paint hall).

Figure 33
Traditional industry and new commercial enterprises around the Titanic Quarter

The types of job that the new Titanic Quarter is bringing to the city are very different to the old heavy manufacturing that used to be on the same site. These new jobs require new skills, new qualifications and new employees, which should help improve the diversity of the city.

Environment

Any building programme which aims to convert a brownfield site to a new urban usage needs to ensure that all industrial pollutants have been removed from the site. In addition, a process of soil remediation was carried out at Titanic Quarter, where land contaminated by the shipbuilding industry was restored. Visually, the area has also been much improved. While contemporary architecture, landscaping and street art have modernised the site, the restoration of its historic buildings, such as the Harland and Wolff Head Quarters, ensure its industrial past is retained.

Transport

The BMAP 2015 notes that newand improved transport links need to be provided to connect the new development to the public transport network. The underlying idea is that someone living in Titanic Quarter could get around without having to purchase a car. New bus services, walkways, bridges and cycle routes have all been implemented (although plans for a light rapid rail system across the city have been shelved). These should all help to reduce congestion and air pollution across the city.

Evaluate the sustainability of this urban planning scheme

A sustainable settlement is one which meets the needs of the present without compromising the ability of future generations to meet their own needs. For a city to remain sustainable it is important that it adopts effective planning policies to help reduce the ecological footprint of the place. Ultimately, any building project where living and working space is increasing is not sustainable. The process of building consumes valuable materials. However, there are measures which can be taken to ensure that the project either fixes previous problems or creates as minimal an ecological footprint as possible.

A sustainable city should aim to:

- manage resources, such as oil and forests, as effectively as possible.
- increase opportunities for recycling and reusing materials.
- ensure that the city grows in ways that minimise damage to wildlife and the countryside.
- source more resources, such as food, from the local area.

Planners can help to influence the sustainability of a city. As a sustainable settlement, the Titanic Quarter has some of the following characteristics:

Society and people

1. The apartments built in Titanic Quarter provide more housing units than individually spaced out houses, giving the development a smaller 'urban footprint'. Generally, compact housing also reduces heating bills, and allows services such as access to water and bin collections to be shared. These factors combine to make the residential development more sustainable.
2. The modern construction is designed to provide a light and attractive environment within which to live and work. If people enjoy their living space, they are more likely to take pride in their area, encouraging a sense of community and safety in the urban landscape. If people feel safe in their environment they are also more likely to walk and drive around the area, improving social links.

Economic

1. Long term employment opportunities are available nearby, for example, in the Belfast Metropolitan College or the Public Record Office for Northern Ireland. This reduces the distance people will have to travel for work and limits the need for personal transport.
2. The area plan incorporates plans for both social and gentrified housing, allowing people with various incomes to live in the area close to their employment.

Environmental

1. The land in the Belfast harbour area was previously used for industry and its regeneration removes much of the old waste and pollution. It is also more sustainable to reuse this land than develop green field sites at the edge of the city.
2. The land is used for a variety of purposes, such as residential, economic and retailing, all in close proximity. This reduces the need for transport by car, as movement around the area is easiest by foot, bus and bicycle.
3. The developers of Titanic Quarter are committed to the reduction of waste and the recycling of waste materials. New initiatives for refuse collection and recycling are being trailed and implemented.
4. New developments in green technologies will potentially allow Titanic Quarter to develop alternative energy sources such as solar and wind power.

Test yourself

1. Describe the main details of the urban planning scheme to redevelop the industrial heart of Belfast. (4)
2. What are the main reasons why planners felt the need to redevelop this part of the inner city? (3)
3. What are the main advantages of living somewhere similar to Titanic Quarter? (3)
4. What are the main disadvantages of living somewhere similar to Titanic Quarter? (3)
5. What role do you think planning has played in the redevelopment of Titanic Quarter? (4)
6. Would it be fair to describe Titanic Quarter as more of a gentrification process than a redevelopment? (3)
7. Draw a Venn diagram with one circle to represent housing, one to represent employment and one to represent the environment. Try to identify two or three ways that each of these methods of urban planning is changing the face of Belfast in Titanic Quarter. (6)
8. Suggest some reasons why the Titanic Quarter development might not be very sustainable. (4)
9. Evaluate the sustainability of an urban planning scheme that you have studied. (What are the good and the bad points of the Titanic Quarter development?) (8)

Check your learning

Now that you have studied Part 3 Urbanisation in LEDCs and MEDCs, return to page 183 and answer the Key Questions for this section.

Organise your revision

Draw a mind map to summarise Unit 2A 'People and Where They Live'.

EXAM PRACTICE QUESTIONS

Some of these questions are from previous CCEA GCSE examination papers and others have been written in the same format to give you practice at answering 'exam style' questions.

Try to answer the questions with as much detail as possible. Also consider the number of marks that each question receives, as this will give you a good indication of the amount of depth that your answer needs.

Resource A

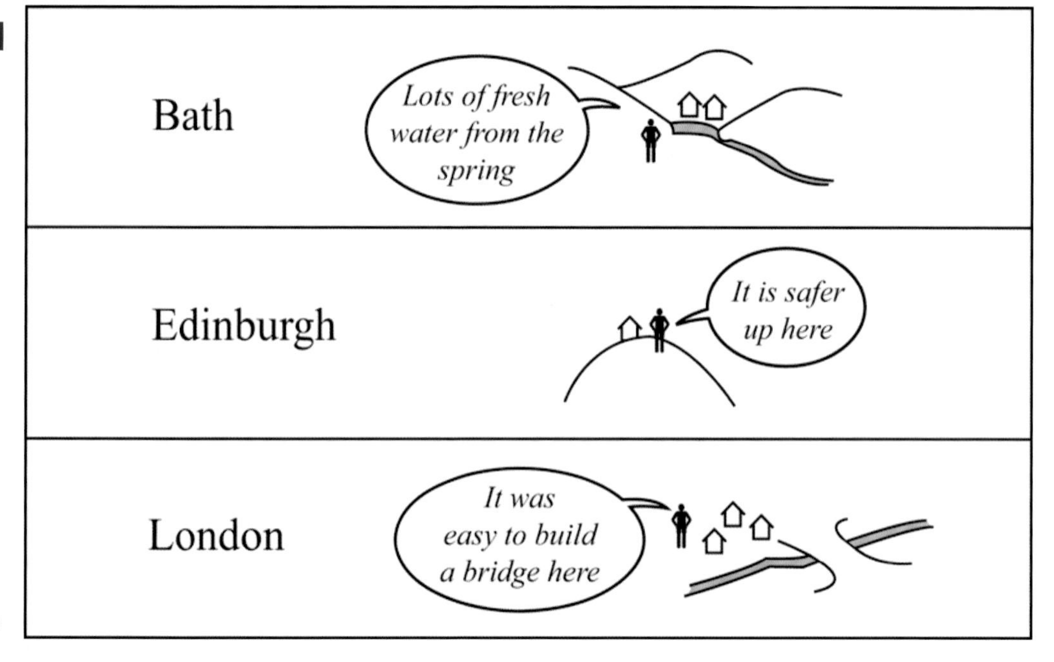

Resource B

SETTLEMENT	TYPES OF SITE
Bath	
Edinburgh	
London	

Resource C

STATEMENT	TRUE OR FALSE
A seaside resort is an island settlement with many coal mines	
Market towns grew up where farmers came to buy and sell goods	
Route centres often develop where many roads meet	True
An industrial town has a lot of factories making things	

Resource D

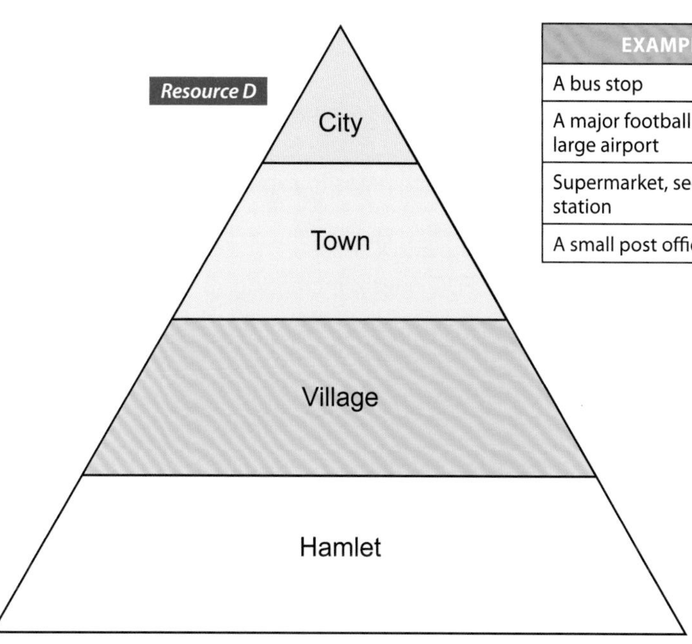

Resource E

EXAMPLE OF SERVICES AVAILABLE	SETTLEMENT TYPE
A bus stop	Hamlet
A major football stadium, hospitals, cathedral and large airport	
Supermarket, secondary school and a small bus station	
A small post office and a general shop	

Why people move to cities in a LEDC

- Better houses and schools
- Poor harvests
- More jobs and better wages
- Poor transport
- Soil erosion

Resource F

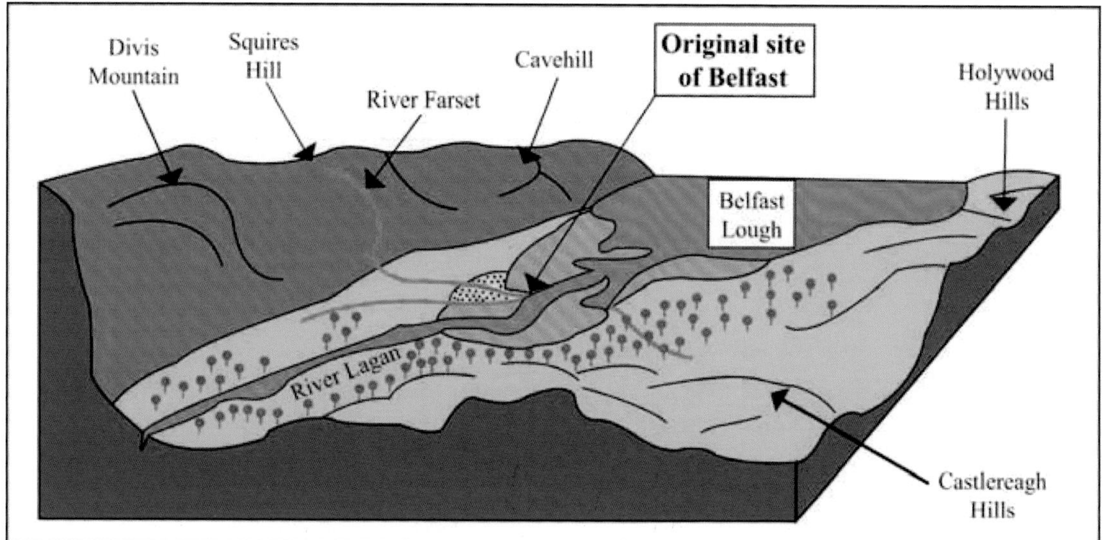

© Adapted from *Higher Ground: A Geography of Northern Ireland for GCSE* by Derek Polley, Colourpoint Books (2001)

Resource G

Resource H

CBD	Rural-Urban fringe
Inner city	Suburban residential

Foundation Tier

1. Study Resource A which shows the site of three settlements. Answer the questions which follow.
   i. State the meaning of the term site. (2)
   ii. Complete the table in Resource B by naming the type of site for each settlement. (3)
   iii. Complete the table in Resource C by stating whether the following statements are true or false. One answer has been completed for you. (3)
2. Study Resource D which shows a geographical hierarchy of settlement types. Answer the questions which follow.
   i. State which settlement type from the diagram has the largest population size. (1)
   ii. Complete the table in Resource E which matches services available to settlement types. One has been completed for you. (3)
3. State the meaning of the term rural-urban fringe. (2)
4. Describe and explain two characteristics of the CBD. (4)
5. Study Resource F which shows some push and pull factors in a LEDC. Answer the questions which follow.
   i. State one push and one pull factor from Resource F which helps to explain why people leave the countryside and move to the city. (2)
   ii. Outline one reason which might stop people from moving. (2)
   iii. Outline one reason why the number of people in LEDC cities is increasing. (3)
6. Many people who move into cities in LEDCs live in shanty towns. Describe and explain one location of a shanty town in a LEDC city you have studied. (1 mark for identifying the city and 4 marks for describing and explaining the location)

Higher Tier

1. Study Resource D which shows a geographical hierarchy of settlements. Answer the questions which follow.
   i. State which settlement type has the greatest population size. (1)
   ii. Describe how and explain why the number of services varies within a settlement hierarchy. (6)
2. Study Resource G which shows the original site of Belfast.
   i. State the meaning of the term site. (2)
   ii. Explain fully why this site would have been attractive to settlers. You should give two reasons in your answer. (6)
3. Study the table in Resource H which names four land-use zones found in cities.
   Choose one of the land zones named and explain its characteristics. (4)
4. State the meaning of the term shanty town. (2)
5. Describe and explain the location and growth of shanty town areas in a LEDC city which you have studied. (7)
6. Many inner city areas have been regenerated and improved. Evaluate the sustainability of an inner city urban planning scheme you have studied. Mention two aspects in your answer. (6)

2B Contrasts in World Development

"You can't get rid of poverty by giving people money"
PJ O'Rouke

We live in a divided world. The rich are getting richer and the poor are getting poorer, and the divide between rich and poor has been increasing at an alarming rate. This contrast is the source of much pain, anger and conflict between people.

This theme takes a look at the differences in development between MEDCs and LEDCs, and looks at some of the issues that are involved in measuring these differences.

The theme is divided up into three parts:

1. **The development gap**
2. **Factors contributing to unequal development**
3. **Sustainable solutions to deal with the problems of unequal development**

PART 1: THE DEVELOPMENT GAP

(a) **Identify the differences in development between MEDCs and LEDCs using social and economic indicators**
Social indicators
Economic indicators

(b) **Assess the effectiveness of social and economic indicators of development in relation to quality of life using the Human Development Index (HDI)**

(c) **The factors that hinder development in LEDCs**
Historical Factors
Environmental Factors
Dependence on primary activities
Debt
Politics

(d) **Describe ONE named strategy that is attempting to reduce the global development gap and explain how it attempts to do so:**
The organisation
The core aims of the strategy
The action taken

Key questions

By the end of this section you will be able to answer the following questions:

1. What is the difference between a MEDC and a LEDC?
2. Write a definition of a NIC.
3. What are the main economic measures of development?
4. What are the main social measures of development?
5. Which is more effective: a social, economic or a composite measure of development?
6. What things have hindered the development of LEDCs?
7. What strategy might be employed to help reduce the global development gap?

Key words

Development
Development gap
Social indicators
Economic indicators
Quality of life
Human Development Index (HDI)
Primary
Secondary activity
Tertiary activity
Quaternary activity
MEDC
LEDC

Identify the differences in development between MEDCs and LEDCs using social and economic indicators

Geographers like to be able to identify and simplify patterns. You will have already started to do this in your Geography course by categorising some countries as MEDCs (More Economically Developed Countries) and some as LEDCs (Less Economically Developed Countries). Yet, no two sources will give you a definitive list of which country sits on which list. Countries such as Brazil, India and China float between the two lists, depending on what economic measures are being used. So, a new category is now recognised to cover this shortfall, the NIC (Newly Industrialising Country).

Even what we mean by International Development is tricky to define. No universally accepted definition exists, but most geographers accept that the study of development looks at the quality of life for humans within a country or area. It considers the wealth that is available to people and how foreign aid, political decisions, healthcare, education, poverty reduction, infrastructure, economics, human rights and the environment all work together to provide a particular quality of life.

For a country to develop, it has to change. Over the last 50 years geographers have noticed that some countries have become richer and have developed their industry more than others.

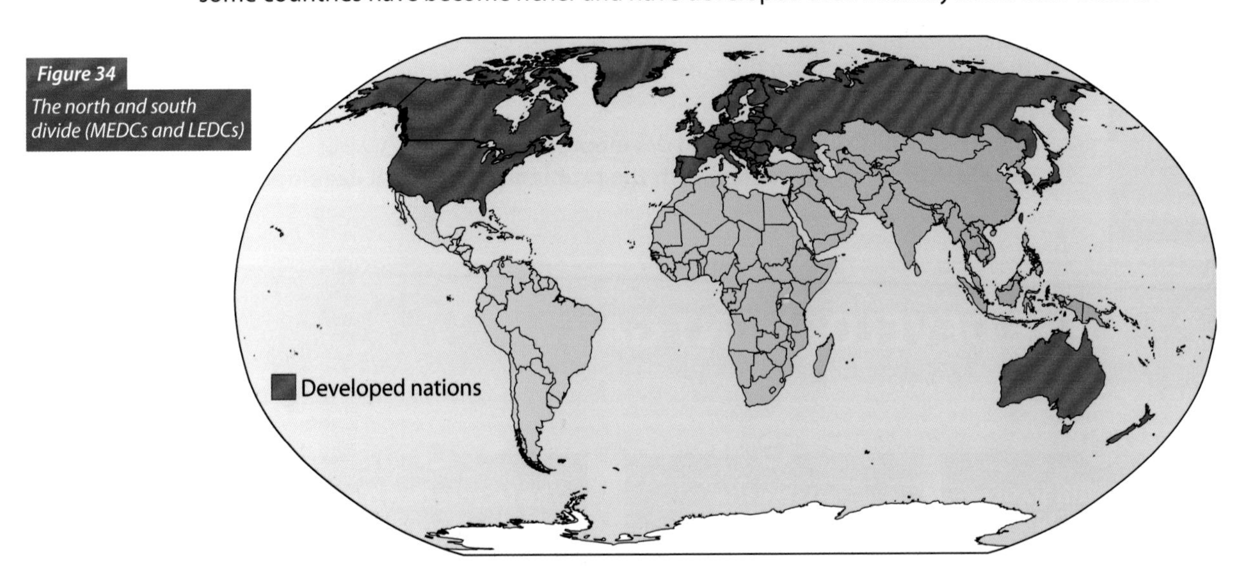

Figure 34
The north and south divide (MEDCs and LEDCs)

No two countries have developed in the same way or at the same rate. In 1980, the Brandt Report divided the world into two halves. 'The north' included all the countries of the industrialised 'developed world' (but also included Australia and New Zealand). 'The south' included the low-income and the middle-income countries of the 'developing world'. Since then, although the designation of the countries has changed a little, the concept behind them remains the same. A **'development gap'** can be identified, which is the difference in economic activity, wealth and social measures between the rich MEDCs and poorer LEDCs. There is a large and widening difference between the quality of life that is experienced by a person living in an MEDC compared with that of someone living in an LEDC. There are many different reasons why this difference exists but much of this is related to the wealth of the country and the interrelationships with debt, trade, population change, impact of HIV/AIDS, political instability and climatic disasters.

The United Nations plays a crucial role in monitoring and enhancing the development of weaker nations. Since the 1960s the General Assembly of the UN has helped to set priorities to deal with particular concerns. At the Millennium Summit in September 2000, world leaders adopted a series of Millennium Development Goals which aimed to remove extreme poverty and hunger, achieve universal primary education, promote gender equality, reduce child mortality, improve maternal health, combat HIV/AIDS and malaria, and ensure environmental sustainability.

The indicators used to measure development

The World Bank uses 298 separate indicators from their dataset to analyse the differences between countries. However, there are a few indicators that are globally accepted which help to show how developed a particular country is.

Social indicators

These are used to assess how well a country is developing in key areas which affect people, such as health, education and diet. They are to do with the quality of life of individuals within a country.

Health

Life expectancy is a health indicator. It is the average lifespan of someone born in that country. It can be affected by wars, natural disasters and disease but generally gives a good indication of how developed a country is. The higher the life expectancy, the more developed the country. For example, people in the UK have a life expectancy of around 78 years, whereas Ugandans have a life expectancy of only 41 years.

Another important health indicator is the **Infant Mortality Rate.** This measures the number of children who die before they reach the age of one (from every 1,000 live births, per year). In the UK this figure is very small (around 6 per 1,000 per year) whereas in Sierra Leone the figure is very high (around 195 per 1,000 per year).

Often birth and death rates are also used as indicators of development. However, this is becoming increasingly complicated to use when comparing the development gap between countries because death rates have fallen in LEDCs and can often be below the death rates for MEDCs.

Another indicator is the **number of patients per doctor.** This measure shows the inequality between people in the north and south as LEDCS usually have poorer healthcare. Remarkably, Cuba leads the world, with 1 doctor per 170 patients. MEDCs are usually around 1 per 380 but LEDCs such as Zambia are 1 per 8,300. Malawi and Tanzania have the least at 1 per 50,000.

Education

One education measure is the percentage of the adult population who are able to both read and write. This tells us what the **Adult Literacy Rate** is. In the UK and other MEDCs you would expect around 99% of adults to be able to read and write. However, in Somalia only around 24% of adults can read and write.

Diet

Increasingly, UN agencies and development organisations are using calorie counts to measure the amount of food that people in different parts of the world are consuming. Generally, wealthy countries have much higher average levels of calorie intake but in poorer countries malnutrition can be more common. For example, the USA's average intake is 3,725 kilocalories per person, while Eritrea is 1,555 kilocalories per person.

Economic indicators

These indicators help us to assess the amount of money or wealth within a country and how the people actually earn that wealth.

Gross National Product (per capita) or Gross National Income

GNP is a very important indicator, as it shows us the total economic value of all of the goods and services that are provided in a country during the course of a year, divided by the number of people who live in the country. The amount is always worked out in US Dollars so that a comparison can be made with other countries. The higher the GNP, the more developed a country is likely to be. Increasingly the term GNI (Gross National Income) is being used to replace the GNP. Both measures are very similar except that for the GNP measurement the indirect business taxes are not deducted.

COUNTRY	RANK	GNI $
Qatar	1	87,030
Luxembourg	2	64,410
Norway	3	62,970
Singapore	4	59,790
Switzerland	5	52,320

COUNTRY	RANK	GNI $
Niger	181	720
Burundi	182	610
Eritrea	183	580
Liberia	184	520
Congo (DR)	185	350

Figure 35

The five richest and five poorest countries as measured by GNI, 2011

Data from World Bank, International Comparison Program database, 2011

Using GNP per capita is still one of the most commonly used measures of development across the world. However, there are a few other economic measures which also help us to understand the contrasts in wealth within countries.

Increasingly GNP is now measured using Purchasing Power Parity (in dollars). This means that 1 US dollar is equivalent to an amount of money that will have the same purchasing power in countries. Therefore the GNP will have been PPP-adjusted. These adjustments can radically alter the way that we look at the wealth of a country. For example, the GNP per capita in India is US$1,704 but when converted to a PPP basis it is US$3,608. Alternatively, in Denmark the GNP per capita is US$62,100 but its PPP is US$37,304.

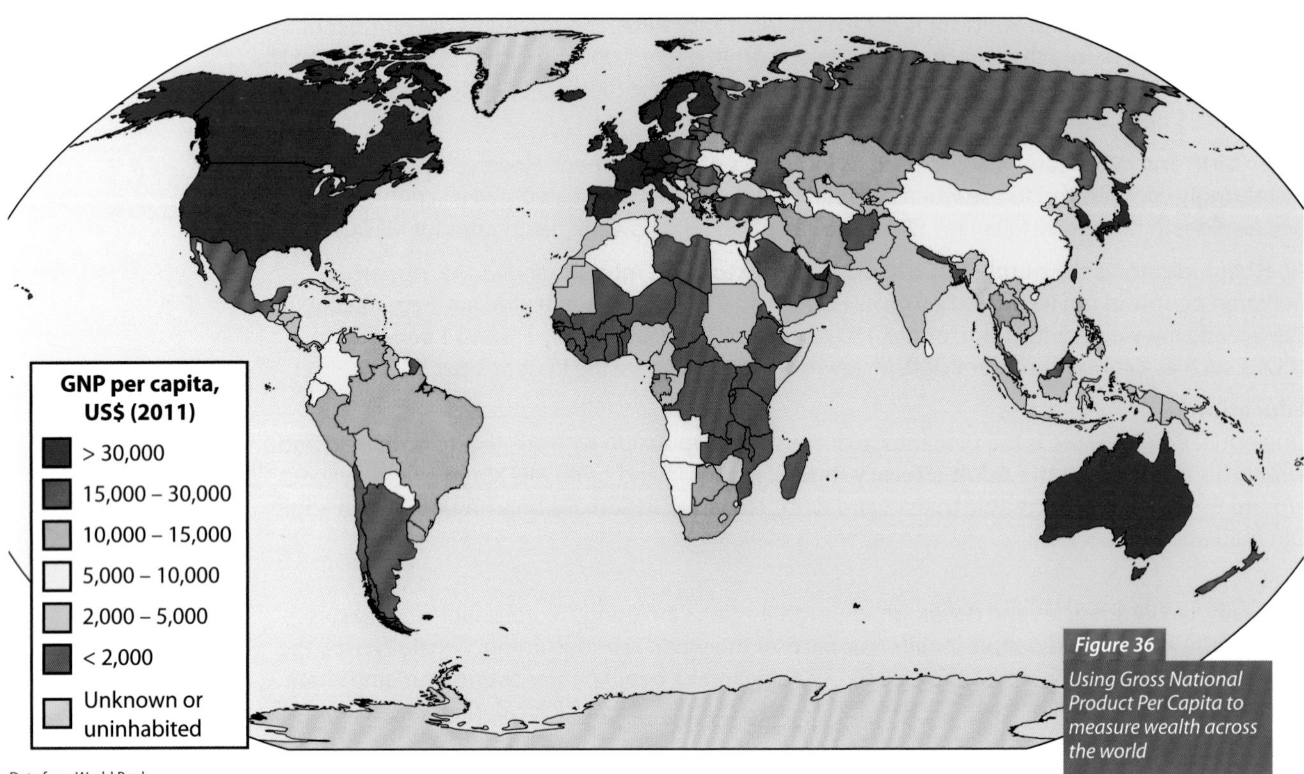

Data from World Bank

Figure 36 *Using Gross National Product Per Capita to measure wealth across the world*

Vehicles per 1,000 people

Sometimes a measure of the number of cars owned by people helps us understand the amount of wealth in a country. For example, Germany has 528 cars per 1,000, Brazil has 79 and China has 8.

Percentage of people employed in Primary activities

Primary activities are those jobs or economic activities where people are involved with collecting and working with raw materials or resources such as farming (agriculture), mining, quarrying and fishing.

Secondary activities are those jobs where people are involved in manufacturing or making something using the raw materials collected in the primary industry. This can be as diverse as heavy manufacturing such as making iron and steel or processing metal and components into cars, or much lighter industries such as making beer, fizzy drinks or bread.

Tertiary activities are those jobs where people provide a service to others, such as doctors, lawyers, teachers and hairdressers.

Quaternary activities is a final category sometimes used to further define people who are involved in the research and development of new products. They mostly feature in the information technology industries.

The percentage breakdown of jobs within each of the employment sectors can indicate development. A rich, more developed country is likely to have more people working in the secondary, tertiary and quaternary sectors. A low number of people working in primary activities such as agriculture is an indication that a country is developed. For example, the UK only has 2% of the population working in agriculture, whereas Vietnam has over 73% of people employed in agriculture.

Newly Industrialised Countries

Newly Industrialised Countries (NICs) is a relatively new category used to describe countries who have not reached MEDC status but who are out-developing their fellow LEDCs. There is some debate about who is entitled to call themselves a NIC. However, many economists are sure that by 2050 China, India, Brazil and Mexico will all have economies to rival the USA. However, in China and India in particular, the immense population of these countries does mean that per capita (per person) income will remain low even if the economy grows.

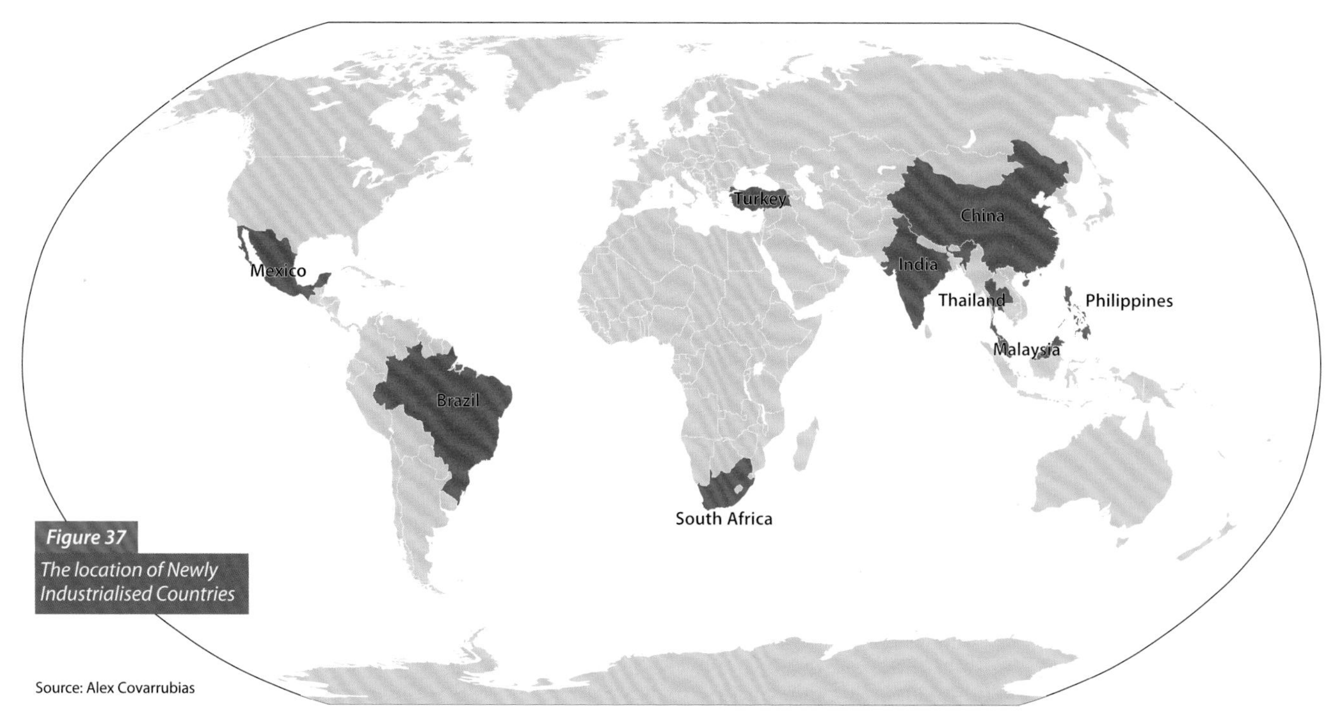

Figure 37
The location of Newly Industrialised Countries

Source: Alex Covarrubias

COUNTRY	GNI (IN BILLIONS OF US$)	GNI PER CAPITA (IN US$)
South Africa	352.038	6,960
Mexico	1060.221	9,240
Brazil	2107.628	10,720
China	6644.327	4,940
India	1746.481	1,410
Malaysia	243.108	8,420
Philippines	209.450	2,210
Thailand	307.129	4,420
Turkey	766.441	10,410

Figure 38
Current NICs in 2011

Data from World Bank national accounts data, 2011

Assess the effectiveness of social and economic indicators of development in relation to quality of life using the Human Development Index (HDI)

As there are so many different measures of development, coming up with a truly reliable measure, which covers all of the possible variations in development is very difficult. As countries change, they develop at different speeds and in different ways. For example, in India and China in recent years more and more people have been investing in very cheap cars, which raises the number of cars faster than the overall development of the country might be growing. As indicators are averages, they can be misleading, and it can be difficult to accurately measure the wealth of a country. Also, there are not just contrasts between countries, but contrasts within countries too.

In response to an overreliance on simple measures, or even on just measurements of wealth, the United Nations started to look for something which would combine some of the major indicators into one easy-to-use measure.

In 1990, the UN started to use the **Human Development Index** (HDI). The very first Human Development Report introduced this new way of measuring development, combining indicators of life expectancy, educational attainment and income into a composite Human Development Index, the HDI. The HDI was celebrated, as it was the first single statistic which contained both social and economic development data. The HDI sets a minimum and maximum for each dimension, called goalposts, and then shows where each country stands in relation to these goalposts as a value between 0 and 1.

Figure 39

Components of the Human Development Index

Source: 2013 Human Development Index, United Nations Development Programme, http://hdr.undp.org/en/statistics/hdi/

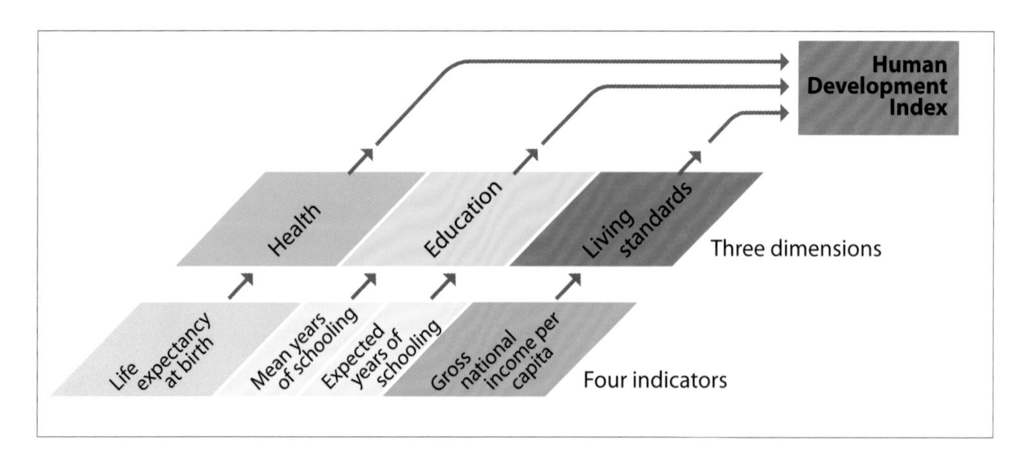

Originally the three measures were for life expectancy, educational attainment (including adult literacy rates and student enrolment rates) and GNP per capita. However, the education component of the HDI is now measured by the mean years of schooling for adults aged 25 years and expected years of schooling for children of school going age.

The life expectancy at birth component of the HDI is calculated using a minimum value of 20 years and a maximum value of 83.2 years. These are the observed maximum values of the indicators from the countries from 1980 to 2010. Therefore a country with a life expectancy at birth of 55 would be 0.554 in the life expectancy index. Overall, the higher the Index score (towards 1) the more developed that a country will be.

For the wealth component, the goalpost for minimum income was set at $163 (PPP) and the maximum of $108,211 (PPP). Some people are concerned that using the HDI still allows too much influence on wealth. However, it does provide a dynamic set of data that tries to take into account the fluctuations within countries every year. Each year the UNDP (the United Nations Development Programme) publishes its Human Development Report, which attempts to measure both equity and sustainability within countries.

Figure 40

Top 15 Countries (Very High Human Development, 2011)

Source: 2011 Human Development Report, United Nations Development Programme.

COUNTRY IN RANK ORDER	HUMAN DEVELOPMENT INDEX (HDI) VALUE	LIFE EXPECTANCY (YEARS)	MEAN YEARS OF SCHOOLING (YEARS)	EXPECTED YEARS OF SCHOOLING (YEARS)	GNI PER CAPITA ($PPP)
1 Norway	0.943	81.1	12.6	17.3	47,557
2 Australia	0.929	81.9	12.0	18.0	34,431
3 Netherlands	0.910	80.7	11.6	16.8	36,402
4 United States	0.910	78.5	12.4	16.0	43,017
5 New Zealand	0.908	80.7	12.5	18.0	23,737
6 Canada	0.908	80.7	12.5	18.0	23,737
7 Ireland	0.908	80.6	11.6	18.0	29,322
8 Liechtenstein	0.905	79.5	10.3	14.7	83,717
9 Germany	0.905	80.4	12.2	15.9	34,854
10 Sweden	0.904	81.4	11.7	15.7	35,837
11 Switzerland	0.903	82.3	11.0	15.6	39,924
12 Japan	0.901	83.4	11.6	15.1	32,295
13 Hong Kong	0.898	82.8	10.0	15.7	44,805
14 Iceland	0.898	81.8	10.4	18.0	29,354
15 Korea (Rep of)	0.897	80.6	11.6	16.9	28,230

COUNTRY IN RANK ORDER	HUMAN DEVELOPMENT INDEX (HDI) VALUE	LIFE EXPECTANCY (YEARS)	MEAN YEARS OF SCHOOLING (YEARS)	EXPECTED YEARS OF SCHOOLING (YEARS)	GNI PER CAPITA ($PPP)
173 Zimbabwe	0.376	51.4	7.2	9.9	376
174 Ethiopia	0.363	59.3	1.5	8.5	971
175 Mali	0.359	51.4	2.0	8.3	1,123
176 Guinea-Bissau	0.353	48.1	2.3	9.1	994
177 Eritrea	0.349	61.6	3.4	4.8	536
178 Guinea	0.344	54.1	1.6	8.6	863
179 Central African Republic	0.343	48.4	3.5	6.6	707
180 Sierra Leone	0.336	47.8	2.9	7.2	737
181 Burkina Faso	0.331	55.4	1.3	6.3	1,141
182 Liberia	0.329	56.8	3.9	11.0	265
183 Chad	0.328	49.6	1.5	7.2	1,105
184 Mozambique	0.322	50.2	1.2	9.2	898
185 Burundi	0.316	50.4	2.7	10.5	368
186 Niger	0.295	54.7	1.4	4.9	641
187 Congo, Democratic Republic	0.286	48.4	3.5	8.2	280

Figure 41

Bottom 15 Countries (Low Human Development, 2011)

Source: 2011 Human Development Report, United Nations Development Programme.

Test yourself

1. Name two social and two economic indicators. (4)
2. What are the main limitations of using just social or just economic indicators? (4)
3. What are the main indicators that are used within the Human Development Index? (3)
4. List some of the problems associated with using the HDI. (3)
5. How effective do you think the HDI is at measuring development? (3)

Use Figures 40 and 41 to answer the following questions:

6. What do you notice about the life expectancy values for the top 15 countries using the HDI? (3)
7. Describe the difference between the top and bottom countries in relation to life expectancy. (4)
8. What do you notice about the differences in education between the countries of the world? (3)
9. Why do you think that the richest countries (in terms of GNI) are not always the ones at the very top of the chart? (3)
10. On a copy of a blank world map, use one colour to highlight the top 15 countries according to the HDI and use a different colour to highlight the bottom 15 countries.
    a. Describe the pattern for the top and bottom countries. (5)
    b. Can you explain why the pattern might be like this? (4)

The factors that hinder development in LEDCs

In many cases around the world, the rich-poor divide between and within countries is increasing. The rich are getting richer and the poor are getting poorer. However, the reasons for this gap are complex. In fact, in many cases, the reason why particular countries have become poor is because of the way that many old colonial powers, such as the UK, have influenced countries far across the globe.

Development is a process of change. Countries are either getting progressively richer or poorer and as a result there are changes in education, social welfare, healthcare and wealth. However, this process of change does not happen at the same rate across the world. Some countries started this process of improvement a long time ago. Others have only recently started to improve their standard of living. In these countries, a range of factors has held them back and has restricted their development. In most cases, the countries that developed first enjoyed their dominance over the less developed countries and attempted to maintain a degree of control over them.

Historical factors

Figure 42

Colonial possessions of world powers, 1914

Source: Dave MacLeod – SHDHS

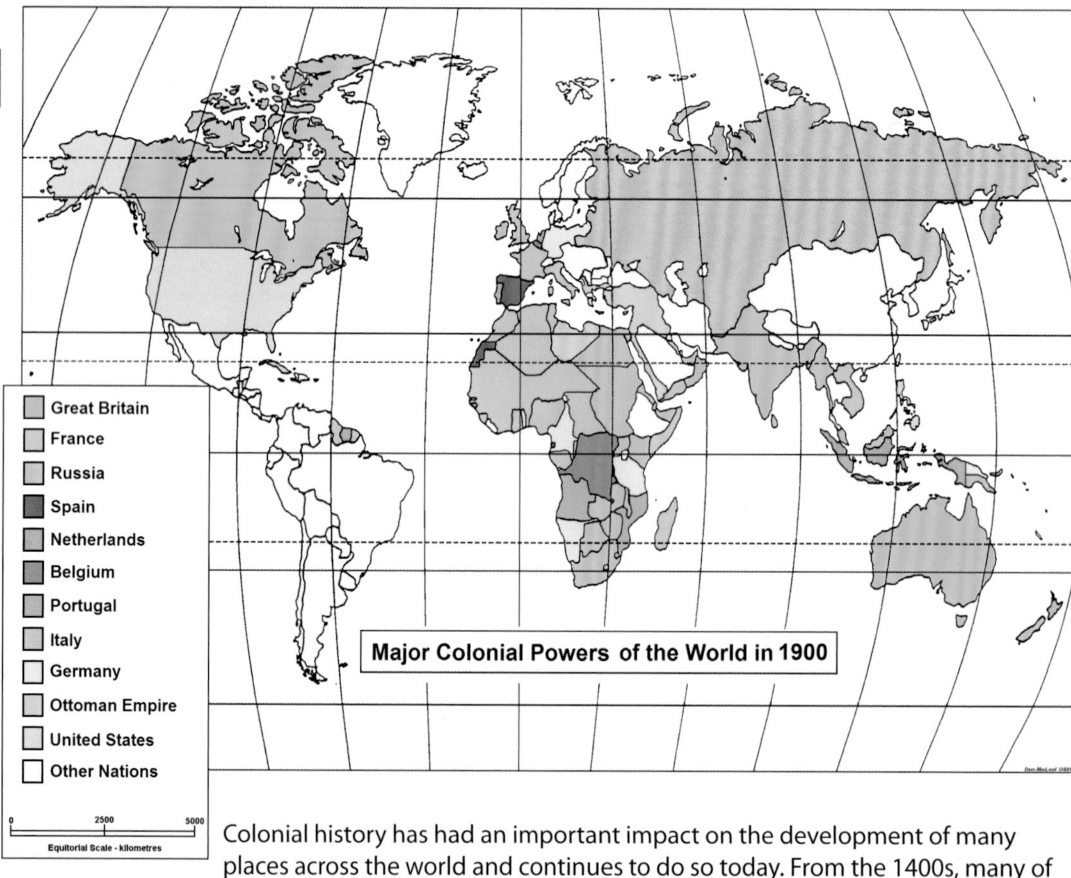

Colonial history has had an important impact on the development of many places across the world and continues to do so today. From the 1400s, many of the European powers had sent various sea expeditions to try and claim new lands and territories for their kings and queens. Initially, Spain and Portugal were able to take territory in Central and South America. France concentrated largely on West Africa and Britain extended an Empire that at its height covered one quarter of the world's landmass, including Canada, Egypt, Kenya, South Africa, India, Australia and New Zealand.

Being part of an Empire or Commonwealth brought advantages to its claimed countries. It was good for trade and business, and European countries such as Britain spent money investing in highly important infrastructure, such as roads, railways, hospitals and schools. However, it also brought disadvantages, mainly that the rulers could inflict their will upon the countries. They could force them to fight in their armies (through two world wars), tax them and take their main assets, such as mineral wealth and land. Most of these colonial lands were ruled centrally by the governments back in Europe but after the Second World War many of these territories were made independent. However, many of the resources and mineral wealth of the country had already been removed by the colonial power.

When the European countries withdrew from their conquered lands and returned political power to the native country, further difficulties arose. The countries sometimes found it difficult to govern themselves effectively, which often led to power struggles, corruption, dictatorships and civil war. This in turn caused instability and countries could not collect taxes effectively, organise education or healthcare, or attract industrial investment which might have helped the country out of poverty.

Environmental factors

The majority of people who live in LEDCs are living far below the poverty line. Environmental issues can cause the already-poor people to struggle further and put them at greater risk than people who live in MEDCs. The following are some of the main environmental factors that can hinder development:

Natural hazards

Although natural hazards occur around the world, their impact is greater in LEDCs, which do not have money to either plan for hazards or cope with their aftermath. Instead of being invested into education or industry, money has to be used for rebuilding and rescue operations.

Extreme climates

Many LEDCs experience extreme climates. They can be extremely cold, wet, hot or dry, making it difficult for people to grow their own food. This can lead to malnutrition and starvation, and leave people more open to the spread of disease. When food yields (the amount of food produced) decline, farmers have fewer crops to sell, which reduces their quality of life. Also, the government will not be able to collect taxes, reducing the money available for healthcare and developing the country.

Figure 43
Typically people in LEDCs try to make a living on the land

Natural resources

Much of the farmland in LEDCs can be marginal, making it difficult for farmers to produce a good crop. Water supplies can also be limited, making it difficult for people to survive. LEDCs might have access to raw materials such as coal, oil or precious metals but many of the rights to these are bought and managed by large multinational companies. The raw materials are often exported rather than used within the country, and are in an unprocessed state, making it difficult to make great amounts of money from these resources.

Dependence on primary activities

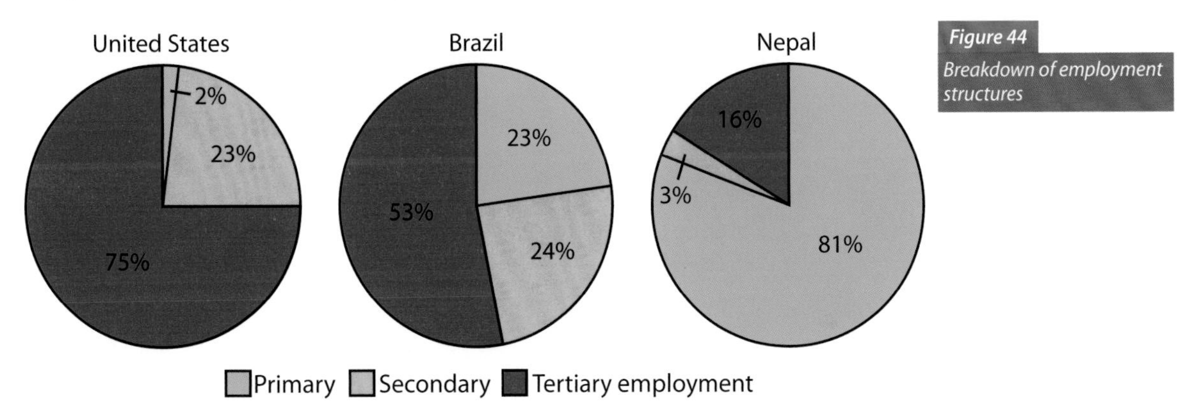

Figure 44
Breakdown of employment structures

Economically, countries that had some sort of mineral wealth or deposits (such as iron ore) or energy resources (such as coal, oil or natural gas) were more likely to develop industrially. This meant that these countries moved from a dependence on Primary activities (such as agriculture) to a dependence on Secondary activities (such as manufacturing).

As time went on, many of the resources that had started the initial Industrial Revolution were depleted and these countries tried to acquire the resources from other countries. Global systems of dependence were created, so that raw materials would be extracted and then sent back to the more industrially developed countries. This created a more global trade system which has allowed large industrial companies such as Ford (from 1903) and Nike (from 1978) to spread worldwide. New overseas factories and processing plants were set up and as companies realised that they could manufacture the same product at a lower cost in LEDCs, they started to move the whole manufacturing process to LEDCs. For example, Nike built factories in Malaysia and China, where profit margins could actually be increased. This change meant that LEDCs would be given the manufacturing stages – the hard labour and often the unskilled or repetitive aspects of the process. Yet, the creative aspects – the product design and sometimes even the final assembly of the product – were kept within the MEDCs.

Debt

As many LEDCs look at their richer counterparts, they realise that there are things that they need if they are going to develop, such as railway lines, hospitals and schools. As the countries cannot afford to raise the much needed cash for development projects, they sometimes borrow money from other countries, international organisations (such as the International Monetary Fund) or even from multinational companies. This money needs to be repaid with interest and sometimes it can take many decades to pay the debt off in full. Any money tied up in paying off debts cannot be used to fund further development projects.

Figure 45

A caricature of Idi Amin, the third President of Uganda. The political landscape of Africa from 1950 to the present has seen many dictators such as this.

Source: Gifted to the U.S. Library of Congress by Edmund S Valtman

Politics

Often the government of a country makes decisions about the jobs, links and trade that it wants to have with other countries and these decisions help to shape the development of that country. Many LEDCs have unstable and corrupt governments, with a few people holding all the power, money and influence. They are rich even though the majority of people remain very poor, and in many cases overseas money and aid never reaches the people, restricting the countries' development. Power struggles within these countries can also lead to civil war and ethnic tensions, and violence that can further cause damage to the already fragile infrastructure. For example, in Saddam Hussein's Iraq much of the money that had been set aside as relief aid for starving people ended up lining Hussein's own pockets and paying for the famous gold taps in his palaces.

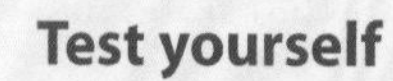

Test yourself

1. Define the term 'development'. (2)
2. Describe how historical factors can hinder development in LEDCs. (3)
3. Describe how environmental factors can hinder development in LEDCs. (3)
4. Explain how a dependence on Primary activities can cause problems for LEDCs. (3)
5. Define the following terms: Primary activities, Secondary activities, Tertiary activities and Quaternary activities. (4)
6. How can debt be an issue for LEDCs? (3)
7. Describe some of the ways that politics can have an impact on LEDCs. (3)

Describe one named strategy that is attempting to reduce the global development gap and explain how it attempts to do so

Many different organisations have been set up that try to help reduce the global development gap. Some of these have been highly successful, some have been very contentious and some have sadly failed. The United Nations has long been seen as a global referee that reports, monitors and attempts to take action where there is need and deprivation. In this study we are going to look at an organisation that took its roots from a group that aimed to make a difference before the new millennium arrived in the year 2000.

The Jubilee Debt Campaign

Background

Jubilee Debt Campaign came into being on 24 March 2001, a successor to the Jubilee 2000 campaign which had sought debt cancellation to mark the beginning of a new millennium. Jubilee 2000 was set up by Martin Dent and Bill Peters and the concept came from the biblical idea of Jubilee, the fiftieth year (quoted in Leviticus) where those enslaved because of debts are freed, lost lands are returned and community which has been torn apart by inequality is restored. Jubilee 2000 aimed to wipe out the £58 billion owed by the world's poorest nations.

The campaign became a global mission to change the world. According to the Guinness Book of World Records, its petition to show support for the aims of the organisation was the largest petition ever signed (with over 24.3 million signatures) and was also the most international (with people from over 166 countries signing).

In their December 2000 report, 'The world will never be the same again', they reviewed the success of their campaign and noted that:

"Rich countries have promised to write off $110bn of debt; and by the end of the year 20 countries are expected to have some received some debt relief. The number of children attending primary school in Uganda has doubled, thanks to $1bn debt cancellation; Bolivia will get $1.3bn of relief and is directing resources released to the poorest municipalities; while Mozambique has a $67m reduction in its debt repayments, allowing it to divert funds to hospitals and housing."

Core aims

Many of the promises that governments made in 2000 regarding cancelling world debt were not followed through, so Jubilee movements in many countries continued the work. The Jubilee Debt Campaign continued their mission in the UK to end all unpayable poor country debts by fair and transparent means.

The JDC is a coalition of charitable and church groups. Their focus is to:

1. Change the UK government policy on debt.
2. Ensure that maximum influence is made on the World Bank and International Monetary Fund (IMF) through the UK government.

In their 2009 publication 'The Return of the Debt Crisis' they noted:

"In the 1960s and 70s rich country governments and their banks lent enormous sums of money to Third World countries, newly liberated from colonialism. This lending was used to buy allies in the Cold War . . . at the end of the 1970s interest rates were increased and scores of Third World countries were caught in a debt crisis. This meant that however much poor governments repaid, the stock of debt never got smaller. And in order to make these repayments, governments were prevented from spending on health, education and economic development. The 1980s and 90s were 'lost decades' for development, when vast sums of money flowed from the global South to the global North, making the poor poorer and the rich richer."

The JDC is based on the work of many different charities and local activists. They are strongly linked to educating people about the issues facing poor countries today and in raising public awareness of the injustices of debt and the importance of debt cancellation. They are a multi-faith project with representatives from six major faiths and many denominations. They also work alongside other organisations such as City Circle, Friends of the Earth, People and Planet, War on Want, and the World Development Movements.

The Jubilee Debt Campaign's basics about debt

The world's poorest countries pay almost $23 million every day to the rich world.

1. Why should we drop the debt?
Debts should be cancelled because they are unjust in terms of their origin, and also because they worsen poverty.

2. How big is the debt of poor countries?
The poorest 48 countries have debts totalling £109.61 billion, whilst for the poorest 128 countries, it is over £2.41 trillion.

3. Where did the debt come from?
Much of the debt of poor countries is left over from the 1970s and often arose through reckless or self-interested lending by the rich world.

4. Hasn't all the debt been cancelled?
No! Thanks to campaigners, debt cancellation has become a reality – but the debt crisis is still with us.

5. Hasn't the UK cancelled all debts?
No! The UK has cancelled many debts, and been active in calling for further cancellation for the poorest – but it still holds billions in debts from other poor countries.

6. What do countries have to do to get cancellation?
There are lots of harmful, undemocratic and irrelevant conditions attached to debt cancellation – we want these strings to be cut.

7. Where do debt payments go?
Poor countries are paying debts to the World Bank, the International Monetary Fund (IMF), other international bodies (all controlled by rich countries), to rich country governments and to companies.

8. Doesn't corruption make debt cancellation pointless?
No. Corruption is a big problem in some places – but funds from debt cancellation do make a difference.

9. How do we prevent further debt crises?
We need economic justice – covering debt, trade, aid, tax, etc and responsible financing in the future.

10. What can I do?
Plenty! Campaigning does make a difference.

Actions taken

Successful campaigns

1. Ending the Vulture Culture

Figure 46
The Drop the Debt campaign in Birmingham, 2008

One type of finance arrangement that the JDC campaigns against is called a 'vulture fund'. This practice is when a private finance company buys up the debt of others. The JDC argues that 'vultures' gain profit from the poverty and misery of others. They make money by buying up defaulted Third World debt very cheaply and then suing the country for the face value of the debt plus interest and compensation. The JDC claims that they make huge profits from very poor countries, often seizing investments and sometimes even aid money. In 2008 two companies took legal action against Liberia and sued it for £20 million in the British High Court in relation to a debt taken out in the 1970s.

The JDC supported a campaign where in spring 2009, Labour MP Sally Keeble raised the issue of 'vulture funds' in the House of Commons. A bill was later introduced which would stop the activities of 'vulture funds' against very poor countries.

2. Cancelling Haiti's debt

Haiti was once described as the 'Pearl of the Antilles'. However, its wealth funded the French Empire and a 'slave debt' made it difficult for Haiti to get out of poverty. In June 2009 Haiti was finally relieved of $1.2 billion of debt. However, following the earthquake of January 2010, Haiti owed an additional $1.2 billion of debt and JDC campaigners were involved in campaigns to cancel Haiti's debts for a second time.

3. Other campaigns

The JDC has been involved in a number of high profile campaigns in recent years. One such campaign was **Lift the lid on bad loans.** This focused on countries that had taken out loans in the past which the JDC argued were "dubious and irresponsible", and suited the lender's needs more than the recipient. In 2006, the **Cut the strings** campaign considered the harmful conditions that they believed to be often attached to debt relief. Many of the world's poorest countries were required by rich country agencies (such as the G8, the IMF and World Bank) to cut public spending and privatise healthcare and other services to secure debt relief. The JDC was also heavily involved in the **Make Poverty History** campaign and **Make Trade Fair** in 2005.

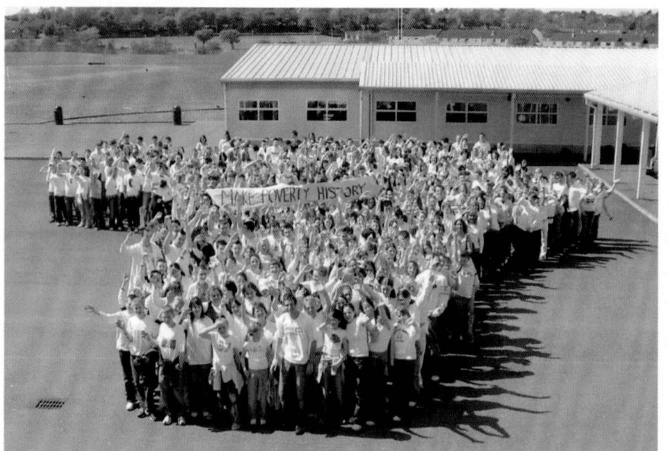

Figure 47
Students from Slemish College raise money and awareness of the MakePovertyHistory campaign in 2005

Continuing campaigns

1. Climate Justice and Climate Debt

Debt does not just restrict countries from spending on social services but it also prevents them from equipping their population with measures to deal with climate change. The JDC are trying to stop the governments from pushing what they believe to be unfair climate loans and instead ensuring that future climate finance is given as grants via the UN.

2. Reclaiming the economy for people

The JDC supported countries using 'debt audits' to examine how responsible the loans to their country were and what impact they had. For example, in a debt audit completed in Ecuador, they found $14 billion of debt had resulted in "incalculable damage" to its people and the environment, and violated the principles of human rights.

3. Setting up a World Debt Court

Debts cancelled one day can be run up the next, unless the lending and wider financial systems are fundamentally changed. There is no protection by insolvency laws for poor countries and new ways are needed to help deal with debt. The JDC propose a world debt court which would allow each borrower and lender to face each other under a neutral arbitrator.

4. Ending UK Dodgy Deals

LEDCs owe hundreds of millions of pounds to the UK. Public money has been used in many cases to support governments who have been involved in human rights abuses or in measures to increase conflict. The JDC has been campaigning for changes to the Export Credits Guarantee Scheme. This would mean that debts would be audited fully, dodgy debts would be cancelled and there would be fewer deals with dictators and arms exporters.

5. Reduction of debt for Zimbabwe, Pakistan and South Sudan

The JDC is committed to seeking resolution and debt audits to evaluate and reduce the debt liability for Zimbabwe, Pakistan and South Sudan.

Test yourself

1. Describe how Jubilee 2000 was set up. (3)
2. How successful was the Jubilee 2000 campaign by December 2000? (3)
3. What are the core aims of the Jubilee Debt Campaign? (3)
4. Why did the JDC feel that the debt crisis had returned? (3)
5. Describe one successful campaign from the JDC. How successful has it been? (6)
6. What are the current campaigns that JDC are involved in? (4)

Check your learning

Now that you have studied Part 1: The Development Gap, return to page 195 and answer the Key Questions for this section.

PART 2: FACTORS CONTRIBUTING TO UNEQUAL DEVELOPMENT

How does globalisation both help and hinder development? (with reference to one case study from an LEDC or NIC)

How does the pattern of world trade create problems for LEDCs?

Key words

Globalisation

Newly Industrialised Country (NIC)

Trade

Key questions

By the end of this section you will be able to answer the following questions:

1. Define 'Globalisation'.
2. Describe two of the key features of Globalisation.
3. How can Multinationals be involved in Globalisation?
4. How can Globalisation help development?
5. In what ways can Globalisation hinder development?

How does globalisation both help and hinder development? (with reference to one case study from an LEDC or NIC)

Globalisation is the process of the world becoming more interconnected and interdependent. People around the world are more connected to each other than ever before, largely due to developments in transport, technology, communications and the Internet. Globalisation can also be seen in trade and business, with goods and services produced in one part of the world becoming increasingly available in other parts of the world. Many companies have businesses, factories and offices in countries across the world, for example, Cadbury, Coca Cola and Ford Motors. These companies are called Multinational Companies (MNCs).

The following are some of the key features of globalisation:

1. Globalisation brings the world's economies closer together than ever before, especially in relation to trade, investment and production.
2. Trade across the world has grown at an exceptionally fast rate and has been the 'engine of growth' in many countries.
3. Global communications and improved transport means that economic decisions taken in one country can quickly have an impact on other countries, with consequences to global finance.
4. With MNCs the profits usually flow back to the global headquarters (usually MEDCS) rather than staying in the manufacturing countries (usually LEDCs). This makes the MEDC rich and the LEDC poor. The wealth and power also tends to be concentrated in the hands of decision-makers in the MEDC.
5. Individual countries are less independent than they used to be. They now rely on important working relationships and contracts with global companies so that they can continue to develop. The impact of decisions made by these companies also affects the individual countries.
6. Globalisation is therefore a process that allows the economies of our world to become more integrated as a global community.

Multinational Companies (MNCs)

MNCs are companies that manufacture goods in more than one country. Many of the biggest multinational companies have offices and factories in every main country across the world. However, some will be more centralised in geographic regions (such as Asia).

In recent years, the main trend for MNCs has been to locate more and more factories in LEDCs. Here the labour costs are lower, allowing the companies to increase their profits. Often the headquarters and more 'high-end' jobs, such as design and research, are retained in the MEDCs. For example, all Apple products are designed in Cupertino, California but are manufactured in China.

Figure 48
The Top Ten Global Companies (2012)

RANK	COMPANY	INDUSTRIAL SECTOR	EMPLOYEES	REVENUE (GROSS INCOME)	HEADQUARTERS
1	Exxon Mobil Corp	Oil and Gas	99,100	$303 billion	Texas, USA
2	Royal Dutch Shell	Oil and Gas	90,000	$293 billion	The Hague, Netherlands and London, UK
3	Walmart	Retail	2.2 million	$278 billion	Arkansas, USA
4	BP	Oil and Gas	83,4000	$241 billion	London, UK
5	Vitol	Commodities	No data	$185 billion	Rotterdam, Netherlands
6	Sinopec	Oil and Gas	400,513	$170 billion	Beijing, China
7	State Grid Corp of China	Electric Utility	1.56 million	$161 billion	Beijing, China
8	Chevron Corp	Oil and Gas	61,189	$158 billion	California, USA
9	ConocoPhillips	Oil and Gas	29,800	$156 billion	Texas, USA
10	PetroChina	Oil and Gas	464,000	$138 billion	Beijing, China

Source: Figures from Wikipedia, http://en.wikipedia.org/wiki/List_of_companies_by_revenue, compiled from individual company reports and Fortune magazine's 'Global 500' annual ranking of the world's largest companies

Simply put, this table shows us that many of the top ten global companies have huge revenue and thousands of employees, many of whom will be based outside the country where the main headquarters are located. In fact, even determining the national 'ownership' of many of these companies is a highly complicated activity.

Walmart is a good example of a company that can be found all over the world, in one guise or another. Its beginnings were humble, with Sam Walton opening the first store in Arkansas in 1962. Now, however, Walmart is the biggest retailer in the world, with over two million employees and 8,000 stores in the USA, Mexico, Japan, Brazil, Canada and the UK (where it is called Asda). If it were a country, Walmart would be the thirty-sixth richest country in the world and would be just above Finland (which has a GDP of $239 billion).

One impact of globalisation that has affected people in Northern Ireland is that manufacturing jobs have been moved out to the LEDCs. Any company that wants to continue to make profit in difficult financial conditions might consider such a move. Take a look at some of the reasons behind this below.

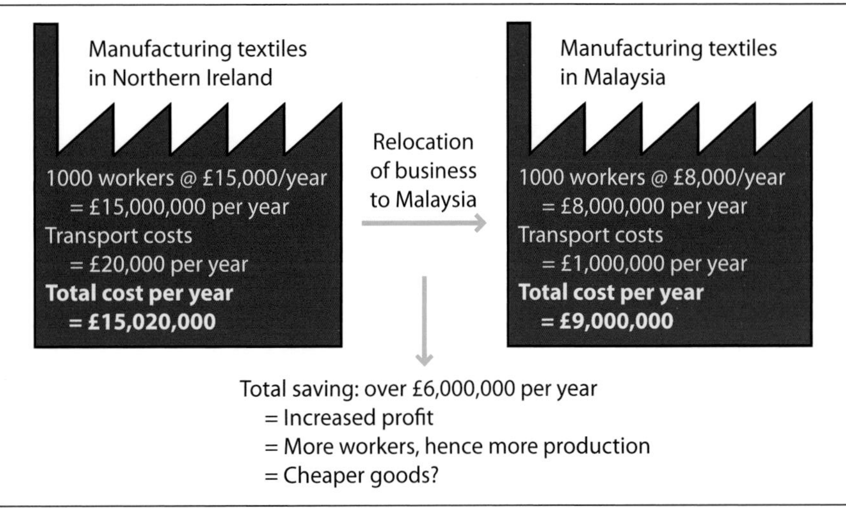

Figure 49
Why might a small company relocate from Northern Ireland to Malaysia?

How globalisation has helped development

1. Globalisation creates jobs in LEDCs, giving employees a reliable source of income in formal jobs.
2. Often MNCs will spend money to try to improve the social conditions and local infrastructure. Governments, keen to attract vital foreign investment, will spend money on improving roads, airports and transportation and communication links.
3. Additional revenue means that foreign investments and money are coming into the economy.
4. People will learn and develop new skills, and often receive better education.
5. New skills, technology and specialist machinery will be brought into a poor country, making it more modern.

How globalisation has hindered development

1. Although jobs are created in LEDCs, the rate of pay tends to be much lower than in MEDCs. For example, a factory worker in the USA might be paid $6 an hour but in China they might only receive $1 an hour.
2. Working conditions are not always good. Employees in LEDCs often have to work long hours in very poor and often unsafe working conditions.
3. In reality the depth of foreign investment is minimal. All profits flow back to the headquarters in the MEDC and the LEDCs do not have the same access to the wealth.
4. There is no guaranteed job security. Many of the jobs in LEDCs are semi-skilled and easily replaced. If a company has chosen to relocate manufacturing once, there is nothing to stop them moving again to a country where more incentives are offered. Often the MNCs will take any machinery, equipment and expertise with them when they go.

Test yourself

1. Write a definition of a MNC. (2)
2. Describe the main reason why a small company might relocate its manufacturing processes from Northern Ireland to Malaysia. (3)
3. What considerations might cause the company NOT to move its operation from Northern Ireland to Malaysia? (3)
4. What benefits would such a move bring to people in the LEDC? (3)
5. What potential disadvantages would this bring to people in the LEDC? (3)
6. Why do you think Walmart can be described as a successful MNC? (3)

Further research

Find out further information about ONE of the companies listed in the table in Figure 48. For example, where do they have offices and factories located? (8)

CASE STUDY

A globalisation case study (from an LEDC or NIC) Nike™ and China

In January 2012, global sportswear company Nike announced plans for the development of a new centralised campus for Nike's employees in Shanghai, China. A new 60,000 square foot headquarters is to be built at The Springs in Shanghai's Yangpu District. The campus is located between the city's two major airports, is on the major transit hub and adjacent to a 33-acre ecological park.

Figure 50

Global sports shops are increasing business all over the world

Nike is based in Oregon, USA and describes itself as being "the world's leading designer, marketer and distributor of authentic athletic footwear, apparel, equipment and accessories for a wide variety of sorts and fitness activities." It was founded in 1964 and has since grown to around 700 shops worldwide, with offices in 45 countries outside the USA. In addition, Nike has 600 contract factories that employ more than 800,000 workers in 46 countries.

Nike made the decision in the 1980s to concentrate on design and marketing and sub-contract production to factories in LEDCs (mostly in Asia). Nike has been working in China for 30 years and at present has 170 factories (with over 272,000 workers). In the 1980s and 1990s, some of the practices that Nike used to reduce costs led to allegations of poor business ethics. Nike could take their business anywhere in the world, if the price was right and this meant that sub-contractors all over Asia were competing with each other. While workers in South Korea were getting their unions to argue for a pay rise to $2.40 a hour, the company moved production to China, where it could make deals to produce the same products for a tenth of the price.

The advantages to Nike of locating their factories in China:

- **Cheap labour** – Wages are low in China (sometimes as little as 50p per hour).
- **Cheap land and services** – Land for factories is less expensive than in the USA, as are electricity and transport costs.
- **Lower taxes** – The government has given incentives and charges lower taxes and rates to attract MNCs such as Nike.
- **Fewer worker rights** – There is much less protection for workers in China than in MEDCs. Employees are not allowed to join unions, take paid holidays or receive sick pay and can be made redundant much more easily than employees in the USA. This saves the company money and gives them more flexibility in the hiring and firing of staff in line with manufacturing demands.
- **Fewer environmental laws** – The Chinese government does not apply the same environmental regulations as those enforced in many MEDCs. This means less money needs to be spent on meeting government standards.
- **Access to a global market** – Nike factories in China are located well away from the majority market for Nike products (in Europe and the USA). However, Nike can transport their goods in bulk which keeps the cost of transport down. In addition, new markets are being developed all the time and since the 2008 Beijing Olympics, China is Nike's second largest market after the USA.

Around 1 in 3 pairs of Nike shoes are made in China. In recent years, an increased focus has been placed on global companies that use labour in LEDCs. According to Nike's 2008 Corporate Responsibility report, "improving conditions for workers throughout our global supply chain continues to be one of our greatest priorities."

How globalisation has helped development in China

The advantages to China of Nike locating their factories in China:

- Investment by Nike brings jobs and often of a better standard of employment than local jobs. This helps to improve the standard of living of many people who work for Nike.
- As the global significance of Nike's relationship with China increases, more and more foreign investments and money will flow into China, further helping the Chinese economy. This is evident in the most recent announcements by Nike to invest in superstores and head offices in Shanghai.

- Facilitating global companies such as Nike has led to improvements in the services and infrastructure within China. Global companies require global transport and communication links and this has raised standards within China.
- Nike claims that it has been instrumental in improving working conditions for employees within China. It spends around $10 million a year implementing its Code of Conduct in factories, enhancing regulations for fire safety, air quality, minimum wages and overtime limits. It has also employed a team of 100 factory inspectors who travel around monitoring conditions.
- In some instances other MNCs who are looking for new production bases will see where Nike has put its factories and consider similar locations for their own factories. This will multiply the impact and amount of work for people in the country and is known as the 'multiplier effect'.
- For employees who do well in the company there might be increased opportunities to work overseas.

How globalisation has hindered development in China

The disadvantages to China of MNCs locating their factories in China:

- Wages are still quite low on a global scale.
- Many of the jobs are low skilled and do not require high levels of education. It is difficult for people to improve themselves.
- Working conditions in Chinese factories, although improving, are not universal and most workers are still working and living in conditions that are much worse than the global norm.
- Some people have likened working for other, less scrupulous, MNCs to economic slavery. In some cases there has been potential for child labour in factories.
- Much of the profits generated by the products manufactured by MNCs in China do not go back into the Chinese economy. For example, in 2012 Nike made profits of $10.4 billion.
- There is always the possibility that a MNC could pull out of China and take its jobs to another country that will give the company a better deal.

Test yourself

1. Why do you think that Nike locates more of its factories in LEDCs than in MEDCs? (3)
2. What are the main advantages for people in China if MNCs such as Nike choose to locate there? (3)
3. What are the main disadvantages for people in China if MNCs such as Nike choose to locate there? (3)
4. When you balance both the advantages and disadvantages, do you think that China should continue to support MNC industry or attempt to remove links with companies such as Nike? (6)

CASE STUDY

How does the pattern of world trade create problems for LEDCs?

There is no country in the world that is self-sufficient and able to produce its own raw materials, goods and services. Countries need each other. **Interdependence** is when a relationship is built up between countries to exchange goods and services. **Trade** is the flow and movement of these goods and services between producers and consumers. Good trade links are essential for the development of any country. **Imports** refer to the products bought and brought into a country. **Exports** are the products sold and sent to another country.

The balance of trade

The balance of trade is the difference between the cost of imports and the value of exports. Some countries can earn a lot of profit from the exports that they make and they do not need to import much back into the country. These countries will have a trade surplus and they will become rich and develop quickly. Other countries will earn less from their exports and their imports will cost a lot more. This means that they have a trade deficit and they will become poorer.

In general, LEDCs make most of their money through the export of primary materials. The problem is that this makes the LEDCs highly susceptible to price changes on these raw materials. If prices go down, extraction can become unprofitable. Usually, LEDCs do not have the industrial expertise or financial capability to process these materials themselves, so enlist the help of MNCs, who extract the raw materials and send them to MEDCs for further processing. MEDCs have more money to invest in manufacturing processes and they tend to keep the technical jobs for themselves. For example, oil which might be extracted by BP in Nigeria is kept in its crude form and shipped to the UK, where it will be processed in oil refineries and turned into plastic, petrol and other oil-based products.

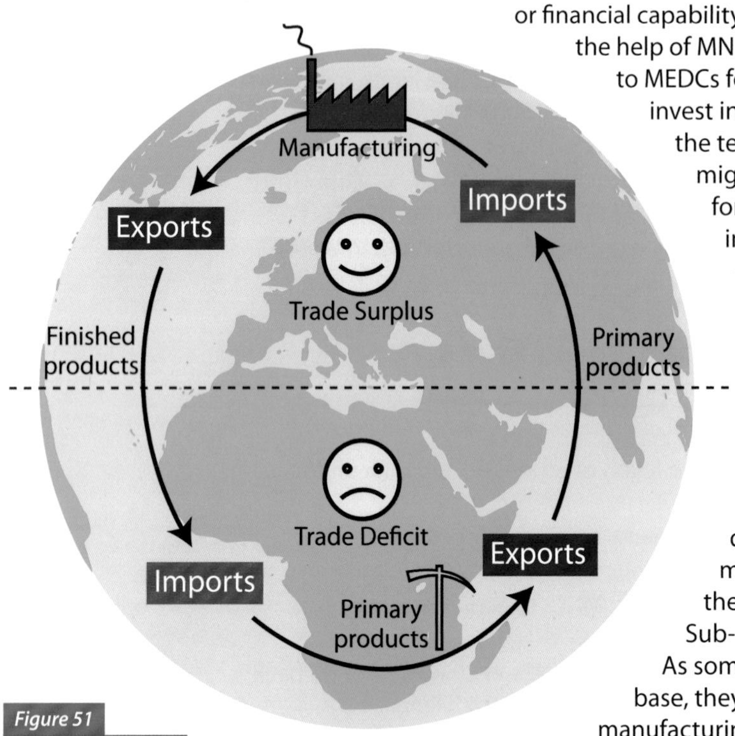

Figure 51
World trade pattern

For many years this has meant that LEDCs could not generate the same amount of money as MEDCs, as the finished manufactured products are more valuable than the primary materials. However, globalisation has led to LEDCs assembling more complicated electronic products, and this balance has started to shift. The distinctions between countries are getting more and more difficult to identify. The IMF noted in 2010 that the US and the EU countries only accounted for 50% of Sub-Saharan Africa's exports (down from 78% in 1990). As some of the NICs have expanded their manufacturing base, they are now taking more materials and carrying out manufacturing processes that formerly were the preserve of the MEDCs. For example, China now has 17% of all exports and there is a noticeable trend of Chinese companies setting up links within Africa. Brazil and India collectively account for an additional 9% of African exports.

Figure 52
The trade balance for the top six richest countries in the world

COUNTRY	TOTAL INTERNATIONAL TRADE ($ billion)	EXPORTS ($ billion)	IMPORTS ($ billion)	TRADE BALANCE (Exports minus Imports, $ billion)
1. USA	3,225	1,289 (3)	1,936 (1)	–647
2. China	2,908	1,506 (1)	1,327 (2)	+179
3. Germany	2,408	1,337 (2)	1,099 (3)	+238
4. Japan	1,404	765 (4)	639 (4)	+126
5. France	1,107	517 (5)	590 (5)	–73
6. UK	971	410 (9)	561 (6)	–151

Figures from The World Factbook, 2011. Washington, DC: Central Intelligence Agency, 2011

Trade issues for LEDCs

1. Reliance on Primary products

Often LEDCs are reliant on the export of one or two primary products to help them generate foreign currency. For example, 90% of Zambia's exports come from copper, while sugar makes up to 90% of the exports from Mauritius. It is very risky for a country to rely on one product, especially agricultural products, as a bad harvest can mean that vital export earnings are lost due to slight changes to the weather. Prices for primary products also tend to fluctuate more than prices for secondary manufacturing goods. Countries that want to develop and improve their trade balance need to try to diversify their exports.

2. Tariffs

The customs duties that are attached to imports are called tariffs. Tariffs give an advantage to manufacturers of locally produced goods over similar goods which are imported. They also help to raise revenues for governments. These tariffs must be paid by any goods coming into an EU country. For example, Palm Oil from Indonesia will face an 8% tariff or 16% if it is refined. This will raise its selling price within the EU and the Indonesian producers are likely to sell less and make less profit.

NAFTA	North American Free Trade Agreement
CAIS	Central American Integration System
UNASUR	Union of South American Nations
EU	European Union
AU	African Union
AL	Arab League
SCO	Shanghai Cooperation Organisation
SAARC	South Asian Association for Regional Cooperation
ASEAN	Association of Southeast Asian Nations
PIF	Pacific Island Forum

Figure 53
World trading blocs

3. Trading blocs

Most countries around the world today are part of at least one trade bloc. This allows imports and exports to be made without incurring tariffs. It encourages 'free' trading and helps to protect the prices of goods produced within the bloc. For example, the UK and Spain are both members of the European Union (EU), which means that goods produced in Spain can be sold tariff-free in the UK and vice-versa. The issue with these trading blocs is that generally the MEDCs have isolated themselves in a few blocs – the European Union (EU) and North American Free Trade Association (NAFTA) – and most of the LEDCs are in their own blocs – the Union of South American Nations (UNASUR), the South Asian Association for Regional Cooperation (SAARC), the Arab League (AL) and the African Union (AU). This means that there is very little reduction of trade tariffs between LEDCs and MEDCs. This allows MEDCs to continue to protect their industries but makes it difficult for LEDCs to establish themselves.

Coffee from Colombia

Many LEDCs find that it is difficult to sell agricultural/primary products to MEDCs. Coffee is a good example of a cash crop. A cash crop is an agricultural crop that is grown for sale for profit. The UK imports 90% of its coffee in the form of beans, which are then processed and the majority are turned into instant powder.

The trade issues

1. Reliance on primary products

The country of Colombia has long been famous for its rich aromatic Arabica coffee. Columbia is able to produce around eight million bags of coffee every year and export about seven million (60 kg) bags of green unprocessed coffee and 635,000 bags of processed coffee. Coffee earns the Colombian economy nearly $2 billion per year and accounts for 5% of all commodities exported from the country. Towards the end of 2011, coffee prices had fallen by 7.4% while other agricultural products had also dropped in price, such as cotton (by 40%) and cocoa (by 28%).

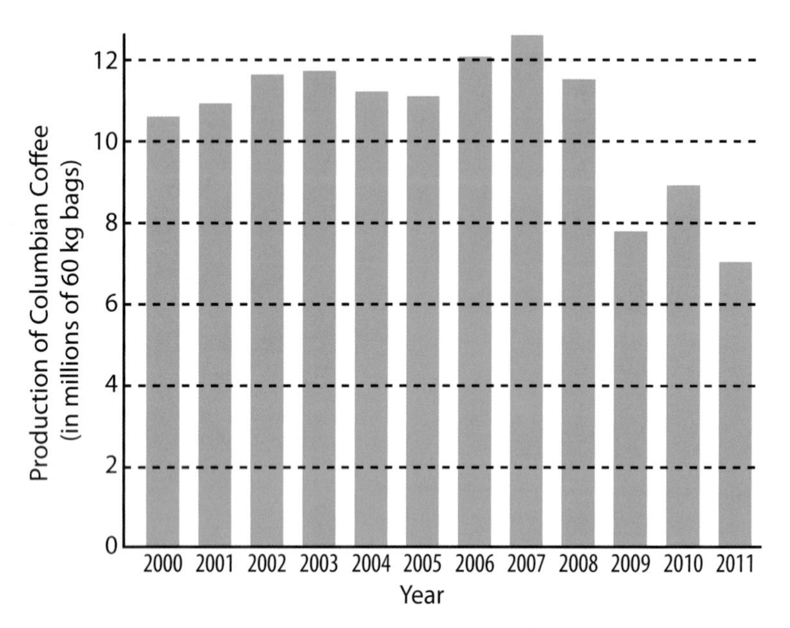

Data from National Federation of Coffee Growers of Colombia

Figure 54 *Production of Columbian coffee (in millions of 60 kg bags)*

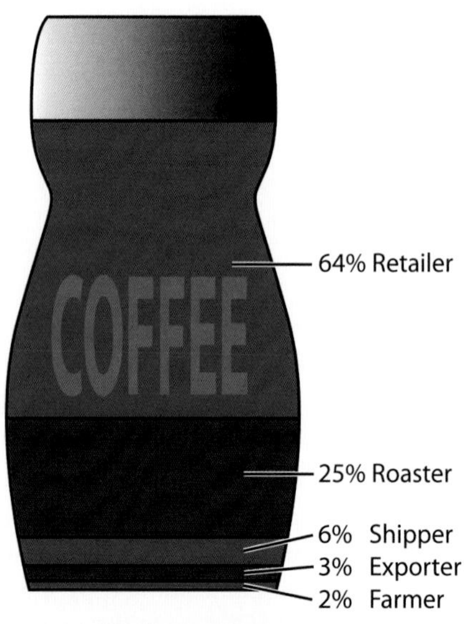

Figure 55 *Share from coffee sales*

2. Tariffs

Although there is no tariff if exporting raw green coffee beans to the EU, a tariff of up to 10% can be applied for processed coffee. This is to prevent LEDCs from processing the coffee themselves. Colombia does not make as much from exporting the beans as the company who roast the beans (in the UK).

3. Trading blocs

Many countries are keen to grow cash crops such as tea and coffee so that they can help to reduce their debts. However, this can sometimes lead to overproduction, which will make the price fall and farmers will struggle to make money. Trading blocs will protect traders within a bloc but trying to move goods from one bloc to another raises the costs to the producer.

4. Large MNC

If coffee growers in Colombia wish to negotiate a better trading deal, they face a large MNC, Nescafé. Nescafé could decide to move production to another country if Columbia's coffee prices increase more than they would like. Nescafé is based in an MEDC (Switzerland), which means that most of the profits will not go to back to the LEDCs.

Test yourself

1. Describe how the balance of trade works. (4)
2. What is the difference between an import and an export? (2)
3. With reference to Figure 52, what are the top three countries in terms of exports only? (3)
4. How can a LEDC's reliance on primary products cause trade problems? (3)
5. How can tariffs and trading blocs influence trade from LEDCs? (4)
6. How can the coffee trade be used as a good example of the trading issues that affect LEDCs? (8)

Check your learning

Now that you have studied Part 2: Factors Contributing to Unequal Development, return to page 207 and answer the Key Questions for this section.

PART 3: SUSTAINABLE SOLUTIONS TO DEAL WITH THE PROBLEMS OF UNEQUAL DEVELOPMENT

(a)
A sustainable project that uses appropriate technology (from a LEDC)
Describe and explain how it uses technology to progress with:
Economic improvements
Environmental improvements
Social improvements

(b)
Evaluate the success of the appropriate technology project

(c)
What is Fairtrade? What are the advantages that it brings to LEDCs?

(d)
What is Aid? Evaluate how aid brings both benefits and problems to LEDCs.
Long and Short term aid
Bilateral, Multilateral, Voluntary and Tied Aid

Key questions

By the end of this section you will be able to answer the following questions:

1. What does 'sustainable' mean?
2. Define the term 'appropriate technology'.
3. What are the economic, environmental and social improvements of an appropriate technology project?
4. What is Fairtrade?
5. What is aid? Describe its benefits and its problems.

Key words

Appropriate technology
Fairtrade
Aid
Bilateral aid
Multilateral aid
Voluntary aid
Tied aid

A sustainable project that uses appropriate technology (in an LEDC)

As we have seen already, in recent years people have become increasingly aware of the widening gap between MEDCs and LEDCs. Many want to ensure that this gap does not continue to widen and that new solutions are sought to allow LEDCs and NICs to progress and develop.

Sustainable development is "development that meets the needs of the present without compromising the ability of future generations to meet their own needs". Therefore, if the solutions to unequal development are going to last, they need to be sensitive to the needs, skills and resources of the people in LEDCs and NICs.

Any project that is defined as sustainable must:

- improve quality of life.
- improve living standards.
- encourage economic development.
- give future generations a higher chance of survival.
- not harm the environment.
- use appropriate technology.

Appropriate technology

Appropriate technology is technology that is suited to the needs, skills, knowledge, resources and wealth of local people, in the environment in which they live. The term was first used by EF Schumacher in 1973 to describe the use of technology that would result in fewer negative impacts on the environment and society.

Often in LEDCs, hi-tech solutions to problems are inappropriate for the inhabitants as they do not have easy access to energy sources or replacement parts. Appropriate technology will provide an innovation that is a suitable for the local people.

Appropriate technology will:

- avoid high capital investment costs by not using expensive machinery and sources of energy.
- usually be labour intensive, as many people in LEDCs are unemployed and there is little value in replacing people with machines.
- be suited to the skill levels of local people.
- attempt to develop local industries and farming methods rather than trying to replace them.
- use local resources as far as possible and will try to source renewable resources.
- help to conserve the local environment.

Case Study: The development of a borehole and hand pump in Uganda (an LEDC)

What is the problem?

In some parts of Uganda people are struggling to survive. The land can be arid in places, few services are available and the majority of people are extremely poor. A lot of the population is nomadic and always on the move. Supplying these people with clean water is a huge challenge.

Figure 56

A Ugandan child collects dirty water from a stream

Courtesy of Fields of Life

The quantity (amount) and quality (standard) of water in Uganda is poor. Families usually have to walk around 48 minutes each day just to fetch water. Few people have access to safe water and water-borne diseases are rife.

Water quantity (amount) is poor due to the following:

- The area is arid in parts.
- The local population is increasing rapidly, which puts more pressure on this valuable resource.
- Farmers are increasingly using water for irrigation.
- As industry in the area starts to develop, more water is needed.
- As living standards improve, more people want to buy technologies that use more water, such as flushing toilets and washing machines.

Water quality (standard) is poor due to the following:

- The local people do not have money to treat their water with chemicals or water purification tablets.
- There is a lack of basic sanitation. Often polluted sewage mixes with drinking water.
- Disease can spread quickly in the hot climate. As water sources are shared, this means that whole families, tribes and villages can quickly become very ill. For example, diarrhoea caused by drinking dirty or contaminated water kills more people in Uganda than anything else (including HIV/AIDS and malaria). This is sometimes called dysentery.

Water borne diseases are thought to cause around 2 million deaths worldwide every year and the UN High Commissioner for Refugees (UNHCR) estimates that 1.1 billion people lack clean drinking water. Many people, especially in rural areas, are suffering from unnecessary water-related diseases such as diarrhoea, bilharzia and typhoid. People do their laundry and bathe in the same places that they water their animals and get their drinking water. Some do not have latrines and defecation takes place in fields, bushes or along drainage ditches. In 2002, the UN stressed the fundamental right that people have for access to clean water for domestic and personal use. With this in mind, many different groups have been looking for solutions that would allow more people access to this vital resource.

Solutions from Fields of Life in Uganda

The Fields of Life trust is a charitable organisation that has been working in Uganda since 1995. It was established when Irishman Rev Trevor Stevenson set up an agricultural project near Kampala and has since been involved in building schools, providing health education and income generating projects across East Africa.

Trevor explains:

"For years I have travelled to Uganda and driving through the rural villages I was appalled watching mostly young children and women fetching water. It is bad enough that they have to carry water for miles on their head, but it is what is inside those jerry cans that we, at Fields of Life, cannot accept any longer. The water is infected with a myriad of diseases and women and children are exhausted from carrying heavy loads. The next day they are sick from drinking the dirty contaminated water but if the family is to survive they have to fetch and drink the same dirty water again the next day."

Fields of Life are committed to supplying safe, clean water, especially to vulnerable and marginalised groups, mainly women and children. Thanks to their drilling rigs, which can drill from 100 m to 200 m into the ground, Fields of Life now have the capacity to drill 10 wells per month. This provides clean water for 10,000 people and one million over 10 years with 1,000 wells.

However, there has been a certain amount of controversy surrounding the sustainability of hand pumps in some parts of Uganda and decisions about whether there are sufficient long term gains from using them. Unfortunately without maintenance most hand pumps have a limited working life of as little as 3–5 years. Some of the difficulties of using hand pumps have concerned:

- cost
- technology
- maintenance
- responsibility of upkeep
- availability of spare parts
- community involvement
- organisation
- education
- hygiene

Figure 57

Fresh, clean water from a sustainable source can make a big difference

Courtesy of Fields of Life

In 2007, only 40% of the 350,000 pumps installed in Africa were still in use. To combat this, between 1981 and 1991, the UNDP and World Bank initiated a project called the 'Rural Water Supply Handpumps Project'. The Village Level Operation and Maintenance (VLOM) concept was introduced as part of this. This aimed to enable village workers to maintain their own hand pumps, to ensure that spare parts were being manufactured within the country and were readily available, and to ensure that pumps could last for lengthy periods.

Why is this appropriate technology?

The initial investment into a village is made through donations to Fields of Life. This provides the necessary materials to complete the construction of a well, which currently costs around £3,300 to drill. The technology that is left in each village is appropriate for the needs of the local people:

- **Cost:** They do not have to pay any money for the construction of the well. However, water user committees, made up of members of the local community, are set up in each village, which can collect small amounts of money for a maintenance fund.
- **Technology:** A hand pump is installed at each borehole rather than a mechanical pump. Mechanical pumps are more complicated pieces of technology that might require more moving parts and maintenance. Electric and battery operated pumps are not used for the same reason.
- **Maintenance:** A water user committee takes responsibility for the operation and maintenance of the borehole.
- **Responsibility of upkeep:** Often, some local people will be trained to troubleshoot simple problems with the well.
- **Community involvement and organisation:** The Fields of Life team prepare the local community in advance of digging any well. Land development agreements are made with local landowners to ensure that free access is granted to any water source.

- **Education and hygiene:** Fields of Life provide some basic training about water use and sanitation. They educate on how to store water efficiently, why to avoid stagnant pools of water and how to dig long drop toilets to ensure drinking water and sewage do not mix. They also implemented a new health educational initiative in 2012 amongst schools.

What are the economic improvements?

Time: Now that people (children and women in particular) do not have to spend long periods of time looking for and carrying water for the family, they have more time for other things. This means that children are able to go to school, making them better educated, more employable and improving their future. Fields of Life often build schools and wells in the same villages. Women are also released to other tasks, for example, starting a small business, which could help support the family and raise standards of living.

Industry: A regular supply of water might allow villagers to engage in small business ventures. For example, they could make tea and drinks for tourists, or the water could be used in cooking and baking, which could be sold in local market places.

Farming: Water can be used for irrigation of vegetable patches and even for drinking water for animals. However, this needs to be monitored carefully and animal troughs should be positioned well away from the well (often outside the village). This means that even in the dry season villagers can produce crops and look after their animals.

Health benefits: Clean water means that people will suffer less from illness and disease. This means that people will have more energy and time to work and have a great opportunity to improve their quality of life.

Figure 58

A local water source means less walking for women and children

Courtesy of Fields of Life

What are the environmental improvements?

Environmental impact: Wells and boreholes generally do not cause any environmental damage, as long as they are positioned in an appropriate place. The only real environmental issue is the construction of the metal pump and concrete run-off plate.

Animals: A source of water means that animals can be kept alive through dry seasons. This reduces dead animal carcasses which can be a source of pollution and disease.

Pollution: Hand pumps do not produce pollution as they do not need diesel or petrol generators to produce electricity. In many cases the pump will actually encourage villagers to take a renewed interest in keeping their village clean and tidy.

What are the social improvements?

Shared responsibility: As the well is left with the local villagers to manage, there is a new sense of shared responsibility to ensure that the well is always working and that there are no pollutants leaking into the ground water.

A new community spirit: This shared responsibility can empower people, renew community spirit and give the people a new hope for a better life.

Feeling better: The water comes from deep underground, which means that it is always clean (and has come through the natural filters of soil and rock). Less disease means that people feel better about life.

Empowerment: As women often spend hours each day carrying water from sources to the home, this allows many women an opportunity to be involved in other activities, perhaps even starting to earn money. In addition, incidences of violence to women (including rape) are reduced as it is normally women who are walking long distances to collect the water, making them vulnerable to attack. Education programmes also help to empower people as they lead to more employment options.

Evaluate the success of appropriate technology

When looking at the impact of appropriate technology on a community, such as hand pump projects in Uganda, it is important to ask 'how sustainable can measures like this be'?

Evaluation exercises

Using the information in this section and the questions below, write up an evaluation of the success of this project. Use the following framework to help you:

Appropriate technology strategy:

- Location and organisation
- Why is this considered appropriate?
- What is GOOD about this strategy?
    - Economic strengths
    - Environmental strengths
    - Social strengths
- What is BAD about this strategy?
    - Economic problems
    - Environmental problems
    - Social problems
- How successful has the project been?

Environmental sustainability:

- Is the strategy good or damaging to the environment?
- Does it help protect wildlife and animal habitats?
- How do these management strategies impact the environment?
- Do they cause pollution?
- How might animals and natural habitats have been affected by the management strategies?
- Do you think the management strategies are largely positive or negative?

Social sustainability

- How does this management strategy impact on the people? How does it change their life?
- Is the strategy something that will add to people's lives? Will it allow them to live close to each other without risk of death or injury?
- How does it affect what they normally do?
    - Does it affect their family?
    - Does it effect how and where they live?
    - Does it affect food supply?
    - Does it affect how they travel around and get from one place to another?
- Do you think that the management strategies are largely positive or negative?

Economic sustainability

- Does the strategy have expensive construction or maintainance costs or is it just a one off cost?
- How do these management strategies impact people and money?
- Has the building of new managements strategies brought prosperity or poverty to local people?
- Does it affect jobs?
- Have the strategies changed the traditional way of life in the areas?
- Will the strategy last a long time?
- Do you think the management strategies are largely positive or negative?

How successful has the project been?

- Why is appropriate technology better than hi-technology projects? In a world where technology seems to be able to solve every problem, it is sometimes difficult for us to realise that technology is not always the most obvious and best solution when working in LEDCs.
- Take a look at the table below which helps to summarise some of the differences between the two approaches.

APPROPRIATE TECHNOLOGY	HI-TECHNOLOGY
• Smaller scale • Cheaper equipment • Small demand for energy/electricity • Large demand for labour (LEDCs have high unemployment so much available) • Uses local resources • Involves traditional skills where appropriate	• Large scale • Expensive equipment • Big demand for energy/electricity • Employs few people • Often needs imported materials • Requires that people get training in new skills

Test yourself

1. What is 'sustainable development'? (2)
2. What is 'appropriate technology'? (2)
3. With reference to the appropriate technology case study:
   a. Describe the problem that the case addressed. (2)
   b. Describe the solution that was created to address the problem. (6)
   c. Explain why this is classed as 'appropriate technology'. (3)
   d. What are the economic benefits it brings? (3)
   e. What are the social benefits it brings? (3)
   f. Describe the environmental benefits it brings. (3)

Fairtrade and the advantages it brings to LEDCs

The world is not based on providing a fair system of trade, where producers can get a fair deal on all that they sell. For many years MEDC producers and retailers have gotten a much better deal than producers in LEDCs.

In recent years, there has been a global movement known as Fairtrade which tries to make some aspects of trade fairer. Fairtrade is a strategy used to provide an organised approach to help producers in LEDCs gain better trading conditions and promote sustainability. It gives workers who were previously under pressure from the usual trade routes an alternative option to help them get a fair price for their efforts. It is keen to ensure that producers get paid a higher price and actively promote higher social and environmental standards.

Fairtrade products have begun to be traded and marketed through MEDC supply chains. These products have either been imported by Fairtrade organisations or the suppliers have applied for 'product certification', where the products have been seen to comply with Fairtrade production guidelines.

The Fairtrade Mark

The Fairtrade Mark is an international label which certifies that products were sourced from producers in developing countries and meet certain standards. Fairtrade labelling was started in the Netherlands in the 1980s, with The Max Havelaar Foundation launching the first label in 1988 on coffee from Mexico. In the UK, the Fairtrade Foundation was set up in 1992 and the mark was used for the first time in 1994.

Why is Fairtrade good for producers?

1. Fair and stable prices

For Fairtrade products, buyers have to pay the Fairtrade minimum price. This price is good for producers as it aims to cover the costs of sustainable production for the producer. This means that when market prices fall below a sustainable level, farmers do not lose out and when the market price is higher than the Fairtrade minimum price, the buyer must pay the market price.

2. Fairtrade for development

Producers are also paid a Fairtrade premium, which is an extra amount paid beyond the price of the goods that the producer can spend on whatever they want. However, it can only be used for the social and economic benefit of workers and not company bosses. It is often invested in farm improvements to help increase crop yield, quality, processing facilities, education or healthcare.

3. Empowering small-scale farmers

Fairtrade was set up to help empower the small-scale farmer. In some products, Fairtrade only certifies small farmer organisations, for example, coffee, cocoa, cotton and rice. Fairtrade aims to ensure that farmers can benefit from better working conditions and aims to protect the environment in a sustainable manner.

Fairtrade has also brought about community development in some areas. In Colombia, members of the COSURCA coffee cooperative have prevented the cultivation of more than 16,000 acres of coca and poppy used for producing drugs. In Guatemala, measures such as this are also allowing coffee producers to send local children to college for the first time.

Is Fairtrade always the best deal?

The Fairtrade Foundation does not monitor how much extra retailers charge for Fairtrade goods, so it can be difficult to work out how much actually reaches the producers. Some have reported that British café chains were passing on only 1% of the extra charge to the producer. Additionally, there is no solid evidence available to show that Fairtrade farmers get higher prices on average.

Some people are also concerned about the ethics of big business such as Wal-Mart, McDonalds and Starbucks, which use Fairtrade coffee but have large profit margins. This is sometimes at the expense of the smaller, independent retailers that often give a higher percentage of profit back to the producer.

The rising popularity of Fairtrade coffee in the UK

Coffee is a great example of a product where Fairtrade has made a real difference. Back in 1998, many coffee producers were struggling to produce coffee and make enough money to survive. Coffee prices were at an all-time low but since then the impact of Fairtrade has meant that coffee growers now get a greater percentage of the final cost of the product.

By 2008 the International Fairtrade Foundation measured nearly $5 billion of global trade which was a 22% year on year increase. Figure 59 shows the sales figures for Fairtrade coffee in the UK from 1998 to 2009. One company called Cafédirect has been at the forefront of making sure that the producers of coffee get their fair share of the profits.

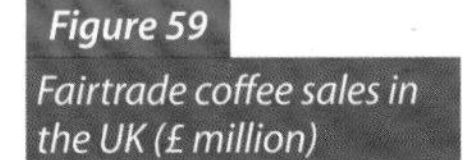

Figure 59
Fairtrade coffee sales in the UK (£ million)

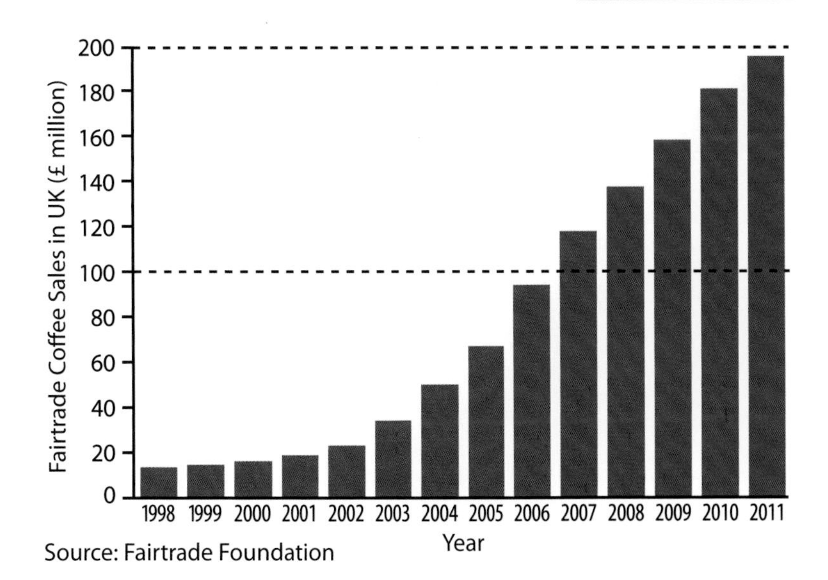

Source: Fairtrade Foundation

Cafédirect Factfile

What they stand for

Superior Quality. They buy the highest quality coffee and work directly with smallholder growers.

Who they work with

Cafédirect work with 38 producer organisations in 13 countries and they say they make a positive impact in the lives of over 1.8 million people.

What makes them different

Cafédirect believe they go further than other hot drinks companies that claim to be Fairtrade:

- They "consider a fair price for the crop to be a minimal starting point, not the end goal".
- They only buy direct from growers and not middlemen at auction.
- They share the profits of final product sales with growers and have invested over 50% of profits in the growers' businesses and communities to date.
- They are keen to help fund growers to buy the machinery that will allow them to process the crop, rather than just exporting the raw material, which will enable the growers to increase their profit margin.
- The growers are made part of the business: they are brought into a cooperative and they own shares in the business. They also have actual growers on the board of directors to have a voice in business decisions.

Their awards

In the last five years they have won 22 awards for great taste and been awarded for their work as an ethical company by *The Guardian*, European Commission and the United Nations.

www.cafedirect.co.uk

Figure 60
Cafédirect ensure that their producers get a fair share of any profits

Test yourself

1. With reference to Figure 59, describe the amount of Fairtrade coffee sales in the UK. (4)
2. Can you explain why the sales of Fairtrade coffee may have increased? (3)
3. Describe three advantages of Fairtrade for LEDCs. (3)
4. What advantages might Fairtrade bring to an MEDC? (3)
5. Why might Fairtrade not always be seen as being truly successful? (3)
6. Evaluate how successful you think Fairtrade has been in dealing with unequal development. (5)

What is aid? Evaluate how aid brings both benefits and problems to LEDCs.

Long and short term aid

Aid is the process of one country or organisation giving resources to another country. The type of aid given depends on the circumstances and might include money, expertise (such as doctors, aid workers or rescue specialists) or goods (such as food, water, shelter or tools). The givers of aid tend to be in MEDCs and the recipients of aid tend to be in LEDCs.

Countries such as the UK feel a social responsibility to continue to change lives and help the world's poorest people. In 2011 the UK International Development Secretary, Andrew Mitchell, announced that by 2015 the UK Department for International Development would:

1. Secure schooling for 11 million children, which is more than they educated in the UK but at 2.5% of the cost.
2. Vaccinate more children against preventable diseases than there are people in the whole of England.
3. Provide access to safe drinking water and improved sanitation to more people than there are in Scotland, Wales and Northern Ireland.
4. Save the lives of 50,000 women in pregnancy and childbirth.
5. Stop 250,000 newborn babies dying needlessly.
6. Support 13 countries to hold freer and fairer elections.
7. Help 10 million more women get access to modern family planning.

Short term aid

Short term aid is often in response to an emergency and is usually linked to a particular need. For example, following natural disasters, LEDCs often need food, water and shelter so that people can survive the immediate aftermath of the event.

Long term aid

Long term aid is often a more far-reaching, sustainable type of help, given over a long period of time. Often it will take a long time for any impact to be noted, as the help programme attempts to improve an aspect of life in an LEDC (socially, economically, educationally or environmentally). The aim with long term aid is to enable the LEDC to sustain itself further in the future and to have less need for help from external agencies.

Bilateral, multilateral, voluntary and tied aid

There are a variety of different types of aid:

1. Bilateral

Bilateral aid is the type of help which is given directly from one government to another government. Bilateral means that two countries are involved and the aid is usually tied. The UK government see this as an opportunity to increase access to the basics (clean water, sanitation, health care and education); try to prevent conflict and climate change; and try to put into place the building blocks of wealth creation (property rights, effective public services, stability and the rule of law). The UK government estimates that by 2015 it will have spent close to £3.5 billion per year in bilateral development projects – a total expenditure of nearly £15 billion. Currently, the two largest priorities for the UK bilateral development funds are Pakistan (getting £1.4 billion until 2015) and Ethiopia (getting £1.3 billion until 2015). Afghanistan is another key country and has been allocated £178 million per year for specific development projects. This will result in £710 million being spent in Afghanistan by 2015.

2. Multilateral

This is aid that is given from national governments to world/international organisations such as the World Bank, the World Health Organisation (WHO) or the United Nations (UN). This money is then distributed to the various development projects which need it around the world.

In 2011 the UK government carried out a comprehensive review of the spending on multilateral agencies to assess the 'value for money' which each investment represents. Some multilateral agencies, such as UNICEF, the GAVI Alliance (the Global Alliance for Vaccines and Immunisation), and Global Fund to Fight AIDS, Tuberculosis and Malaria, are seen as doing well and helping to meet the UK development objectives. For example, since 2000 GAVI have prevented 3.4 million deaths and vaccinated 213 million children. Since 2004 the Global Fund to fight AIDS, Tuberculosis and Malaria have distributed 160 million insecticide treated nets, treated 7.7 million new Tuberculosis cases and treated 3 million HIV positive people with antiretroviral drugs. However, others were seen as being fragmented, with administration and coordination problems and weak or ineffective practices on the ground.

Source: Rick Scavetta, U.S. Army Africa

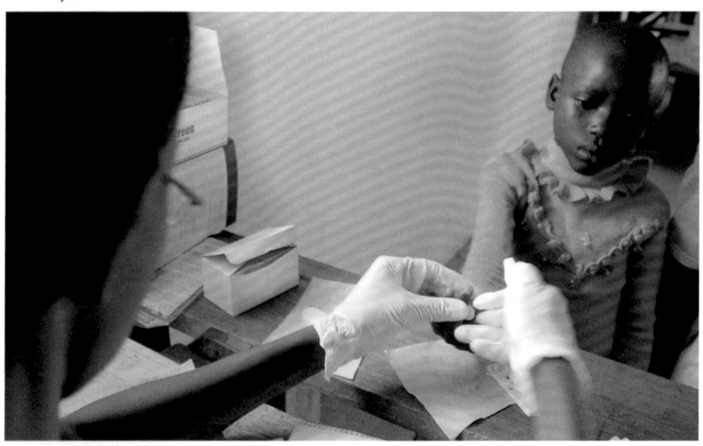

Figure 61

An example of multilateral aid: vaccinations as part of World Malaria Day 2010 in Kenya

The UK Department for International Development notes that it gives out around £3.5 billion to multilateral organisations each year. About one fifth of this is part of EU obligations through the European Commission. From the remainder, 30% goes to the World Bank, 25% goes to global funds, 15% to the European Development fund and 9% to the UN. Often this type of aid is 'conditional' as the use of aid will be determined by the organisation rather than the government which donated the money.

3. Voluntary

Sometimes this is referred to as charitable aid. Many charities are called NGOs (non-governmental organisations), are based in MEDCs and have a particular interest in a part of the world, medical issue or cause. These are usually funded by the public and often organise their own aid programmes and projects on a small scale within countries, on their own or alongside partner organisations. This type of aid makes up a surprisingly small percentage of all aid given (around 6%).

Charities such as Christian Aid, Comic Relief and Oxfam are all Voluntary organisations that work directly on development projects. Often this type of aid is short term and responds to a particular emergency situation. The advantages are that this type of aid does help to save lives and provides people with emergency shelter, food and medicine. However, the aid often runs out quickly after the initial emergency is over and no real solutions are found to the original causes of poverty in the country.

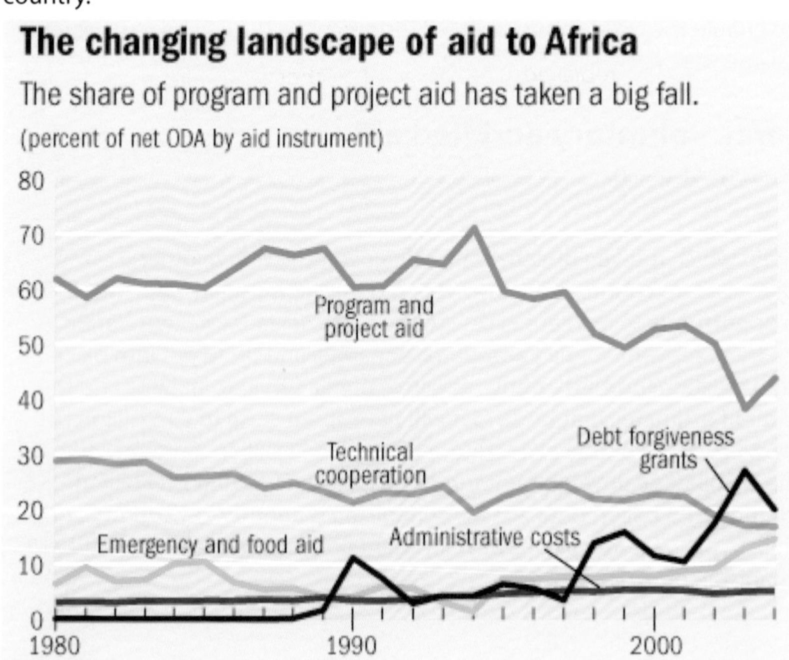

Figure 62

The changing landscape of aid to Africa

Source: Chart 3 (The changing landscape of aid to Africa) from The International Monetary Fund (IMF)'s Finance and Development Magazine, Volume 43, Number 4, December 2006 edition. Based on data from OECD-DAC database, © OECD.

4. Tied

The big issue with much of the aid given to countries is that it is 'tied'. This means that the money given by the donor government is specified for particular projects. For example, a government might give £20 million to a country but specify that the money should be used for building hospitals, schools or roads. Often the donor country also benefits economically from the aid. In the

Comic relief is a UK based charity which was set up in 1985 in response to the famine that was devastating Ethiopia. Their idea was simple: get British comedians to help raise money for people in desperate need. From that the bi-annual Red Nose Day and Sport Relief have continued to raise money and the profile of the organisation so that they can achieve their aim of "a just world free from poverty". By the end of 2012 they had raised over £800 million and worked on development projects in over 70 countries.

most extreme cases of tied aid, the government might even indicate that only firms from the donor country can be used to construct the project, meaning that money, jobs and profits will return to the donor country.

How does aid bring benefits and problems to LEDCs?

It is not uncommon for the leaders of LEDCs to complain that aid can do more harm than good to a country. In recent years the leaders of Rwanda and Niger have both challenged the way that richer countries often attach conditions to aid. In 2005, the President of Niger accused aid agencies of exaggerating his country's food crisis for their own gain. Niger was in the midst of a drought which caused a nationwide food crisis. Yet, the President became concerned that the aid which was coming into the country was not addressing the underlying problem – world food prices.

In 2002 during the southern Africa food crisis, Zambia banned aid because it was worried about getting genetically-modified US maize. The US ambassador to Zambia was livid and spoke out: "Leaders who refuse to let their people have food, should be put in the dock for the most serious crimes against humanity".

The key issue is whether aid is actually good or bad for a country. Does it make a difference and is it always appropriate for the needs of the people?

The benefits of aid

Many economists call for an increase in the amount of overseas assistance. In 2000, the 189 member states of the UN signed a declaration in which they said, "we will spare no effort to free our fellow men, women, and children from the abject and dehumanising conditions of extreme poverty to which more than a billion of them are currently subjected". They also promised to grant more generous development assistance.

Some economists with the IMF (International Monetary Fund) argue that a doubling of aid to Africa could have a very strong, positive effect on growth, saving a lot of lives and taking a number of poor people out of poverty.

Figure 63
Trucks on the way to Haiti to supply humanitarian aid after the earthquake in 2010

In December 2011, at the fourth High Level Forum on Aid Effectiveness in Busan, Korea, policy-makers and practitioners met to review the success and challenges to make the £76–95 billion of aid that is delivered globally even more effective. The main thrust of these discussions was to ensure that aid was as transparent as possible, that tied aid would be reduced and commitments would be put into place to ensure that more aid was actually spent in recipient countries.

Some of the key benefits of aid are:

- Much short term and humanitarian aid can help to save lives following a natural disaster or war. In some cases, charities have been able to make a difference by working alongside partner organisations to help to improve education, agriculture or health care.
- As short term aid is brought in as actual goods by charities, there is less chance of it being sold to pay for political corruption.
- Aid can help to improve the standard of living of people in recipient countries.

- If aid is used properly and focused effectively, it will allow a country to develop and improve in a sustainable manner.
- Some governments (such as the UK) and some global organisations have been able to take audits of the value for money that they are getting for the aid they give. They want to make sure that money is making a difference and changing lives. Many countries have responded by cutting back the amount of multilateral or tied aid that they use.

The problems with aid

The assumption is often made that if you pour money into an economy through foreign aid, it will bring about an increase in output and growth. However, many estimates suggest that less than 20% of aid actually reaches the people who need it directly. Two-thirds of the world's poorest people live in 10 countries that together receive less than a third of official development assistance. People live in rural areas where there is little infrastructure in place to support the movement of aid resources beyond the urban areas. Alex de Waal (Director of the World Peace Foundation) wrote: "Aid is essentially a Western, Anglo-Saxon model of charitable endeavour that's being imposed on the rest of the world", continuing, "Major problems are solved not through charitable action but through political action".

One of the biggest criticisms of humanitarian organisations is that they 'waste' too much money on staff costs and running overheads. Charitable causes have become big business and many people argue that the continued 'professionalisation' of aid agencies has destroyed the 'spirit of volunteerism'.

Some of the key problems with aid are:

- Aid does not always reach the poorest people who really need it. Political corruption and poor administration systems in countries means that aid fails to reach the people.
- Many LEDCs have come to depend on the regular flow of aid. This means that if the flow ever ceases (for example, during a global recession) it could cause problems for the people in the LEDC.
- It can be difficult for countries to become independent if they rely on other countries for handouts.
- If the aid is given as low interest loans, this creates a new dependency relationship between the two countries and leads to debt.
- Tied aid puts a lot of conditions on LEDCs, which can make receiving aid less desirable.
- Aid can undermine local producers. As new food supplies come into the country this can have a negative impact on the prices of produce grown and sold within the country. People will also prefer to 'free food' to food they have to pay for.

Who knows what shape aid will take and what role charities and MEDC governments will have in the future, as LEDCs continue to develop and seek independence.

Check your learning

Now that you have studied Part 3 Sustainable Solutions to deal with the problems of Unequal Development, return to page 215 and answer the Key Questions for this section.

Organise your revision

Draw a mind map to summarise Unit 2B 'Contrasts in World Development'.

Test yourself

1. What is the meaning of the term aid? (2)
2. What is the difference between short term and long term aid? (2)
3. What are the four main types of aid and how do they differ? (8)
4. What is the difference between aid given by charities such as Oxfam and governments? (4)
5. Explain two benefits and two problems of aid being received by LEDCs. (8)

EXAM PRACTICE QUESTIONS

Some of these questions are from previous CCEA GCSE examination papers and others have been written in the same format to give you practice at answering 'exam style' questions.

Try to answer the questions with as much detail as possible. Also consider the number of marks that each question receives, as this will give you a good indication of the amount of depth that your answer needs.

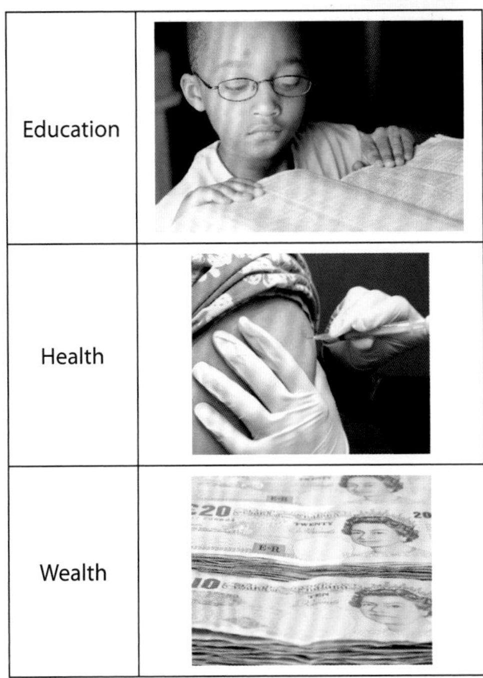

Resource B

SUCCESS	STATEMENTS	FAILURE
Yes	In Asia, living standards have improved as new industries have been set up	
	There are still many people who are short of food and money in LEDCs	
	Many children under 5 years are still dying from diseases due to dirty water	
	90% of the world's children are now attending primary schools	

Resource C

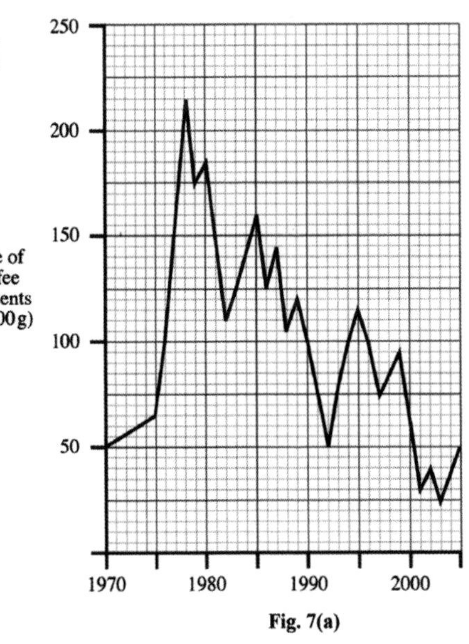

Fig. 7(a)

Resource D

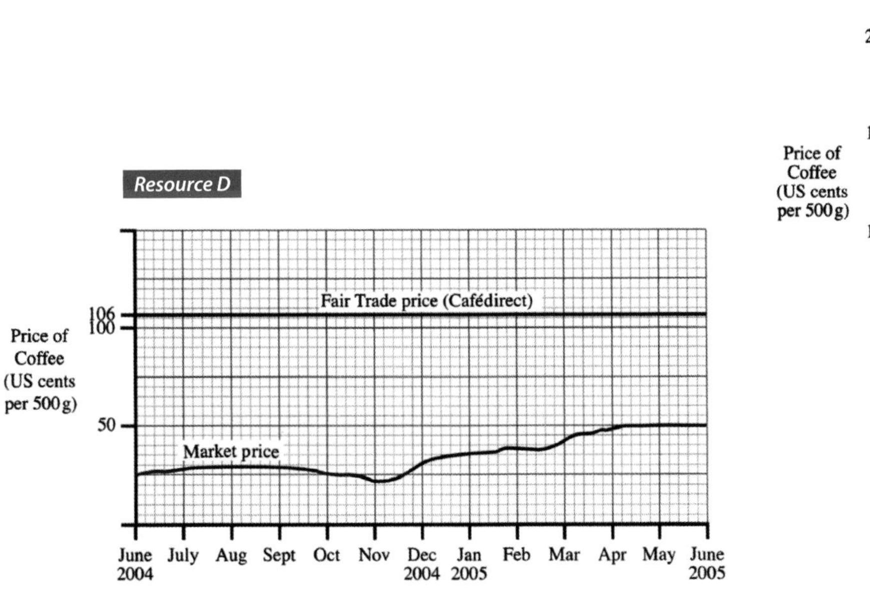

Foundation Tier

1. The Human Development Index (HDI) takes into account three different measures of development, as shown in Resource A.
   Explain why the HDI is a good indicator of development. (3)
2. MEDCs are trying to help LEDCs to develop.
   i. Name one strategy they are using to help the LEDCs to develop. (1)
   ii. Outline two advantages this strategy could have for LEDCs. (4)
3. Describe one advantage Fairtrade brings to LEDCs. (3)
4. Name one way (apart from Fairtrade) in which MEDCs can help LEDCs to develop. (1)
5. Giving information about one sustainable project you have studied, explain how appropriate technology can be used to improve conditions in a LEDC. (4)
6. Complete the table In Resource B to show if MEDCs are successful in helping LEDCs to develop. Write yes to show success or no to show failure. One has been completed for you. (3)

Questions and resources from CCEA GCSE Geography Papers
June 2010 (Q5), June 2011 (Q1, 3, 4, 6; Resources A, B), Specimen 2010 (Q2), ©CCEA

Higher Tier

1. Explain why the Human Development Index (HDI) is a more effective indicator of development than an economic indicator such as energy consumption. (3)
2. Trade Patterns can help or hinder a country's development. Study Resources C & D which shows changes in the market price of coffee.
   i. Describe the changes in the price of coffee on the world market shown in Resource C. (3)
   ii. Explain how the Fairtrade price helped coffee growers in LEDCs from June 2004 to May 2005 using Resource D. (3)
3. Discuss how globalisation can both help and hinder development. Refer to one case study country within your answer. (8)
4. Fairtrade is one sustainable method of reducing the gap in development. Describe one advantage Fairtrade brings to LEDCs. (3)
5. Outline one international strategy that aims to reduce the development gap between countries and evaluate its effectiveness. (8)
6. Giving information about one sustainable project you have studied, explain how appropriate technology can be used to improve conditions in a LEDC. (5)

Questions and resources from CCEA GCSE Geography Papers
June 2010 (Q3, 6), June 2011 (Q1, 4), Specimen 2010 (Q2, 5; Resources C&D), ©CCEA

Managing Our Resources

As the global population continues to rise at a fast rate and LEDCs continue to develop steadily, more and more pressure is being applied to the world's resources. Worldwide many people have decided that action needs to be taken to protect the precious resources that we do have and to look for new opportunities for energy, transport and tourism.

"We are just statistics, born to consume resources"
Horace

This theme takes a look at the different resources that are available and the measures that can be taken to manage them more carefully. It also takes a closer look at what strategies can be used to protect the environment and how aspects of how we view tourism can be made to become more sustainable.

This unit is divided up into four parts:

1. The impact of our increasing use of resources on the environment
2. Increasing demand for resources in LEDCs and MEDCs
3. Managing waste to protect our environment
4. Sustainable tourism to preserve the environment

PART 1: THE IMPACT OF OUR INCREASING USE OF RESOURCES ON THE ENVIRONMENT

The human impact on the environment in terms of carbon footprints

Identify and evaluate measures to manage traffic in a sustainable manner (using one case study of a city within the EU)

Key questions

By the end of this section you will be able to answer the following questions:

1. State the meaning of the term 'carbon footprint'.
2. What is the impact of a globally increasing carbon footprint on the environment?
3. For a European city that you have studied, evaluate two traffic control measures that have been put into place.

Key words

Resource
Renewable resource
Non-renewable/finite resource
Resource depletion
Carbon footprint

The human impact on the environment in terms of carbon footprints

The earth provides us with many different valuable resources which allow us to survive and add value to our lives. Natural resources are substances obtained from the environment that help us to satisfy human needs.

Every day people need a wide variety of resources such as water to drink, air to breathe, food (from vegetation and animals) for energy, heat, light and a cooking source. In Unit 1B, whilst looking at the issues behind climate change, we also considered how using global resources was affecting our planet. Many of these resources can be classified as being either non-renewable or renewable. A non-renewable resource is one which can only be used once. It is usually burnt or manufactured into something else and therefore cannot be used again. A renewable resource is one which is naturally replaced and can be used repeatedly.

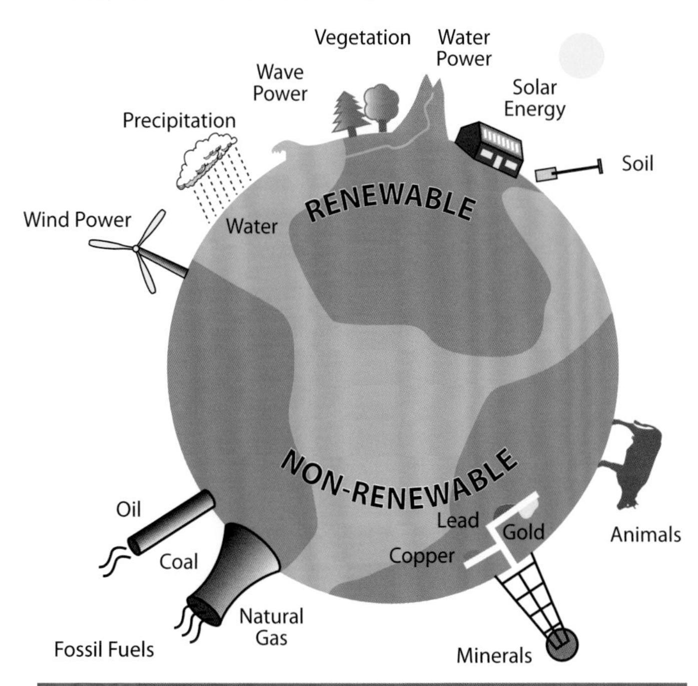

Figure 64 *The main renewable and non-renewable resources for our planet*

We have already noted that the global population is increasing at a very fast rate. This increase means that more people need to share the same amount of resources and, as the majority of resources we use are non-renewable/finite, they are going to run out faster. In addition, more and more people are starting to improve their standard of living and have more disposable income to spend on luxury goods. This leads to increased spending on cars and electronic devices such as mobile phones, laptops and MP3 players, all of which require increased energy supplies to function.

More people plus more electricity 'needs' mean that there is a greater demand for power supply. Scientists have estimated that in 2010 alone there was a 5% increase in the amount of energy consumed in the world. In particular, countries such as China and India are continuing their demand for energy at a rapid rate. This means that new sources of energy are needed. Figure 65 shows how energy consumption has been changing between 1990 and 2011. Many of the MEDCs have maintained a small but steady increase in consumption. However, the large increase in consumption in places such as the Middle East, China and India has created big issues for energy supply.

Figure 65
Energy consumption across the globe

	ENERGY USE IN TERRA WATT HOUR (1,000 KILOWATT HOURS)* *1 kwh is equal to 1,000 watt hours		
	1990	2011	Difference
USA	2712	3852	1140
China	534	4079	3545
India	198	828	630
UK	284	325	41
EU countries	2260	2869	609
Middle East	208	757	549
Latin America	501	1151	650
Africa	253	573	320
The World	**10096**	**19016**	**8920**

Figures from Enerdata yearbook 2012, Enerdata, www.enerdata.net

The majority of energy resources at present continue to come from non-renewable sources. Around 86% of global energy sources in 2011 were non-renewable, with around 13% coming from renewables. In recent years, we have started to suffer from the effects of resource depletion. This means that some of the raw materials that we have come to depend on are running out and becoming exhausted.

Figure 66
Global energy by power source in 2010

POWER SOURCE	%	
Oil	32.4%	Non Renewable Energy = 86.8%
Coal	27.3%	
Gas	21.4%	
Nuclear	5.7%	
Hydro	2.3%	Renewable Energy = 13.2%
Biofuels and waste	10.0%	
Others (includes geothermal, solar, wind and heat)	0.9%	

Figures from Key Word Energy Statistics © OECD/IEA 2012, p6

Test yourself

Answer the following questions with reference to Figure 65 concerning energy consumption across the globe.

1. What three areas consumed the most energy in 1990? (3)
2. What three areas consumed the most energy in 2011? (3)
3. What three areas showed the greatest increase in energy consumption from 1990 to 2011? (3)
4. Why do you think the world consumption of energy rose during this time? (3)

Answer the following questions with reference to Figure 66 concerning energy by power source in 2011.

5. What was the most used power source in 2011? (1)
6. Why do you think that non-renewable energy makes up 86% of all energy consumed in 2011? (3)
7. Renewable energy only makes up 13% of all the energy consumed in 2011. Why do you think this is the case? (3)
8. Use the figures from Figure 66 to create a pie chart to represent this data. (3)

The term 'carbon footprint' has been around since the 1990s. It is often defined as being 'the total set of greenhouse gas emissions that are caused by an organisation, event, product or person'. However, it is usually measured by the amount of carbon dioxide and methane emissions within a population.

Figure 67
Carbon footprint

The idea behind the carbon footprint is the recognition that every action people take has an environmental consequence. As people use particular resources, they will produce greenhouse gases through the transportation, storage and presentation of products. For example, a coffee farmer in Colombia will need to use some basic machinery to help harvest his crop; he will need some form of energy to help process, roast and package his coffee beans; he will need to use local transportation to get his coffee to a port; and a ship to transport his coffee to a consumer market in another country. Further energy will be used in the market, to transport, package and present the final product for sale. Each stage in this production and sales process uses up energy, which is gained from fossil fuels.

In February 2011, the Northern Ireland executive approved the 'Northern Ireland Greenhouse Gas Emissions Reduction Action Plan', which took a look at current policies and tried to map out what Northern Ireland emissions would look like in 2025.

Figure 68 shows that Northern Ireland has a relatively high level of emissions per capita compared to the rest of the UK.

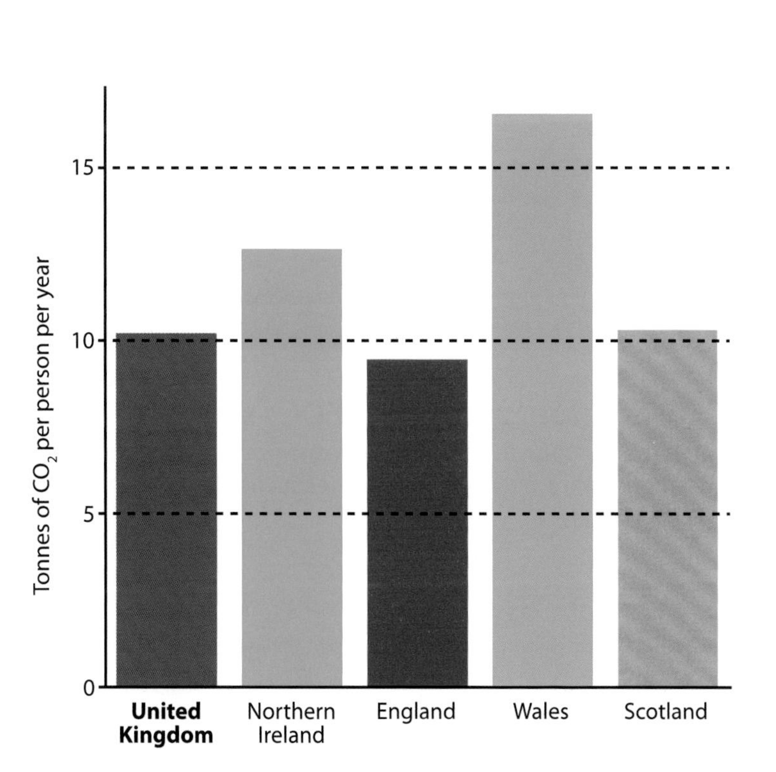

Figure 68
Emissions per Capita

Figures from the Northern Ireland Greenhouse Gas Emissions Reduction Action Plan, February 2011

Some of the main reasons for this include:

1. Northern Ireland has higher transport emissions due to greater levels of rural driving.
2. Northern Ireland has a higher dependency than other areas on oil for home heating systems, as many of the houses cannot be connected to the mains gas grid.
3. Agriculture plays an important part in the economy of Northern Ireland where there is a reliance on agricultural machinery, chemicals and fertilisers.

In addition, the UK government through the Department of Energy and Climate Change is also committed to cutting emissions. The Climate Change Act 2008 along with 'The Carbon Plan' published in December 2011, set out plans for achieving emissions reductions up to 2050. They have introduced a 'Carbon Budget' which is a cap on the total quantity of greenhouse gas emissions emitted in the UK. This means that every tonne of greenhouse gas production emitted between now and 2050 will be counted so that as emissions rise in one sector, they will have to achieve corresponding falls in another sector.

Figure 69
NI Emissions by Sector 1990–2008

SECTOR	EMISSIONS 1990	EMISSIONS 2008	DETAILS
Agriculture	23%	22%	Cattle and livestock Agricultural soils Agricultural engines Field burning
Power	22%	22%	Power Stations
Business and Industrial Process	12%	11%	Cement production Industrial off-road machinery
Public	2%	1%	Railways Public sector combustion
Residential	20%	18%	Domestic combustion Houses & garden machinery
Transport	15%	23%	Railways, aircraft and all other road transport
Waste	7%	4%	Landfill, wastewater handling and waste incineration

Figures from the Northern Ireland Greenhouse Gas Emissions Reduction Action Plan, February 2011

Test yourself

1. How 'green' are you? Do a Google search for a 'carbon footprint calculator' and fill this in to see what sort of carbon footprint your house is making.
   You could try the one at:
   http://www.cooltheworld.com/kidscarboncalculator.php
2. Then, try to analyse your carbon footprint. What is the breakdown of your footprint? How many resources are you using up? How many kilograms or tonnes of carbon dioxide are you producing every year?
3. How do you think that you and your family could reduce your carbon impact?

Why are the carbon footprints in MEDCs so high?

There is little doubt that the carbon footprint of someone living in an MEDC is much higher than in an LEDC. Whilst the amount of carbon used in countries such as China and India is increasing, it has not quite reached EU or USA levels... yet.

The main reasons for high carbon footprints in MEDCs are:

Transport

People in MEDCs travel greater distances than those in LEDCs, both within their country and on holiday from their country.

Higher car ownership

Most families have at least one car in MEDCs and many have more than one. Personal motorised transport is a major source of carbon dioxide in the atmosphere.

Large homes and modern appliances
It takes a lot of energy to heat and to power the homes in MEDCs. People have more disposable income than those in LEDCs, giving them more money to spend on buying technology, which uses a lot of energy.

Products
With a large disposable income, people in MEDCs buy products which use a lot of energy either in their production or their use (for example, clothes, cosmetics and computers).

Food
Much of the food supplied in MEDCs is not produced locally. People in MEDCs eat more meat and exotic fruit and vegetables than those in LEDCs, which have been flown many miles.

Identify and evaluate measures to manage traffic in a sustainable manner (using one case study of a city within the EU)

Case Study: Freiburg, Southern Germany

Figure 70
Rooftop view over Freiburg

Figure 71
Location map of Freiburg

One strategy which could limit the carbon footprint and carbon emissions produced within a country is a sustainable traffic and transport policy. Such a policy allows the authorities to monitor and change the different ways that people access transport across a city.

One city that has adopted a rigorous sustainable traffic management policy in the European Union is the city of Freiburg in Southern Germany.

Introduction to Freiburg
In a 2008 article in *The Observer*, Andrew Purvis took a look at the city of Freiburg and asked, 'Is this the greenest city in the world?'

Freiburg was destroyed by Allied bombers during the Second World War and was then rebuilt on enlightened energy-saving principles. In particular the newly built areas of Vauban and Rieselfeld were developed according to the idea of sustainability. In 1992 the Freiburg council voted a law that would only allow the construction of 'low energy buildings'. They encouraged the use of solar panels for heating and electricity, and pioneered the concept of recycling.

The main employer in the city is the local university and these academics were not happy when plans were presented to build a nuclear power plant in the local area. As a result they were forced to try and solve the problem of having a finite amount of electricity, but a growing population.

Over the last three decades, Freiburg has started to co-ordinate its transport, tripling the number of bicycle trips and public transport (68%), reducing car trips to 32% and reducing CO_2 emissions from transport, even though the population has increased.

Sustainable measures used to manage traffic
Most normal cities with a population of 225,000 find that congestion, pollution and traffic jams are constant issues. In Freiburg, the city council has adopted a number of different strategies which have helped to solve traffic problems.

1. Public Transport
Freiburg operates an excellent public transport system which is operated by VAG Freiburg. The main part of this is an ever expanding number of electric trams, first started in 1983, which work within an extensive network across the city. The trams are an efficient, quick and cheap transport link. The first transferable monthly 'environmental ticket' was introduced in 1984.

Figure 72
The VAG trams that operate an efficient public transport system across the city

2. Pedestrians encouraged

Freiburg has an extensive pedestrian zone in the centre of the city where no cars are allowed.

3. Cycling encouraged

Over 400 miles of cycle paths have been created within the city, with bikes being given priority on roads and at traffic lights. Safe bike storage has been introduced, with more than 6,000 bike parking slots in 2009. Many people cycle everywhere and do not actually own a car. Instead many pay around £500 a year to join a car-sharing club where they can hire a car to move 'big loads' of shopping or for when they go skiing in the mountains.

Figure 73
Bike racks and water canals in the streets show the more sustainable aspects of city planning

4. Car use discouraged

Many of the areas of the city are now car-free. All free parking has been removed and replaced with very high car parking charges. In the Vauban neighbourhood, residents have to sign an agreement that they will never own a car and if they do purchase a car they have to pay £15,262 to the council to park it in a solar garage (multi-storey car park) on the edge of the city.

Evaluation of sustainability in Freiburg

The way that Freiburg has organised its transport means that the integrated transport model is sustainable. Compared to any other city in Germany and beyond, Freiburg has taken more measures to reduce its carbon footprint than anyone else. It continues to push the boundaries of sustainability and has developed green industries and eco-villages. Freiburg describes itself as a 'sustainable city' that is driving down carbon dioxide emissions through regulation, incentives, design, long-term commitment and policy reform.

Positive impacts of sustainable measures

- ***Fewer cars:*** Every day fewer cars come into the centre of the city. It is easier to go by bike or tram and it is more difficult and more expensive to use a car for short journeys and parking. People are encouraged to avoid buying a car, which means there is less personal transport and less carbon emissions. Over the last 10 years carbon dioxide emissions have been reduced by 10% and 35% of the residents choose to live without a car.
- ***Greater use of public transport:*** 68% of trips in the city are made using the tram or the special 'bendy-buses'. The transport system is continually being improved and integrated, with cycle racks at railway and tram stations so that there is a seamless transfer of people from one transport mode to another. The trams are electric and powered from a sustainable source (wind and solar power). Public transport brings reduced carbon emissions.
- ***Reduced congestion:*** Across many cities around the world, congestion means that people are sitting in cars, burning fuel while not moving. As there is now less congestion across the city, this means that any petrol or diesel engines which are used will be more efficient, have shorter journey times and therefore cause less pollution.
- ***Improved environment:*** Streets are not used for cars to the same extent that they once were. This means that there is less air pollution around the city, so people can enjoy the pedestrian areas and sit outside cafes and restaurants. This makes the city look and feel better, and planting programmes have helped to improve the green space and provide more vegetation to absorb the carbon dioxide.

Negative impacts of sustainable measures

- ***Public transport links are limited:*** The trams tend to travel into the city centre, which can sometimes be inconvenient if you want to travel outside or across the city.

- ***Public transport times limited:*** The public transport system works well for those travelling close to the peak times. However, if people work late at night there is a longer gap between trams arriving, which does not suit the needs of people who work unsociable hours.
- ***Overcrowded public transport:*** During peak times the trams and buses can become overcrowded which can be unpleasant and stressful.
- ***Problems for people with car dependent jobs:*** Some people require cars for their jobs. The restrictions and rules for some areas of the city, along with intensive traffic calming measures can cause problems. For example, it can be difficult to get vans and lorries into the city centre for deliveries or to transport shopping and heavy items back home.
- ***Car ownership is still high:*** Although car use has been reduced, many people will still use a car daily, even if they do not enter Freiberg city centre.
- ***People do not like the high prices and restrictions:*** In recent years some have started to question the rules and restrictions that people have to live under. Some think that prices and penalties for having personal transport such as cars are too restrictive and some argue that this is stopping people from choosing to live in Freiberg.

Figure 74
Pedestrian streets are reclaimed for the people, not automobiles

Test yourself

1. What are the main ways in which Freiburg can be considered 'sustainable'? (4)
2. There are a reduced number of cars in Freiburg compared to other similar sized German cities. Explain one positive and one negative impact of this. (4)
3. Explain two other positive impacts of these sustainable transport measures. (4)
4. Explain three other negative impacts of these sustainable transport measures. (6)
5. In what ways can cities like Belfast compare to Freiburg? How would you evaluate the sustainability of Belfast compared to Freiburg? (6)

Check your learning

Now that you have studied Part 1: The Impact of our Increasing use of Resources on the Environment, return to page 229 and answer the Key Questions for this section.

PART 2: INCREASING DEMAND FOR RESOURCES IN LEDCs AND MEDCs

(a)

How population growth and economic development in LEDCs increase the demand for resources and put pressure on people and the environment

Evaluate the benefits and problems of ONE renewable energy source (wind or solar or biofuels) as a sustainable solution
Case Study: Walney Wind Farm, Cumbria (a MEDC case study)

Key questions

By the end of this section you will be able to answer the following questions:

1. Explain how population growth has put pressure on the environment in LEDCs.
2. Name a renewable resource used in an MEDC. Explain one benefit and one problem of using this resource.

Key words

Renewable energy
Solar
Wind
Biofuels

How population growth and economic development in LEDCs increase the demand for resources and put pressure on people and the environment

One of the biggest challenges for LEDCs is that in many cases their population is increasing at a very fast rate. As this population continues to grow, all of the people that need to be fed and need access to water, fuel and other resources puts vast pressure on the resources that are available within the country.

In addition, many of the LEDCs aspire to become NICs. They want to become more developed and are keen to do this through industrialisation, creating products which will be bought by the rich consumers living in the MEDCs. Industrialisation attracts people to the cities, many moving into slums and shanty towns, and urbanisation puts further pressure on the delicate resources.

The more resources that are needed, the more that will have to be collected, processed and manufactured. The 'need' for new mobile phones, computer devices, widescreen TVs and sports equipment puts more and more pressure on global resources. Using these resources remains deeply unsustainable as many are not renewable.

Population growth

As the global population continues to increase, further pressure will be put on global resources. In particular there is a huge demand for the consumption of non-renewable energy. The US Energy Information Administration projects that by 2030 non-OECD (Organisation for Economic Co-operation and Development) Asia will account for 43% of the overall increase in world consumption of solid fuels (eg coal) and liquid fuels (eg oil). They project an increase of 6.5 million barrels per day up to 2015 and then another 8.5 million barrels per day from 2015 to 2030. The world economy continues to be controlled (and lubricated) by oil. In spite of the warnings over the last 20 years that oil reserves are rapidly running out, research into alternative energies has been limited. Many economists argue that if the price of oil rises too fast, it will trigger a global recession.

The 'World Population Balance' group claim that population is not just a matter of numbers. They believe that the environmental impact that energy use can have on particular countries must also be considered. In particular, they think it is important to compare the energy consumed between different countries to the impact in the USA.

Figure 75
Energy consumption of the USA compared to India

	USA	RATIO	INDIA
Land area (km^2)	9,158,960	3:1	2,973,190
Population (2007)	315 million	1:4	1,210 million
Energy Consumption (million barrels of oil)	18.9	5:1	3.4
Energy per capita (barrels of oil per person)	312.14	16:1	19.85

Source: Figures from U.S. Energy Information Administration (2011)

The worry of course is that at some point in the future we could be in the situation where the number of people in the world far exceeds the amount of resources available. This could create some serious problems across the world, push energy prices higher and increase the difference between the rich and poor across the world.

Economic developments in LEDCS

Since the 1980s many LEDCs have taken opportunities to start their own 'industrial revolution'. Meanwhile MEDC energy consumption has continued to increase, fuelled by the growing number of technological devices that are powered by electricity. Many of these devices are designed in MEDCs but are manufactured in the LEDCs, where the labour costs are much lower. The demand for electricity is increasing in both areas of the world, especially in NICs where there is a real push for economic expansion though manufacturing and exports. Figure 76 shows that world energy consumption is projected to grow by 66% between 2008 and 2035. The most rapid growth is for countries outside the OECD, especially the non-OECD Asian countries (Asian LEDCs).

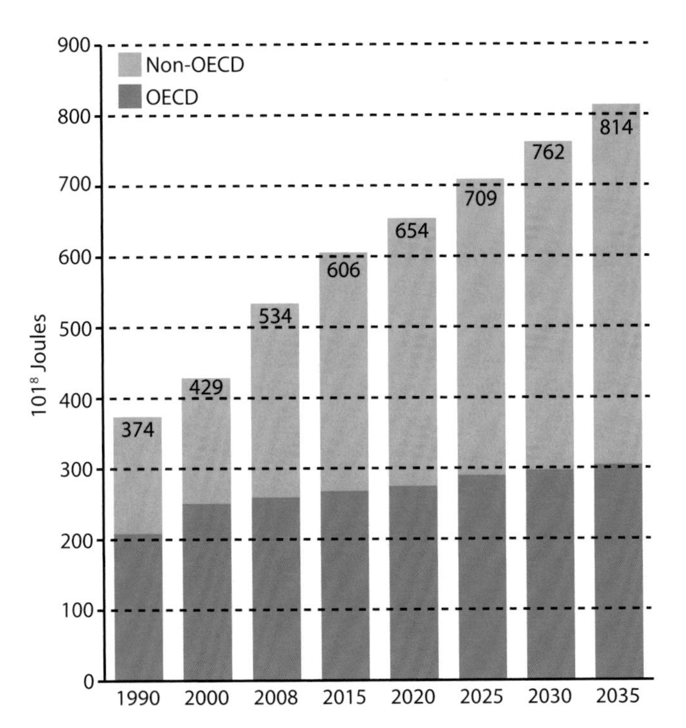

Figure 76

Energy consumption by economic area

Source: Figures from U.S. Energy Information Administration (2011)

> **NOTE:**
>
> ***OECD*** = the 34 countries that have signed up to the Convention on the Organisation for Economic Co-operation and Development (mostly comprised of MEDCs).
>
> ***Non-OECD*** = the other countries that have not signed up (mostly comprised of LEDCs and NICs).

Interestingly, the resource which fuelled the industrial revolution in many countries in the nineteenth and twentieth centuries is still at the core of industrial development in LEDCs in the twenty-first century. Figure 77 shows that coal accounted for nearly half of the increase in global energy used over the last ten years, with the bulk of this growth coming from the power (electricity generation) sector in NICs and LEDCs.

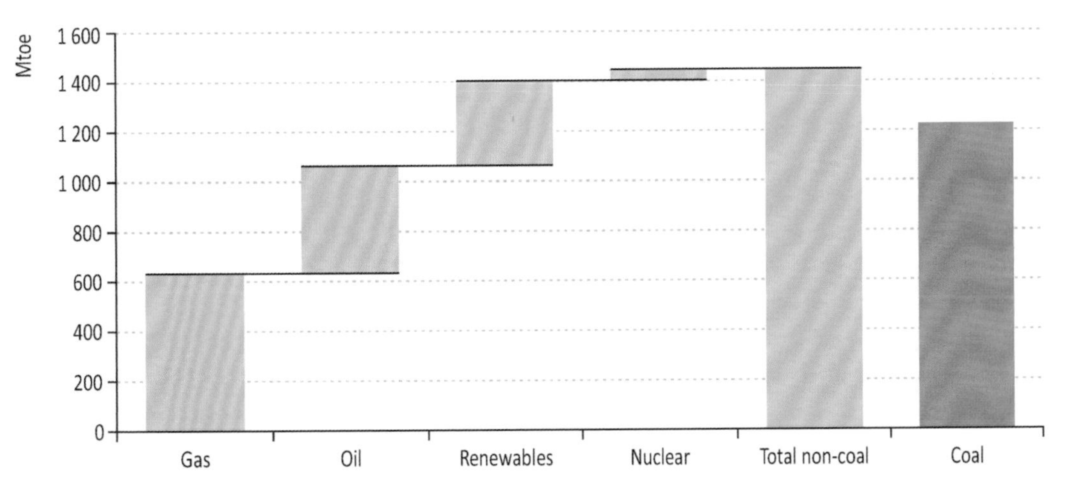

Figure 77

(Left) Incremental world primary energy demand by fuel 2000–2010

Source: World Energy Outlook 2011 © OECD/IEA 2011, fig 10.1, p354

Figure 78

(Below) China's energy consumption

Figures from U.S. Energy Information Administration (1991–2009)

The demand for resources puts pressure on people

Over the last 50 years Newly Industrialised Countries have seen their economy grow. China, for example, is the fastest growing major economy in the world, with growth rates averaging 10% over the last 30 years. It is also currently the world's second largest economy after the United States. As economic growth results in increased manufacturing and production, more energy and resources are needed. This increased demand for resources has started to put pressure on people in a number of different ways.

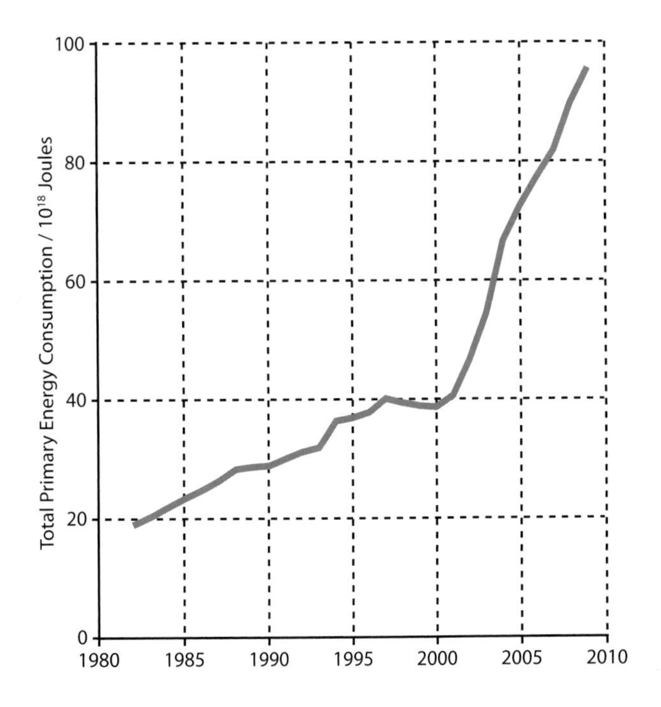

1. Living space

The vast number of people in NICs such as China and India is putting increasing pressure on living space. In China, only 5% of the people live on 64% of the land (in the west) and the remaining 95% of people live on 32% of the land (in the east). The population density there is the third highest in the world (364 people per km²).

People are tightly squeezed into small areas and can often be forced to live in slums or shanty towns. Quality of life can be compromised and life expectancies can be decreased as a result. As more people move into urban areas there is also an increased pressure on resources to support urban living.

2. Food production

India is a rapidly urbanising society but it still retains over 54% of its arable land. However, it is having ever increasing difficulties in growing enough food for its people. By 2040, the gap between land availability and demand will have grown.

China has 20% of the world's population but only 6.44% of the land space. Its current agricultural land per capita ranks sixth from the bottom in the world and much of the agricultural land is under pressure, suffering from soil erosion and pollution.

Figure 79

The increasing amount of dryland areas and resulting degradation across the world

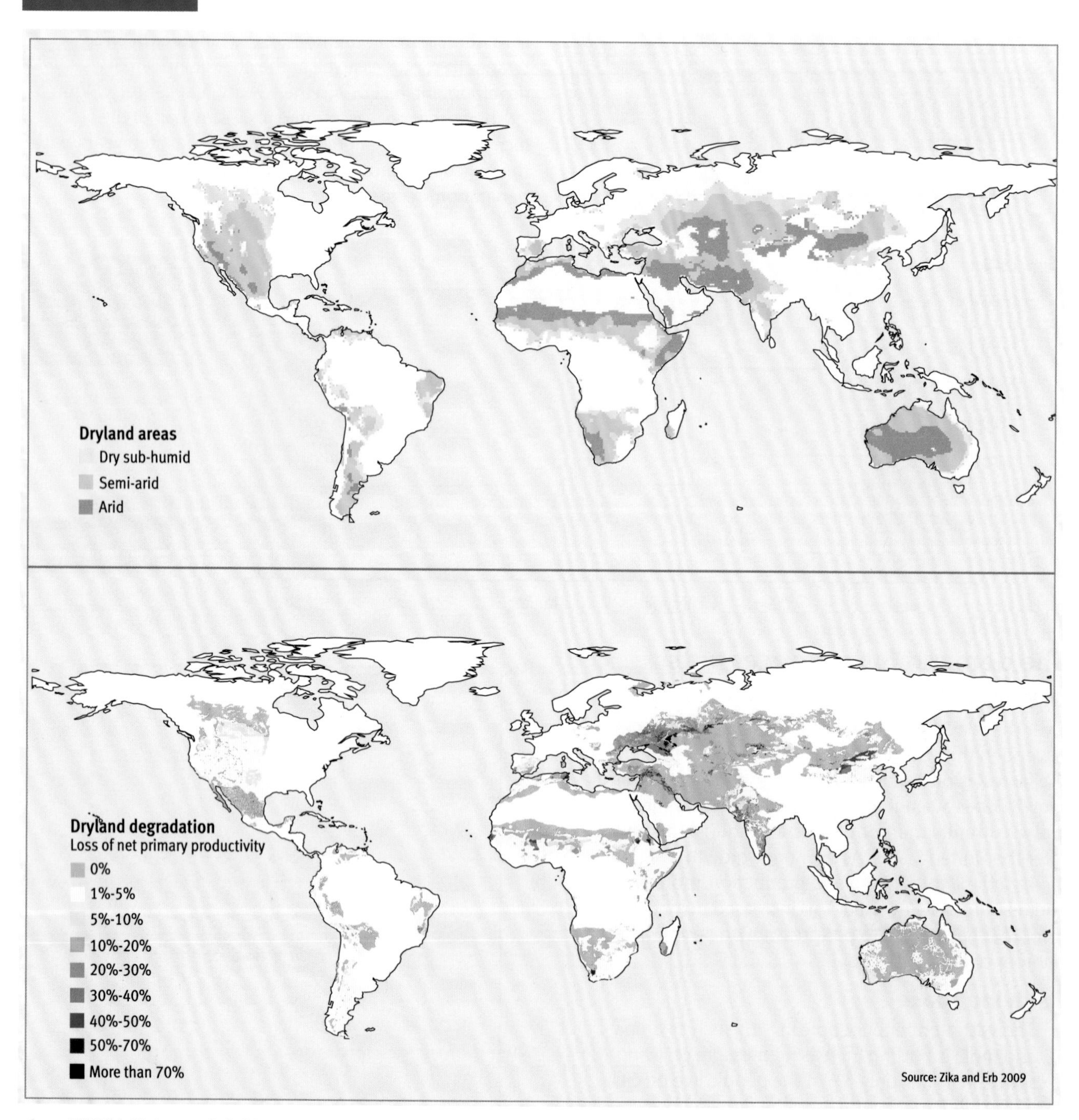

Source: UNEP Global Environment Outlook 5
Zika, Michael, Karl-Heinz Erb, 2009. The global loss of net primary production resulting from human-induced soil degradation in drylands. Ecological Economics 69(2) 310–388. doi: 10.1016/j.ecolecon.2009.06.014

3. Water

Figure 79 shows the ever increasing amount of dryland areas across the world. Much of this affects people in LEDCs. In 2000, about 300 million Africans were living in a water-scarce environment. The UN Environment Programme (UNEP) highlights 10 areas across the continent of Africa where water will be under extreme stress by 2025, affecting nearly 600 million people.* It clearly identifies the increasing population pressure and climate change as the two main factors which are driving this change.

In China, water resources are insufficient and often badly polluted, mostly from sewage, animal and farm waste. Both agriculture and industry suffer from water shortages. Almost the entire rural population of 700 million people has substandard drinking water. Among the people with health problems, 88% can blame their illness on dirty drinking water and 33% of deaths are related to water contamination.

4. Steel

Many LEDCs and NICs' demand for iron and steel is fuelled by building and urbanisation programmes. Scientists estimate that demand from China alone will be around 32 billion tonnes of iron over the next 10 years (from a global iron reserve of 80 billion tonnes!). Increasingly, resources will become more difficult to source and prices will increase. Cost of living for people will continue to rise unless new, alternative sources for materials and energy can be found.

The demand for resources puts pressure on the environment

Many of the recent global meetings to discuss environmental issues (such as the Conferences of the Parties or COP) have noted that the environment continues to be under serious threat from the amount of industrial pollution.

Industry makes up 47% of China's GDP. This means that a vast amount of wealth is created through the manufacturing and export of consumer goods. Major industries in China include iron, steel, aluminium, machinery, textiles, petroleum, cement, chemicals, fertilisers, food processing and automobiles. The pursuit of industry within an economy can have a big impact on the environment.

1. Land

Urban areas are increasing all the time. LEDC cities (such as Sao Paulo) are growing at a fast rate, with people moving from the countryside to urban areas, often into shanty towns and slums. As these cities grow they are taking up more and more of the rural landscape, which is destroying natural environments and animal habitats.

2. Natural resources

Manufacturing, which drives the success of many LEDCs and NICs, requires inexpensive raw materials to turn into saleable products at a minimum cost. Often, the methods used to harvest and collect these raw materials are based on obtaining them as cheaply as possible and environmental concerns are rarely a priority. Open cast mining, large scale tree-felling and gas extraction techniques (such as fracking) can do much to destroy the natural environment. In addition, little money is used to return the environment to normal after the resources have been collected. Between May 2000 and August 2006, Brazil lost nearly 150,000 square kilometres of forest (an area larger than the whole country of Greece).

3. Waste management

Humans today produce more pollution and waste than at any other time in history. Urban residents each produce around 200 kg of rubbish per year. Disposing of this waste, either by burning or landfill, produces air pollution and smog, can spoil the scenery and leave vast areas of land unusable in the future due to radioactive and chemical waste.

* 'Water availability in Africa under stress due to increasing population pressure and climate change', Vital Climate Graphics Africa, UNEP/GRID-Arendal, http://www.grida.no/publications/vg/africa/page/3116.aspx

4. Air pollution and global warming

Increased congestion from burning fossil fuels, wood stoves and from engine exhausts has increased the amount of air pollution in urban areas. In addition, the growth of greenhouse gases in the atmosphere has increased the threat of global warming.

As many LEDCs are investing in industry for the first time, there has been a big increase in air pollution as countries try to develop their economy. This often means that declines in the quality of air and increases to the threat of global warming are happening faster in many of the LEDCs than in MEDCs. According to the WHO, over 750,000 people die every year from air pollution in China and much of this pollution is a direct result of the burning of fossil fuels in the cities.

5. Traffic

As car ownership expands at a fast rate, traffic congestion is increasing and this puts pressure both on people and on the environment. If traffic moves more slowly through the congested arteries of an urban area, more air pollution is created and there is a greater production of greenhouse gases.

Test yourself

1. Explain how population growth in LEDCs has increased pressure on resources. (6)
2. Explain how economic development in LEDCs has increased the pressure on resources. (6)
3. Describe and explain the pressure on the environment due to increased demand in LEDCs. (4)
4. Describe and explain the pressure on people due to increased demand in LEDCs. (4)

Evaluate the benefits and problems of one renewable energy source (wind, solar or biofuels) as a sustainable solution

In Unit 1 (pages 95–96) we looked at some of the renewable sources which provide alternative energy, specifically wind power, solar power and biofuels. In recent years there has been a surge in the amount of renewable energy being developed.

Wind power: has become a popular option in the UK, as it is very cost effective. The UK is now the world's eighth largest producer of wind power, with a total of 321 wind farms using 3,506 wind turbines.

Solar power: is much more costly than wind today but can be very effective when integrated into new buildings.

Biofuel: is not the most popular renewable energy in the UK. However, it is often more important in LEDCs, as many different materials such as chicken litter, straw or energy crops (for example willow) can be used to help fire power stations.

Some other renewable resources found in the UK include:

Hydroelectric power: using large dams was popular in the UK before the 1960s, especially in Scotland. However, social and environmental concerns related to the flooding of vast areas of farmland mean that this is a less favourable option today.

Tidal and wave power: technologies are developing at a fast rate at present but are currently less advanced than wind power.

Case Study: Walney Wind Farm, Cumbria (a MEDC case study)

In February 2012 Walney Wind Farm became the world's biggest off-shore wind farm. The farm is located about 14 km west of Walney Island, off the coast of Cumbria in the Irish Sea. It covers an area of around 73 km^2 and its current capacity is 367 MegaWatts of energy.

In 2004 the energy company Dong energy secured a 50 year lease to develop a wind farm at Walney Island that will be able to supply energy to approximately 320,000 homes. The wind farm grew up in two phases with a total of 51 wind turbines and cost around £1 billion to develop. This cost was met jointly by energy companies, pension funds and equity funds. The farm was officially opened by the Energy Secretary Ed Davey on 9 February 2012. At the opening ceremony he said:

> "Britain has a lot to be proud of in our growing offshore wind sector. Our island's tremendous natural resource, our research base and a proud history of engineering make this the number one destination for investment in offshore wind. And Walney is the newest, biggest and fastest-built jewel in that crown, providing clean power for hundreds of thousands of households."

Figure 80 One of the turbines that forms part of Walney Wind Farm

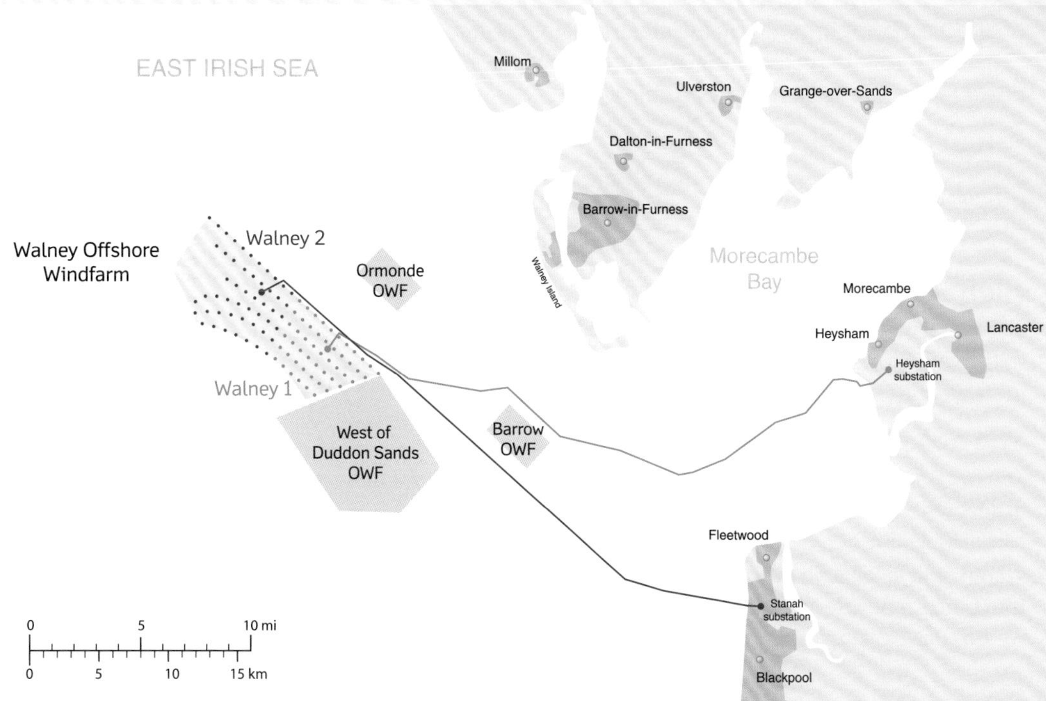

Figure 81 Location of Walney Wind Farm, Cumbria

Benefits of developing wind power

- ### Reduced carbon emissions
  With so much of the UK's energy coming from the burning of fossil fuels, this creates greenhouse gas and carbon emissions. A shift towards renewable energy sources such as wind power means that an inexhaustible, locally available and 'green' fuel can be used.

- ### Sustainable energy source
  Wind power is viewed as one of the most cost-effective and technologically advanced renewables available within the UK. Modern wind turbines are very powerful, reliable and can capture more energy than older versions. A turbine in the UK is likely to be producing power for 70–85% of the year and will have a lifespan of 20 years.

The UK has the best and most geographically diverse wind resources in Europe, and the UK government aims to obtain 10% of the UK's electricity from renewable sources by 2010 and 20% by 2020. The National Grid has estimated that 8,000 MW of wind power might be able to displace about 3,000 MW of conventional energy.

- **Reduced dependence on imported fossil fuels**

  Over the years the UK has become increasingly dependent on the non-renewable energy that is supplied through the burning of fossil fuels. However, any increase in the amount of energy which can be gained from renewable sources will help to reduce the amount of fossil fuels that are needed. As the majority of UK energy is currently imported from other countries this will also help to improve energy security.

- **Value for money**

  The average cost of setting up wind power has been estimated at about 3p per unit more expensive that current electricity prices. Given the recent increase in oil prices, this means that the cost of wind power is actually the same as the cost of producing electricity in the traditional manner.

Problems of developing wind power

- **Harm to wildlife**

  In the early years of wind farm development many marine biologists were concerned that offshore wind farms would have big impacts on the animals living in the sea. One concern was that sea mammals such as seals, porpoises, basking sharks and whales might have their sense of direction compromised by the electric fields created by the electric generators. In some cases sea birds are scared off by the turbine blades and their natural navigation systems may be disrupted by the large banks of turbines.

- **Visual pollution**

  Many people find the wind farms ugly and believe that tourists will be put off coming to an area because of the proximity of the wind farm. However, the contractors argue that it is actually quite difficult to see the wind farm, 14 km offshore, even on a clear day.

- **High set up costs**

  The initial investment of £1 billion is seen to be very expensive. However, the wind farm moved to full capacity production very quickly and the investment will bring a constant return over many years. The turbines should require minimal maintenance.

- **The need for alternative sources of energy**

  The big issue with wind power is that there are some days when the wind will not blow and others when the wind speeds exceed the maximum limits (above 26 knots) for safe use of the turbines. This means that there will be days when no energy is produced and a backup energy system will always be needed to ensure there is not a power blackout. This usually means that backup non-renewable power stations need to be ready to be brought up to full power if weather forecasts predict a lack of wind.

Is this a sustainable solution?

In 2009 the UK government noted that "the goal of sustainable development is to enable all people throughout the world to satisfy their basic needs and enjoy a better quality of life without compromising the quality of life for future generations".

A set of five shared principles was noted to underpin the purpose and framework, making any government policy 'sustainable':

- Living within environmental limits
- Ensuring a strong, healthy and just society
- Achieving a sustainable economy
- Promoting good governance
- Using sound science responsibility

Test yourself

1. Briefly describe three different renewable energy sources. (6)
2. Explain three advantages and three disadvantages of setting up an offshore wind farm in the UK. (6)
3. Why do you think wind power is considered a renewable energy source? (3)
4. Describe in detail two problems with developing wind farms off the UK coast. (4)
5. Evaluate how sustainable wind farms in the UK are. (6)

Check your learning

Now that you have studied Part 2 Increasing demand for resources in LEDCs and MEDCs, return to page 235 and answer the Key Questions for this section.

PART 3: MANAGING WASTE TO PROTECT OUR ENVIRONMENT

(a) Why has waste become such a major issue in the UK?
Shortage of landfill sites
Environmental and health concerns
The need to meet government targets

(b) The waste hierarchy and the concept of 'reduce, reuse and recycle'

(c) Why do we need a range of sustainable waste management approaches?
Case Study: Belfast City Council (one case study of a local government area)

Key questions

By the end of this section you will be able to answer the following questions:

1. Explain the concept of 'reduce, reuse and recycle'.
2. Using one case study, explain why a range of waste management approaches are necessary.
3. What is the waste hierarchy?
4. Why has a shortage of landfill sites made waste a big issue in the UK?

Key words

Waste hierarchy
Landfill site

Why has waste become such a major issue in the UK?

As long as people have been alive, they have been producing waste. According to the 'Wake up to Waste' leaflet produced by the Department of the Environment in Northern Ireland: "Every year in Northern Ireland we produce enough waste to cover the whole country and the amount is rising. The cost of dealing with this waste is also rising." The Rethink Waste NI website (www.rethinkwasteni.org) also notes that "Waste constitutes one of the main areas of expenditure for local councils in Northern Ireland".

Recent Northern Ireland government figures tell us that:

1. Local council waste peaked in 2006 at 1.064 million tonnes, which fell to 1.061 million tonnes in 2007 and 1.017 million tonnes in 2008.
2. Waste that gets recycled or composted has been increasingly steadily from 191,000 tonnes in 2004 to 303,000 tonnes in 2008.
3. The amount of waste collected from the average household is 1.29 tonnes per year.
4. Of all of the local council waste in 2008, 68% of it was sent to landfill and the remainder was mostly Biodegradable Municipal Waste (BMW).

Figures from Rethink Waste NI, www.rethinkwasteni.org

However, the figures for 2010 show some improvement:

1. Composting rates for households are on the increase (up to 42%) and the amount of municipal waste being sent to landfill is decreasing by about 4% per year.
2. The amount of waste sent to landfill was reduced to 59%.
3. The biggest change in household waste in recent years has been the successes in recycling. Of all household waste collected, almost 21% was sent for recycling.

Figures from 'Wasteline', Issue 7, Summer 2011, arc21, www.arc21.org.uk

There are three main issues that are currently affecting the management of waste in Northern Ireland:

Figure 82
Dargan Road Landfill, Belfast

1. Shortage of landfill sites

Over the last 50 years the main method of waste disposal across the UK, and indeed within Northern Ireland, has been the use of landfill sites. However, Northern Ireland is running out of available sites for landfill and this is even more pressing within Belfast. One of the biggest landfill sites in Northern Ireland is the site at the North Foreshore at Dargan Road, just two miles north of the city centre beside the M2 motorway. It came into operation as a landfill in 1958, though the site was expanded for waste disposal in 1973. Landfill operations were halted on the site in 2006 and there are some plans to redevelop the land in the future.

Pressure from the European Union (from 2000) has meant that many local councils across Northern Ireland have been forced to move from a strategy based on landfill to one based on recycling.

2. Environmental and health concerns

The average household in Northern Ireland produces 1.29 tonnes of waste per year. This works out at around 24 kg of waste every week. With fewer landfills allowed for waste management, this means that treating and disposing of this material has become a major problem.

Any waste which goes to landfill will break down and begin to release methane gas. This is a greenhouse gas. A liquid called 'leachate' is also produced and this has the potential to pollute ground water.

3. The need to meet government targets

In 1999 the European Parliament brought out new legislation which divided landfill waste into three main categories:

1. Landfill for hazardous waste
2. Landfill for non-hazardous waste
3. Landfill for inert waste

This new legislation emphasised that waste needed to be treated before being landfilled. The aim of this directive was to prevent or reduce as much waste as possible. Landfill sites across Europe were required to change the way that they operated, complying with national policy and with EU policy. Many countries set their own targets as to how they would transform the way that waste was dealt with.

In Northern Ireland, the Department of the Environment took action (from 1997) and provided direction to transform the waste management practice in Northern Ireland, aiming to use a 'reduce, reuse, recycling, recovery' strategy for the disposal of waste. The EU policy, along with the UK's target for reducing greenhouse gases, as agreed at Kyoto in 1997, led to targets where landfilled Biodegradable Municipal Waste (BMW) would be:

75% of 1995 levels by 2010
50% of 1995 levels by 2013
35% of 1995 levels by 2020

The Northern Ireland Landfill Allowances Scheme was introduced in 2005 and set a maximum amount of BMW that can be disposed of by each council.

The most recent government strategy (2006 and 2011) forms part of the Northern Ireland Waste Management Strategy. The aim of this strategy is to help manage waste and resources effectively. This means trying to minimise its impact on the environment and public health in a way that contributes positively to economic and social development. The core principles of this strategy include:

1. **The waste hierarchy**
2. **The 'polluter pays' principle** – where waste generators should pay the costs of providing services to manage their wastes.
3. **The proximity principle** – where waste needs to be treated or disposed of as close to the point of generation as possible, so minimising the environmental cost of transport.
4. **Self sufficiency** – where people attempt to manage waste within their own borders.
5. **Best available techniques (BAT)** – where the most effective processes and technologies are used to reduce the impact of waste on the environment.
6. **The best practicable environmental option (BPEO)** – where an assessment of waste treatment and management options is used to look at the sustainability of the measure.

Test yourself

1. Describe the current Northern Irish position on waste. (4)
2. Why is there a shortage of landfill space in Northern Ireland? (2)
3. Describe two potential ways that waste can harm the environment. (4)
4. What targets does the Northern Ireland government have to reach? (4)

The waste hierarchy and the concept of 'reduce, reuse and recycle'

The European Union issued a directive in November 2008 which tackled the issue of waste. Article 4 in the Revised Framework Directive set out five steps for dealing with waste and ranked them into their environmental impact or the 'waste hierarchy'. In Northern Ireland new regulations and guidance in relation to waste came into operation in April 2011.

The waste hierarchy

The 'waste hierarchy' ranks waste management options in order of what is best for the environment. The best option is at the top of the pyramid and the least preferred option is at the bottom. Top priority goes to measures that prevent waste in the first place. However, when waste is created, the hierarchy gives priority to preparing it for reuse, then recycling, then recovery and last of all disposal (for example, through landfill or incineration without energy recovery). In other words, the waste hierarchy is a tool that helps local governments to work out what is the best way to deal with waste. This means that strategies such as avoiding waste generation, reducing the quantity

Figure 83
The waste hierarchy

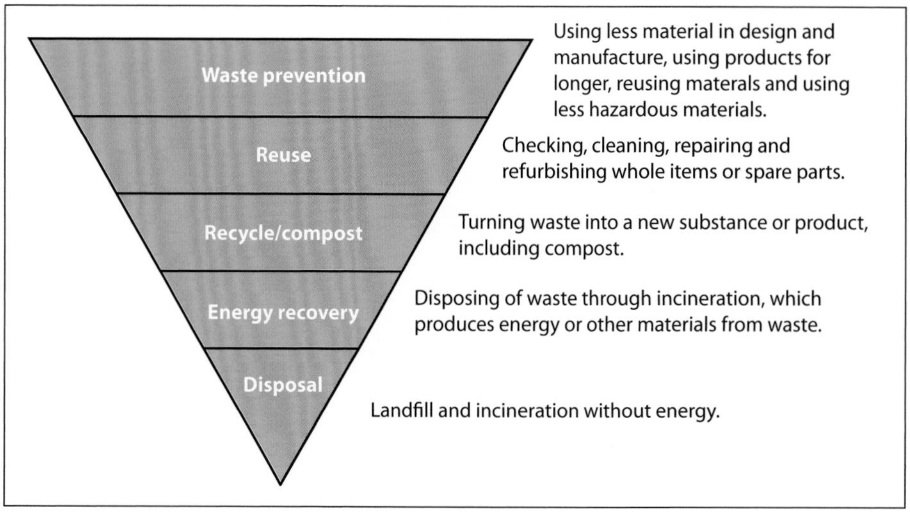

Waste prevention – Using less material in design and manufacture, using products for longer, reusing materals and using less hazardous materials.
Reuse – Checking, cleaning, repairing and refurbishing whole items or spare parts.
Recycle/compost – Turning waste into a new substance or product, including compost.
Energy recovery – Disposing of waste through incineration, which produces energy or other materials from waste.
Disposal – Landfill and incineration without energy.

and nature of waste, and an emphasis on reusing products before they enter the waste stream, are all important to local councils in Northern Ireland.

The main problem in Northern Ireland is that the majority of waste still goes to landfill. Landfill not only takes up more valuable land space, but it also causes water, air and soil pollution. Greenhouse gases such as carbon dioxide and methane are both discharged into the atmosphere and chemicals and pesticides can be washed into the groundwater, causing health problems for animals and people.

Reduce, reuse, recycle

The waste hierarchy refers directly to the idea of 'reduce, reuse, recycle'. Some people have come to refer to this as the '3Rs'. In recent years this has become the backbone of making people environmentally aware. People are encouraged to engage in activities which will have an environmental impact on each of these areas.

Figure 84

Landfill: the least sustainable waste solution

Reduce

People are encouraged to buy less and to reduce the amount of energy that they actually use by turning lights off, taking shorter showers and sharing lifts to work and school. People are also encouraged to be more involved in composting and to buy products that are low energy.

According to the Rethink Waste NI website: "the challenge to rethink waste starts at home. Northern Ireland must get to grips with waste generated at home, as municipal waste remains one of the main environmental challenges." The site explains how we can make efforts to reduce our impact by taking part in the European Week for Waste Reduction, ensuring that we compost garden and food waste (using brown bins), reducing what we buy and the amount of packaging that we use, and through the introduction of the carrier bag levy (April 2013).

Figure 85

A government tax on reusable shopping bags is encouraging people to reuse them

Reuse

This is when different materials are used again, without making them into new products. People are encouraged to use plastic food containers, refillable containers, travel mugs and flasks, which can be washed and reused, instead of using cling-film or aluminium foil, single-fill containers and paper cups, which can be used only once. They are also encouraged to use plastic bags more than once or to buy reusable fabric shopping bags. Marks and Spencer used to give out 460 million food carrier bags annually and were able to drop usage by 81% following the introduction of a 5p charge for bags. Unwanted items can be sold online, at car boot sales or given away to charity shops. Gatherings where people swop unwanted clothes for new ones are also becoming popular.

Figure 86

Easy access to recycling bins encourages people to recycle more

Recycle

This is when waste materials are separated into component parts that can be incorporated into new products. It is slightly different from 'reuse', as energy is used to change the physical properties of the material, turning them into an alternative final product. Paper, glass, mobile phones, batteries, drinks containers, fluorescent tubes, keys, videotapes, metals and electrical equipment can all be manufactured into something else such as egg cartons and fleece jackets.

Upcycling is where waste materials are converted into items that are more expensive than the original article. Downcycling is where the waste materials are converted into items which are less expensive than the original articles. According to Rethink Waste NI "nearly two-thirds of all household rubbish can be recycled, saving energy and raw materials and avoiding waste going to landfill."

Test yourself

1. What is the waste hierarchy? (2)
2. Why do you think disposal is less 'green' than reuse? (3)
3. Explain the difference between the different concepts of reduce, reuse, recycle. (4)
4. Describe some of the Rethink Waste NI waste reduction strategies. (4)

Why do we need a range of sustainable waste management approaches?

Case Study: Belfast City Council

(one case study of a local government area)

All local governments within Northern Ireland, across the UK and as part of the EU are involved in measures to reduce waste and greenhouse emissions. This means that local governments need to be directly involved in ensuring that residents are taking an active role in each stage of the waste hierarchy.

Belfast City Council is the largest district council in Northern Ireland and serves around 270,000 people. In recent years the council has had to develop a comprehensive strategy for waste and recycling within its confines. Belfast City Council has joined up with 10 other councils in the east of Northern Ireland to form arc21, which is one of three waste management groups in Northern Ireland. Arc21 was established in 2003 in order to ensure closer co-operation between councils.

Why does Belfast need a waste management strategy?

1. Belfast produces a large amount of waste

The average house produces around 1.29 tonnes of waste every year. In Belfast, around 130,000 tonnes of waste is sent to landfill every single year. If this amount is not reduced in the future, the Council will be forced to pay heavy fines (as much as £4 million in 2012), which will continue to rise each year that the council fail to meet their targets. Although around 29% of all waste in Belfast is currently recycled, there are plans to increase this to 50% by 2020. Every household has access to a recycling service but recycling alone will not allow the Council to reach its targets.

2. There is a shortage of landfill space

Large landfill sites such as Dargan Road can no longer be used as the only solution for dealing with waste. New solutions and new strategies need to be introduced to reduce the amount of material which is sent into the remaining landfill sites, as no new landfill sites have been identified.

3. Environmental and health concerns

Many of the people who live near to the landfill sites have argued for years that they should be closed. They have been concerned about the smell and the safety of such sites. In recent years there have been gas explosions at a few landfill sites across Northern Ireland.

Dargan Road site

BELFAST

Figure 87
Location of Dargan Road landfill site

4. EU laws and targets

Currently Belfast has not been able to hit the EU targets for waste management and the targets are even harder to attain by 2020. Further measures need to be taken to ensure that the Council does not have to pay large fines.

What methods are being used to manage the waste?

1. Recycling bins

The Belfast City Council has invested money into supplying residents with a variety of bins. All households can dispose of their waste properly by selecting the correct bin. Most residents will have access to the following:

Black bin

This is for general waste that is not suitable for recycling.

Blue bin

This is for items that can be recycled such as paper, brochures, aerosols, plastic bottles, cardboard, cans and aluminium foil.

Brown bin

This is for items that are usually organic and biodegradable such as food waste, grass cuttings, plants, leaves and shredded paper.

Kerbside recycling box

Some people get this instead of a blue bin. It is collected by Bryson recycling and is for paper, cardboard, plastic bottles, glass, cans, foil, textiles, tools and batteries.

Food waste caddy

This is for any house which also has a brown bin and allows people to collect leftover scraps of food that can then be collected in a biodegradable bag and put in the brown bin.

Home composter

This is used to recycle garden waste and can contain grass, branches, food scraps, vegetable peelings, teabags, coffee grounds, cardboard and shredded paper.

2. Reduction in black bin collections

Belfast currently recycles only 31% of its waste but this needs to increase to 60% by 2020 to meet EU targets. The strategy encourages residents to recycle more by building new recycling centres across the city and inserting new bottle and textile banks in accessible locations such as supermarket car parks. It has also reduced black bin collections from once a week to once every two weeks, encouraging people to use their recycling bins and centres.

3. Education and advertising

In the last few years a comprehensive education and advertising campaign has been ongoing within the Belfast City Council area and also regionally across Northern Ireland. Waste management officers are keen to engage with all aspects of society from schools to churches to local community groups. Recent TV advertisements have also attempted to educate people about how to recycle more and enjoy a 'green' lifestyle. The Northern Ireland curriculum in schools was also changed to ensure that students have to deal with waste issues in both Primary and Secondary school. Schools are also encouraged to take part in 'Eco-schools' campaigns.

Figure 88

Sign at the entrance to Dargan Road Landfill site

4. Waste treatment facility

In June 2009, Belfast City Council decided to give arc21 permission to build a new waste treatment facility. The site will house a Mechanical Biological Treatment (MBT) facility. This will sort and compact waste before it is sent on to a landfill or an Energy from Waste (EfW) plant. This plant will help to meet the EU targets in relation to the amount of waste that people in Belfast send to landfill.

The MBT facility will take waste from the black (general waste) bin, with the aim of recovering any material that can be recycled before sending the rest to a landfill. The plant uses machinery to sort through the waste and any remaining waste is biologically treated, which reduces the total amount of waste by one third.

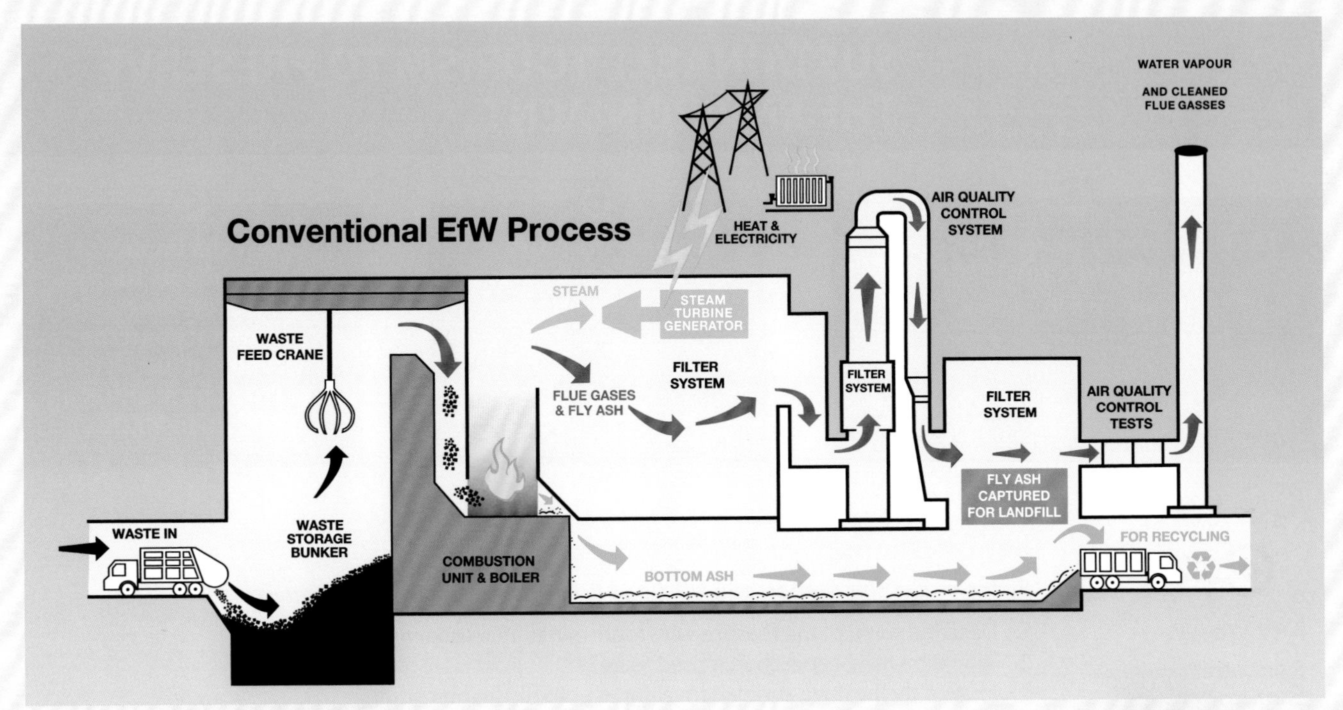

Source: arc21, www.arc21.org.uk

Figure 89

Diagram explaining how an Energy from Waste (EfW) plant works

Energy from Waste (EfW)

Many of the councils and waste management groups such as arc21 argue that one MBT unit in Belfast is not enough to ensure that Northern Ireland will meet EU targets. They express the need for at least one more EfW plant in order to:

- help to prevent a UK energy deficit.
- be a cost effective solution during the economic downturn.
- contribute towards a more local fuel supply.
- be compatible with effective recycling.
- be good for the environment and public health, as it produces no harmful emissions.

EfW generates energy from waste through the use of an incinerator. Although councils and waste management groups state that EfW plants produce no harmful emissions, some people still have concerns that they might be harmful to both the environment and people.

The current proposals for Belfast are to locate both a MBT and an EfW plant on the site of the Hightown Quarry near Mallusk. The Becon Consortium is developing plans for these sites. For up to date information visit the Becon Consortium website www.becon.co.uk.

Test yourself

1. Describe in detail why Belfast needs a waste management strategy. Try to choose at least three different reasons. (3 × 2 = 6 marks)
2. How have Belfast City Council been using different bins to encourage recycling? (3)
3. What impact might a new waste treatment facility have on reducing landfill in Belfast? (3)
4. Evaluate the impact that an EfW plant might have on Belfast. (6)

Check your learning

Now that you have studied Part 3: Managing Waste to Protect Our Environment, return to page 243 and answer the Key Questions for this section.

PART 4: SUSTAINABLE TOURISM TO PRESERVE THE ENVIRONMENT

The reasons why tourism has grown globally since the 1960s
Increased leisure time
Increased disposable income
Cheaper travel
Increased health and wealth of pensioners

(b)

Evaluate the positive and negative impacts of tourism
Cultural
Economic
Environmental

(c)

Assess the impact of one sustainable tourism project on the local community and the environment (using ONE case study from either an LEDC or an MEDC)

Key words

Ecotourism
Green tourism

Key questions

By the end of this section you will be able to answer the following questions:

1. Describe some of the reasons why tourism has increased in recent years.
2. Why is travel cheaper than it used to be?
3. Explain the positive and negative impact of tourism on culture.
4. Evaluate the positive and negative impact of tourism on the environment.

The reasons why tourism has grown globally since the 1960s

People have always liked to travel the world, to explore, to find new places, to enjoy the sun. The concept of travelling to a place for rest and relaxation is not a new one. In fact the Romans pioneered the concept of travel purely for tourism. One definition of tourism is:

> "Tourism is the temporary, short term movement of people to destinations outside the places where they normally live and work, and activities during their stay at these destinations; it includes movement for all purposes, as well as day visits or excursions".
>
> (Source: The Tourism Society, http://www.tourismsociety.org/)

In recent years there has been a big increase in the tourist industry. From the 1960s to the present tourism has expanded rapidly to become a global phenomenon.

Figure 90
Key events in UK tourism history

1949	First overseas flying 'package holiday' is sold by Vladimir Raitz of Horizon.
1965	Lord Thomson creates the Thomson Travel Group.
1969	Development of Tourism Act sets up the England, Wales and Scotland Tourist Boards, plus the British Tourist Authority (BTA).
1970	Introduction of the Boeing 747 'jumbo' jet.
1974	The UK's No 1 tour operator Clarksons goes into liquidation.
1986	The number of UK package holidaymakers tops 10 million for the first time.
1994	The Channel Tunnel is opened.
1998	25.7 million visitors come to the UK, spending more than £12 billion.
2007	A record 32.8 million visitors come to the UK, spending £15.9 billion (and staying an average of 8 nights per visit).

Up until the 1960s tourism was a more of a localised industry. From 1850–1960 there was little opportunity for people to travel much beyond their home country. Those who could afford to take holidays (usually for one week in the summer each year) might travel to local resorts that were a short train journey away. In Northern Ireland, many people from Belfast would travel north to Portrush or south to Newcastle. These resorts were modest, with few amenities or services and visitors would rarely venture beyond their hotels, spending most of their time strolling along specially made paths or eating in the hotel restaurant.

Following the Second World War things changed remarkably and the 1960s were a real turning point. In particular there were four key changes which widened the opportunities for travel and allowed people to travel increasingly further distances.

1. Increased leisure time

In recent years working hours have changed, with the average worker spending 37 hours at work compared to around 50 hours in the 1950s. There has also been an increase in the amount of holiday entitlement and bank holidays. School holidays now shape the times when people can take holidays, making these the peak and most expensive times to travel from the UK.

The way that many households are now organised also encourages people to have more leisure time. Houses are full of 'labour-saving' devices such as dishwashers, washing machines and microwaves, which give people more time to relax or participate in their hobbies.

2. Increased disposable income

Disposable income is the amount of money that households have left over when tax, housing and the basics of life have been paid out. since the 1970s In the UK, people's disposable income has increased since the 1970s. This means that people have more money for short breaks, domestic holidays and overseas travel.

In the last few years many tourism commentators have noted that there has been a 'slow down' in the travel and tourism industry. The national economy in a country can have a big impact on the amount of disposable income that people have. If the economy is under pressure, people will have less disposable income and will therefore go on fewer holidays or not go as far away as previously. If the economy is booming, people should have more disposable income and will therefore go on more holidays and travel further.

3. Cheaper travel

• Increased car ownership

Following the Second World War, individuals were able to enjoy greater freedom within their leisure time and could travel to tourist resorts using the increasing number of private cars. In 1950 there were 2.3 million private cars in the UK, by 1970 this has increased to 11 million and by 1999, there were over 24 million vehicles. As more and more people had a personal motor car, they could use this to explore different domestic and foreign tourism resorts. Camping and caravanning became increasingly popular and people had much greater flexibility as to when and where they could travel.

• Increased use of aeroplanes for travel

Aeroplanes were used extensively during the Second World War. Following the war there was a rapid advance in aircraft technology and the introduction of fast jets (such as the Boeing 707 in 1958). Air travel became an affordable method of travel and throughout the 1960s there was a surge in demand for cheap air flights. The 'package holiday' was born.

• The rise of low cost air travel

Even up to the year 2000, air travel was often the most expensive part of the holiday budget. 'Low cost airlines' such as easyJet and Ryanair made flights more affordable to the traveller. As costs were reduced, people were able to afford a greater number of both short haul and long haul flights.

Figure 91
International travel is more accessible today than ever

4. Increased health and wealth of pensioners

Pensioners today are much wealthier than they used to be and modern medical care ensures that many of them are still active, fit and able to travel the world.

Until recently, many people were also able to take retirement packages early. As life expectancies have risen and with many people retiring early, tourists in the 'grey market' have around 20–25 years to spend time and money travelling the world.

As a result of all of these changes, the UK tourist is much more likely to go overseas for a holiday than ever before. In 2002 over 60 million visits were made abroad (this is three times as many as in 1982). By 2012, the total number of visits abroad by UK residents had slipped back to 56 million, with 43.5 million going to Europe and 3.5 million to North America. Traditionally, UK travellers go to Spain and France more than any other country.

Test yourself

1. Why do you think the UK tourist industry changed so much during the 1960s? (3)
2. Define the term 'tourism'. (2)
3. Explain why transport costs have made international travel more accessible for people. (4)
4. Of all the different reasons for the changes to the tourism industry since the 1960s, which do you think had the biggest impact? (4)

Evaluate the positive and negative impacts of tourism

Since the explosion of the tourist industry from the 1960s, many national governments have realised the potential for tourism to bring large amounts of money into the country. In the UK, tourism accounts for around 9% of the total GDP. This means that tourism has been a very important area for development in recent years. This increasing importance and development of tourism has meant that a number of positive and negative impacts have started to appear.

Cultural

Positive Impacts

Many tourists want to travel to extreme parts of the world to engage with new cultures and find out more about them.

Figure 92

A local woman selling traditional Peruvian clothes and fabrics in Sacred Valley, Peru

- People from rich and poor countries can mix and learn from each other in new ways.
- Often when tourists come to a particular area, this triggers a revitalisation of neglected areas and causes the government to invest money in repairing and maintaining them. New community facilities are often built up and local houses, churches and old buildings sometimes receive attention too.
- An increase in tourism can cause a rebirth of local arts, crafts and customs, and locals sometimes engage in producing traditional goods for the tourist market, bringing cash into the community.
- Community tourism can bring benefits to local communities who can host tourists in their village, managing the scheme communally and sharing the profits.

Negative impacts

When groups of people from different countries and cultures come into contact with each other, there can be problems. Some believe that cultural impacts can be both greater and harder to fix than many of the environmental issues.

- Traditional activities such as farming can be lost as young people seek to move into the more attractive, modern jobs in travel and tourism. People think there is a better chance of a future and more pay in service jobs.
- The behaviour of tourists can distort local customs and tourists might insult the culture, values and religious beliefs of the host community.

- Religious codes and practices might be changed to adapt to the needs of visitors, for example, Sunday opening of facilities.
- Local languages may be lost through under-use.
- Traditional crafts might be lost in favour of mass-produced items which tourists will take home as souvenirs.
- Crime, including public disturbance, drunken behaviour and burglaries may increase. The drunken behaviour of tourists on holiday (for example, in parts of Ibiza and Majorca) can cause noise and vandalism problems for local residents. Crime can be a problem in LEDCs as poorer, local people sometimes try to steal possessions from richer tourists.

Tourism Concern note that Community Tourism should:

1. be run with the involvement and consent of local communities.
2. give a fair share of profits back to the local community.
3. involve communities rather than individuals.
4. be environmentally sustainable.
5. respect traditional culture and social structures.
6. have mechanisms to help communities cope with the impact of western tourists.
7. keep groups small to minimise the cultural and environmental impact.
8. brief tourists before the trip on appropriate behaviour.
9. not make local people perform inappropriate ceremonies.
10. leave communities alone if they do not want tourism (people should have the right to say 'no' to tourism).

Source: Tourism Concern, http://www.tourismconcern.org.uk/

Economic

Positive impacts

Many governments around the world see potential earnings as the most important reason for encouraging tourism. Tourism has many possibilities for job creation and producing income for the people who own and operate the various facilities that are aimed at tourist use.

- ***Employment Creation:*** Increasing tourist numbers mean that more people need to be employed to cater for their needs. The United Nations World Tourism Organisation (UNWTO) estimates that approximately 385 million people around the world are employed in the travel and tourism industry, which is about 11% of the total workforce. Much of the work is direct employment, with jobs in transport, travel companies, tourist attractions, tour operators, hotels and other types of accommodation. Other work indirectly engages with tourism, for example, general product manufacturers (such as shoe and food suppliers) sell to the general population and to tourists.

- ***Foreign Currency:*** Foreign currency coming into a country can help the economy to grow and in LEDCs and NICs it can even help stabilise the economy. Tourism is estimated to generate an output every year of around US$ 7.1 trillion and makes up 12% of the global GDP.

- ***Income for commercial organisations:*** If more people are using services within the country, this will improve its performance and profits. For example, in Northern Ireland in 2011 the number of GB and overseas visitors through the first nine months showed a 6% increase (from 1.09 million visitors in 2010 to 1.17 million visitors in 2011) and the total spend went up 20% (from £242 million to £291 million).

- ***Improvements to infrastructure:*** If a country attempts to improve its transport, utilities and communications to attract tourists, this will also benefit inhabitants. For example, the London 2012 Olympics was a hugely successful event that brought large numbers of visitors to London. One legacy was the improvements to transport and to the accommodation and services available in east London.

In 2012 there was a renewed focus on marketing Northern Ireland and increasing visitor numbers. The Northern Ireland Tourist Board hoped that 2012 would "be the tipping point for Northern Ireland and a real chance for change perceptions".

The extended impact of the London Olympics in July/August, the Irish Open at Royal Portrush, the opening of Titanic Belfast and the opening of the new Giant's Causeway Visitor centre have all combined to build into a special advertising campaign called 'NI2012 – Our Time, Our Place'. The Northern Ireland Department of Enterprise, Trade and Investment Tourism Strategy 2010–2020 specifically aims to increase visitor numbers and revenue within the Northern Ireland economy, and to increase the number of additional jobs that will be created by any increase in this sector.

The 2012 campaign aimed to increase both visitor numbers and the value of the tourism industry within the Northern Ireland economy, "giving an incredible opportunity to change the world's perception of Northern Ireland." The hope is that visitor numbers and revenues would continue to build towards 2020.

Opportunity	2012	2013	2014	2015	Total
Additional Visitors	150,000	333,000	200,000	150,000	833,000
Additional Revenue (GBP)	24 million	52 million	34 million	30 million	£140 million
Created Jobs	612	1,326	867	765	3,570

Figures from 'Why NI 2012?', Northern Ireland Tourist Board, www.nitb.com

Negative impacts

- ***Services:*** The services in many tourist resorts are designed and priced to cater for the tourist, and locals may not be able to afford the inflated prices. Also, the emphasis will be on supplying the tourist and local people might have less access to water services, entertainment and transport.
- ***Extra charges:*** Often any local infrastructure changes and development programmes have to be paid for by the local people, through increased rates or tax bills.
- ***Land and house prices:*** Some tourists buy second homes in their favourite holiday destinations, increasing the demand and price for housing. Developers may also buy up land for holiday home sites. This can make it difficult for first time buyers to afford houses in their local area.
- ***Character changes:*** Rural areas, such as the Lake District, may find that they lose the traditional characteristics that they once had as increasing numbers of tourists come to stay in the area. Traditional services and shops could be replaced with services that cater for the needs of the tourist rather than the local. Retail outlets might cater for walkers and cafes and restaurants may supply food for the tourist. Gift shops might predominate and local people could find it increasingly difficult to get their daily requirements in their village and may have to travel further to use more modern shops and supermarkets.
- ***Loss of traditional employment:*** As tourism brings new job opportunities, more people will leave traditional employment to work in the tertiary/service sector.
- ***Problems of keeping income:*** Often much of the income generated by the tourist does not stay in the area or benefit it. For example if a guest stayed in an international hotel chain, some of the money earned would make its way to the hotel chain's headquarters. This is known as leakage.

Environmental

Positive impacts

Although tourism is often blamed for the environment damage it can cause, sometimes tourism can be a positive force for environmental change:

- ***Sustainable tourism:*** Some organisations are keen to involve tourists in sustainable campaigns, where visitors can actively participate in helping the area to be maintained and protected. Pressure groups such as Tourism Concern campaign for tourists to be more sustainable and ethical as they visit places.
- ***Awareness:*** Tourism to remote places such as the tropical rainforests of South America and to the savanna grasslands of East Africa can help expose the environmental problems and activities which are taking place here.
- ***Improvements:*** The development of tourism can bring much needed improvement to derelict areas and tidy up and clean waterways.

Figure 93
Tourism can damage fragile environments

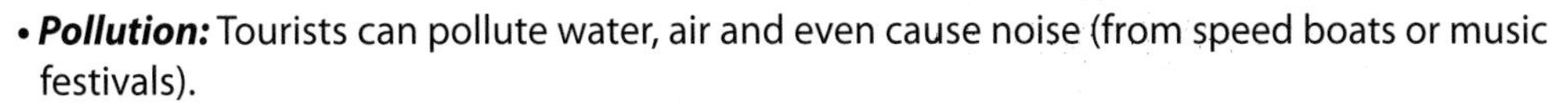

Negative impacts

Unfortunately, the environmental impacts that tourists bring to an area often outweigh the positives.

- ***Damaging fragile environments:*** On a global scale, environmental issues such as global warming, acid rain and large scale deforestation have had a large impact on very sensitive environments. Coral reefs have been dangerously exposed and fragile mountain environments have been destroyed with litter.
- ***Soil erosion:*** Erosion of soil, rock and vegetation can be caused by walkers, horse riders, mountain bikers and (motor) bikers.
- ***Congestion and overcrowding:*** Too many people visiting areas such as the Peak District National Park (for example on Bank Holidays) at any one time causes overcrowding and the traffic leads to an increase in carbon emissions.
- ***Pollution:*** Tourists can pollute water, air and even cause noise (from speed boats or music festivals).
- ***Loss of habitats for animals and plants:*** As more land is used for tourist accommodation, there are fewer places for animals and plants to grow.
- ***Pressure on water supply:*** As tourism increase, so does the amount of water used, from drinking water and toilet flushing to irrigating golf courses.

Figure 94
The impact tourism can have on an area

Impact	Positive	Negative
Economic	*Jobs created* Hotel managers, restaurant waiters, souvenir shop assistants and tour guides. *Money brought into the country* Tourists bring valuable cash into the country, which is spent in the local economy.	*Jobs are seasonal* At off peak times (in the winter) there are fewer jobs for local people. Also, in economic recessions people do not travel as far and visitor numbers will fall which decreases employment. *Tour companies benefit more than locals* The big international travel companies such as Thomas Cook and Kuoni make more money than the local people and guides.
Environmental	*Wildlife is protected and National Parks are created* Tourists are attracted to places of natural beauty such as Indian Ocean beaches and African game parks. Local councils will ensure that these are protected so that tourists keep coming. *Improved historical buildings* Local councils will maintain old buildings as tourist attractions. *Improved services and facilities* Councils will ensure that the areas used by tourists have good roads, airports and leisure facilities which benefit local people too.	*Wildlife is disturbed* The large numbers of tourists will disturb and damage natural habitats. For example, too many jeeps on safari will cause soil erosion. *Loss of land and water resources* The best land will be used for tourists. Water will be used for swimming pools and for tourist comforts whilst local farmers may struggle to get enough water for their crops and animals.
Cultural	*Tourists learn about different cultures* Tourists and local people can learn from each other.	*Crime* If many wealthy tourists visit an area, some locals might be tempted to steal cameras or money from them. *Insulting the culture* Tourists may not understand local culture and their behaviour could insult local values or beliefs. Local people can feel undervalued as groups of tourists come to see how the locals live. This can create resentment.

Test yourself

1. Describe and explain two positive cultural impacts of tourism on an area. (4)
2. Describe and explain two negative cultural impacts of tourism on an area. (4)
3. Describe some of the most important recommendations from 'Tourism Concern' to ensure that tourism involves the local communities. (5)
4. Describe some of the economic benefits that tourism can bring to a place. (3)
5. Why do some local people feel that the economic cost is just too great to pay for tourism? (3)
6. Are there any environmental positives for tourism? (3)

Assess the impact of one sustainable tourism project on the local community and the environment

Ecotourism is often described as tourism that:

- is 'environmentally sound'.
- protects natural environments, wildlife and resources.
- is socially appropriate and respectful of local culture.
- does not damage local communities.
- provides economic benefits for local people.
- leads to sustainable tourism.

The UN World Tourism Organisation defines sustainable tourism as:

"Tourism that takes full account of its current and future economic, social and environmental impacts addressing the needs of visitors, the industry, the environment and host communities"

Figure 95

Location of Mara Intrepids Camp, Maasai Mara Nature Reserve, Kenya

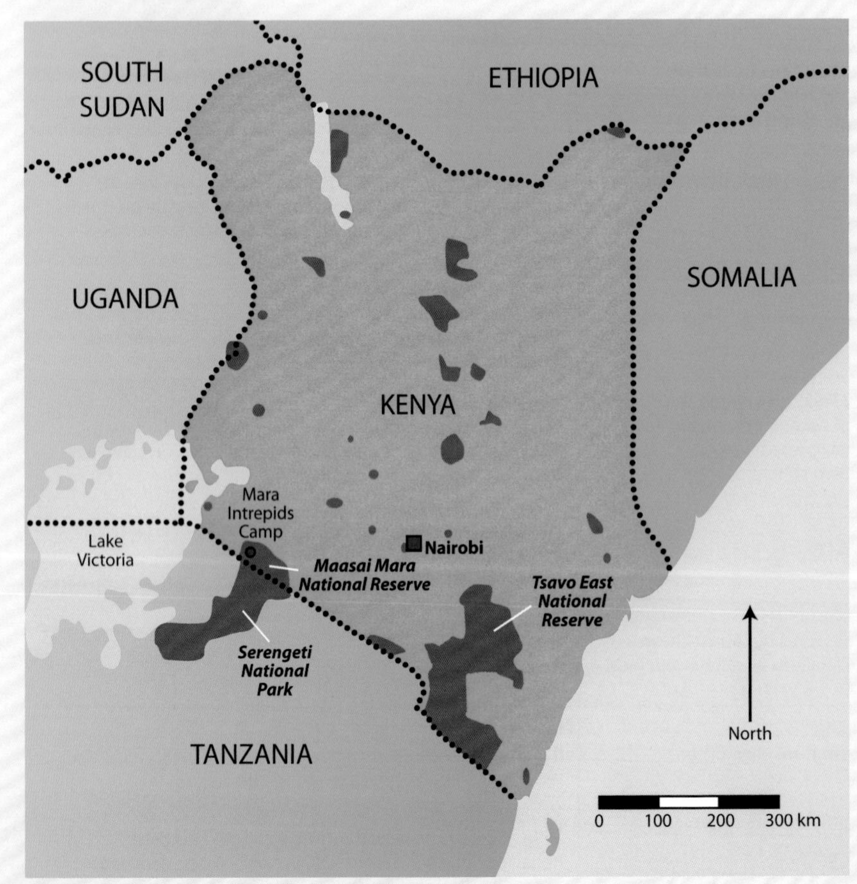

Case Study: Kenya (a LEDC case study)

Kenya was one of the first African countries to embrace mass tourism. Its natural features, wildlife reserves, beautiful coasts, coral reefs and white sandy beaches on the Indian Ocean are the main attractions.

Most of the tourists come from Europe and stay in coastal resorts but also spend some time on safari to an inland wildlife reserve. In 1997 Kenya earned over $450 million from tourism and it continued to grow. By 2010 the highest number of arrivals was recorded at 1.1 million visitors. This was a 15% growth on 2009 and brought in around 73.68 billion Kenyan Shillings (which is around $887 million).

The best known of the wildlife reserves is the Maasai Mara National Reserve, which borders the Serengeti National Park in Tanzania. In 1997 there were around 133,000 visitors to the reserve. The numbers peaked in 2006 at 316,500 but in 2009 went down to 158,000.

Figure 96
Safari drives allow tourists to get close to the animals

One of Kenya's most popular hotel chains is Heritage Hotels Ltd. They have hotels in various national parks and locations across the country and win International Tourism awards on a regular basis.

The Mara Intrepids Camp is found on the savanna grasslands of the Maasai Mara National Reserve, with the river Talek flowing past. The hotel website describes it as follows:

"The Mara Intrepids is a short drive from the Mara River, where more than a million wildebeest and zebra make their perilous migration crossing every July and August. For the remainder of the year, the camp offers some of the world's finest game viewing, with large populations of plains games, elephants, rhinos, buffalos and all of the big cats. In 2004, some 75% of the footage of the BBC's famous 'Big Cat Diary' was shot in the vicinity of the camp. Mara Intrepids is 298 km and a six-hour drive from Nairobi, while our all-weather airstrip enables guests to be at the camp within 45 minutes of leaving the capital".

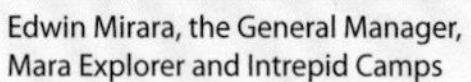

Edwin Mirara, the General Manager, Mara Explorer and Intrepid Camps

Figure 97
Sustainable tourism can also be luxury tourism

The camp itself is made up of 30 luxury tents which are spread across a riverside. Guests can choose from a Luxury tent (with large four poster bed, reproduction furniture, ensuite bathroom, hot shower and flush toilets) or a Luxury Family tent (with one queen bed and two twin beds, both with ensuite, hot shower and flush toilets). Both tents include airstrip transfers, three guided safari drives into the game reserve each day, a morning bush walk, all meals and an evening cultural talk.

What makes this 'Sustainable Tourism'?

The Mara Intrepids Camp has received the second highest Ecotourism award in Kenya – a silver award.

Social and economic impacts

- ***Jobs:*** Many of the staff at the camp are local Maasai people. This creates much needed employment and a good income.
- ***Staff training and opportunities:*** The Camp are very aware of the risks of injury for game drivers and walking safari guides, so professional training and first aid training is provided.
- ***Cultural visits:*** The camp organises a series of cultural visits for tourists, which helps to build a rapport with the local communities. They also organise regular cultural talks.
- ***Community Development Projects:*** The Mara Intrepids Camp supports local community projects.
- ***Community Action:*** In June 2010 Heritage Hotels teamed up with MEAK (Medical and Educational Aid to Kenya) to set up a local 'Eye Mission to the Mara'. Many Maasai people suffer from difficult eye infections that can lead to blindness. 193 patients were diagnosed with minor conditions and 11 had cataract operations. Many were treated for conjunctival infections and 300 bottles of eye drops and 11 pairs of reading glasses were prescribed.
- ***Supports education:*** The facility (and its guests) supports a local primary school in Talek. It provides reading and writing materials, and assists students with registering for their Kenya Certificate of Primary Education (KCPE) exams.

Environmental Impact

- ***Wildlife conservation:*** Guests are educated about wildlife conservation and can obtain literature about how wildlife is protected. Guides are trained not to disturb the animals and jeep numbers are kept to a minimum to help reduce noise pollution and stop the erosion of the soil by tracks.
- ***Water:*** Grey water (from washing) and black water (from toilets and sewage) is filtered before being released. Laundry is done manually to minimise water use and all water is metered.
- ***Electricity:*** The majority of electricity is from a generator but char dust (compacted coffee husks) is used to heat water. All appliances and lights in the camp are low energy.
- ***Waste:*** All kitchen waste is composted and is used in an organic vegetable garden. Other waste is recycled (into separate bins) and any remainder is incinerated. Any waste which cannot be managed on-site, such as plastic, glasses and cartons, are taken back to Nairobi for proper management.
- ***Food:*** The camp has an eco-garden where it grows its own vegetables.
- ***Buildings:*** Many of the camp buildings and structures are temporary and are designed to blend into the environment. Path lights are covered with local materials and painted green.

Test yourself

1. Why is tourism so important to countries such as Kenya? (3)
2. What are the main attractions to places such as the Mara Intrepids camp in Maasai Mara? (3)
3. Describe in detail how the camp tries to be as sustainable as possible. (6)
4. Evaluate the sustainability of the camp. Can any tourist resort claim to be sustainable? (6)

Organise your revision

Draw a mind map to summarise Unit 2C 'Managing our Resources'.

Check your learning

Now that you have studied Part 4: Sustainable Tourism to Preserve the Environment, return to page 250 and answer the Key Questions for this section.

EXAM PRACTICE QUESTIONS

Some of these questions are from previous CCEA GCSE examination papers and others have been written in the same format to give you practice at answering 'exam style' questions.

Try to answer the questions with as much detail as possible. Also consider the number of marks that each question receives, as this will give you a good indication of the amount of depth that your answer needs.

Resource A

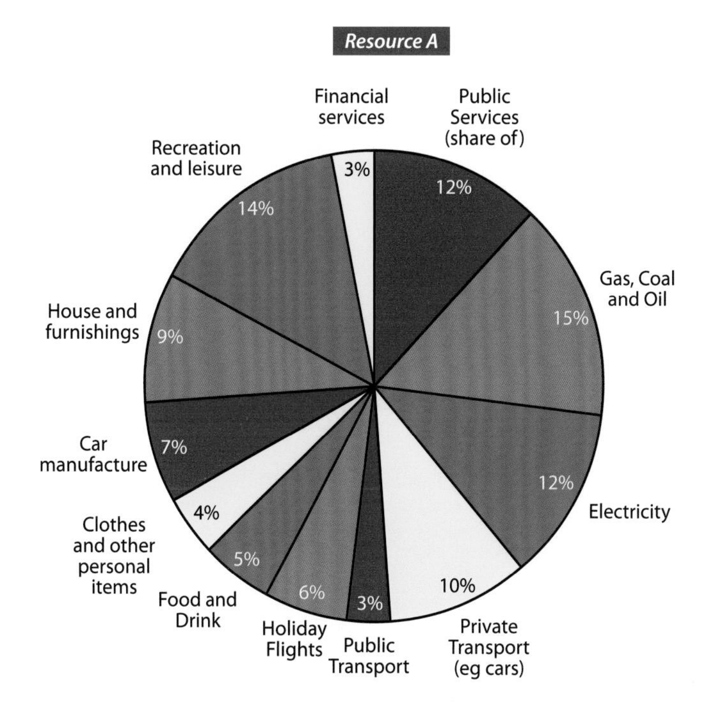

Resource B

Number of cars

Home heating systems

Electricity production

Resource C

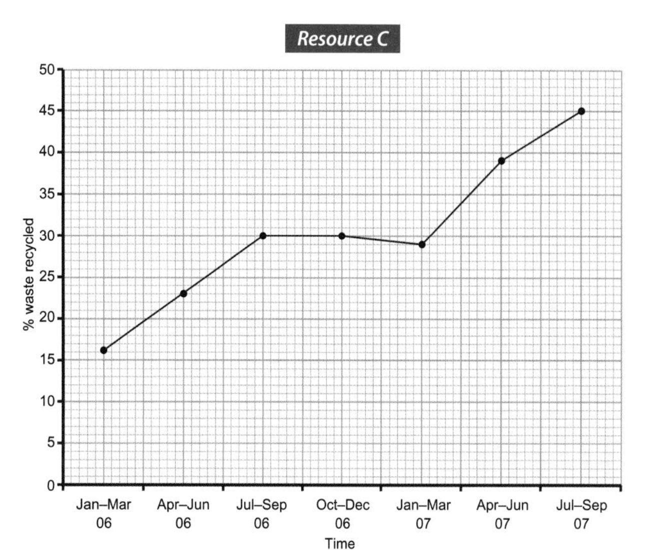

Resource D

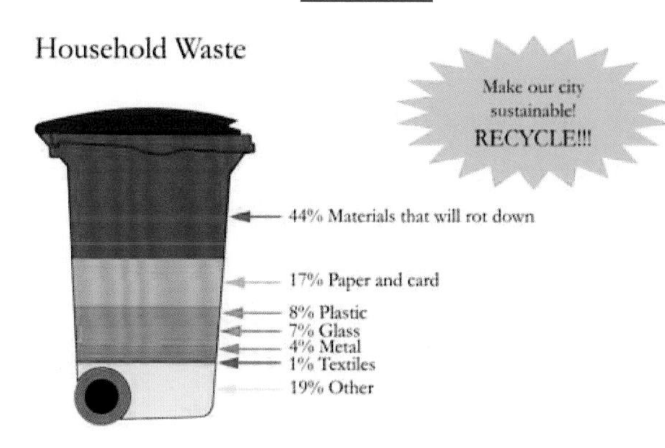

Resource E

COUNTRY	Number of tourists (millions)
FRANCE	80
UNITED STATES	60
SPAIN	55
CHINA	53
ITALY	42

Resource F

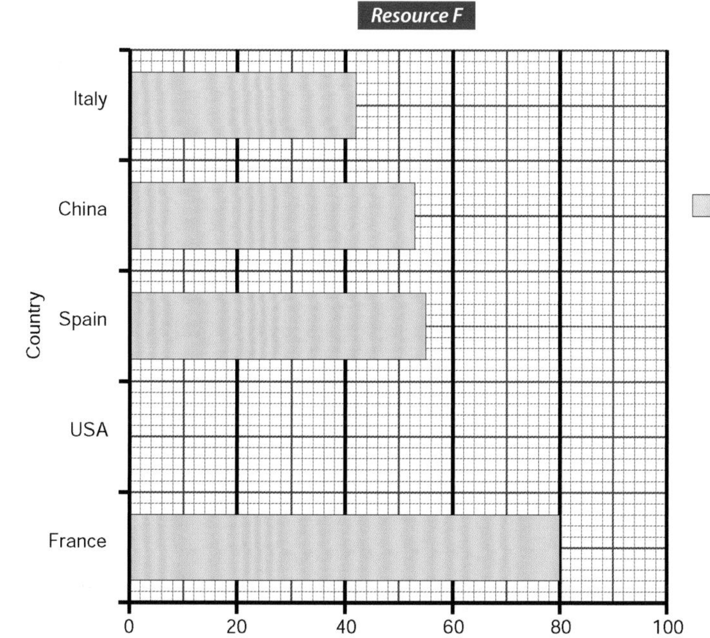

Resource G

CULTURAL IMPACT	STATEMENT	ECONOMIC IMPACT
	Jobs are created in the area. ⟶	
	Tourism helps us see how other people live.	
	Tourists may disrespect local customs.	
	Money gets invested into an area	

Foundation Tier

1. Study Resource A which shows the breakdown of a typical person's carbon footprint.

   a. Which of the following graphical techniques is used to display this data? (1)

   - Scatter graph
   - Pie chart
   - Bar chart

   b. State the percentage that public transport contributes to a person's carbon footprint. (1)

   c. Define the term carbon footprint. (2)

2. Study Resource B which shows images relating to the use of oil.

   a. Explain why the use of oil varies across the world. Use one example from Resource B in your answer. (3)

   b. Which of the following types of resource is oil an example of? (1)

   - Non-renewable
   - Renewable
   - Reusable

   c. State one renewable energy source in a MEDC. Explain one problem that it caused.

   - Name of MEDC (1)
   - Renewable energy source (1)
   - Problem (3)

3. Study the graph in Resource C which shows the recycling rates for a Council District (Omagh) in Northern Ireland.

   a. What is the correct answer from the statements below: (3)

   - The graph is an example of a line/bar chart
   - Recycling rates have increased/decreased over time
   - The highest % of waste recycled is 56%/ 45%

   b. From the list below identify the two non-sustainable forms of waste management that this council could use. (2)

   - Composting
   - Incineration
   - Landfill
   - Reuse products

4. Study Resource D which shows average household waste

   a. Name two waste items that could be recycled. (2)

   b. State whether the following sentences are true or false. (3)

   - A sustainable city would recycle most of its waste.
   - More glass than plastic is dumped by households.
   - In a sustainable city most people travel to work in cars.

5. Study Resource E which shows the countries most visited by tourists in 2009. Make a copy of Resource F.

   a. Using Resource E to help you, complete the number of tourists who visit the USA on your copy of Resource F. (2)

   b. Study the Resource F and choose the correct word (from the bold) to complete each sentence: (3)

   - Spain is the second/third most popular destination in the world.
   - France is the most popular destination as it has few/many tourist attractions.
   - The three European countries receive 177/187 million tourists each year.

   c. Explain one negative environmental impact of tourism. You should refer to a place in your answer. (3)

6. Study Resource G which shows statements about the impacts tourism can have on an area.

   a. Make a copy of Resource G and draw an arrow to match up each statement to its impact. (3)

   b. Explain why tourism has grown rapidly since 1950. Only one reason is needed. (3)

Questions and resources from CCEA GCSE Geography Papers
June 2010 (Q2, 3, 6; Resources B, C, G), June 2011 (Q1, 5; Resource A, E, F), Specimen 2010 (Q4; Resource D), ©CCEA

Higher Tier

1. Study Resource A which shows the breakdown of a typical person's carbon footprint.

   a. State the name of the graphical technique used to display this data. (1)

   b. State the percentage to which public and private transport contributes to a person's carbon footprint on the diagram. (1)

   c. State the meaning of the term carbon footprint. (2)

2. a. State the meaning of the term carbon footprint. (2)

   b. Explain one way in which carbon footprints can be reduced. (3)

3. Some human activities have to be managed to protect our environment.

   For a named MEDC city, describe and explain how one traffic control measure contributes to the sustainable development of the city. (6)

4. Study Resource C which shows recycling rates for a Council District (Omagh) in Northern Ireland.

   a. Describe the trend in recycling rates shown in Resource C. (3)

   b. State one other sustainable (not recycling) and one non-sustainable approach to waste management that this council could use. (2)

5. Using one case study from either a LEDC or a MEDC explain how a sustainable tourism project benefits both the environment and the local community. (5)

Questions and resources from CCEA GCSE Geography Papers
June 2010 (Q2, 4; Resource C), June 2011 (Q1, 5; Resource A), Specimen 2010 (Q3), ©CCEA

Unit 3:

Fieldwork Report (Controlled Assessment)

3 Unit 3 Fieldwork Report (Controlled Assessment)

The CCEA GCSE specification for Geography states that "candidates must complete a report of no more than 2,000 words based on primary data collection". Students have to work on a piece of fieldwork which they write up into a geographical report. If they show knowledge and understanding they can walk into the exam with up to 25% of the GCSE Geography grade already gained.

It is impossible in this short chapter to go into sufficient depth to support every different type of coursework that could be assessed. Instead, this section takes a look at some of the basics of each of the different sections involved in the presentation of the final report.

This section is divided up into five parts:

3A What is involved in the Controlled Assessment?

3B Research and data collection stage

3C Analysis and Evaluation of findings stage

3D Tips and guidance on writing up your fieldwork report

3E Focus on Controlled Assessment for a river

3A: WHAT IS INVOLVED IN THE CONTROLLED ASSESSMENT?

For many years, GCSE Geography has included the compulsory completion of a piece of coursework that involves some primary data collection out in the field, which is written up into a fieldwork report.

Figure 1
Learning to read the landscape features overlooking the Giant's Causeway

Figure 2
Drawing a field sketch on Ballintoy beach

Centres chose one task from six options provided by CCEA. The tasks are available for one year only and can be found on the CCEA website www.ccea.org.uk/geography.

CCEA states that Controlled Assessment tests the ability of the students' knowledge and skills to:

- identify, analyse and evaluate relevant geographical questions and issues.
- investigate using geographical skills, including enquiry skills.
- extract and interpret information from a range of different sources, including field observations, maps (including an OS map of the study area obtained from a digital source) drawings, photographs (ground, aerial and satellite imagery), diagrams and tables.
- evaluate methods of collecting data and suggest improvements to the investigation.

Your role is to complete ONE Controlled Assessment task. Although the fieldwork can be carried out with other people in your geography class, you still have to complete your own report, which needs to be all of your own work. Any time that you use another source for maps, information or quotes, you must make sure that you acknowledge and reference the material carefully.

Once you have completed the write up of your fieldwork report, your geography teacher will mark your work and a selection of completed reports are then moderated by the exam board.

All of the information presented in tables with blue headings is information that teachers use to help to mark the final report. This information is invaluable when deciding what to put into your write up.

The following are some examples of the titles used in recent years:

YEAR OF SUBMISSION	TASK	PLACE WITHIN THE SPECIFICATION
May 2015	1. A study of population change at the local level.	Unit 2, Theme A, Part 1
	2. An investigation of changing characteristics along a chosen stretch of river.	Unit 1, Theme A, Part 2
	3. An investigation of attitudes towards sustainable waste management.	Unit 2, Theme C, Part 3
	4. A study of coastal erosion.	Unit 1, Theme A, Part 3 and 5
	5. An investigation of the location of land-use zones along a transect in an urban area.	Unit 2, Theme A, Part 2 and 3
	6. An investigation of attitudes towards the use of alternative sources of energy, eg wind power/solar power.	Unit 2, Theme C, Part 2 Unit 1, Theme B, Part 3
May 2014	1. An investigation of change within a CBD (Central Business District).	Unit 2, Theme A, Part 2
	2. An investigation of attitudes towards sustainable waste management.	Unit 2, Theme C, Part 3
	3. An investigation of migration at a local level.	Unit 2, Theme A, Part 1
	4. An investigation of how local weather conditions compare to national weather.	Unit 1, Theme B, Part 1
	5. A study of how river features change downstream.	Unit 1, Theme A, Part 2
	6. An investigation of sustainable coastal management.	Unit 1, Theme A. Part 5
May 2013	1. An investigation of river features/ characteristics along the long profile of a chosen river.	Unit 1, Theme A, Part 2
	2. An investigation of the weather conditions and change associated with the passage of a frontal depression at the local level.	Unit 1, Theme B, Part 2
	3. An investigation of the attitudes towards and use of alternative sources of energy.	Unit 2, Theme C, Part 2 Unit 1, Theme B, Part 3
	4. An investigation of the sphere of influence of a chosen settlement.	Unit 2, Theme A, Part 2
	5. A study of the attitudes towards aid.	Unit 2, Theme B, Part 3
	6. An investigation of tourism at the local level.	Unit 2, Theme C, Part 4

Once your teacher has selected which title you are going to complete, you can start to plan your fieldwork and organise how you are going to collect your primary data.

3B: RESEARCH AND DATA COLLECTION STAGE

RESEARCH AND DATA COLLECTION	
	Maximum mark (out of 70)
(a) Introduction and methodology	10
(b) Data processing and presentation	10

At each stage of the Controlled Assessment procedure, teachers have to enforce different levels of control. The level of control for the Research and Data Collection stage is limited. This means that students are allowed to research and prepare for this section without the direct supervision of the teacher. You can use this information to research and plan the early stages of your report.

Each stage also has clearly defined time limits. Students must not exceed a time allocation of 12 hours to prepare and then write up their Introduction and Methodology, and Data Processing and Presentation sections.

Figure 3
Conducting a sand dune survey at Magilligan Point, Co Londonderry

(A) Introduction and Methodology (10 marks)

Assessment Criteria

(A) INTRODUCTION AND METHODOLOGY	
Candidates should be assessed on their ability to complete a written report: • The overall aim of the investigation and its theoretical context based on the learning outcomes stated in the specification. • Spatial context, to include: *a:* one regional map showing the location of the investigation in context. *b:* an OS map of the location of the data collection using GIS. • A list of objectives or hypotheses to be tested. • Methods of data collection described.	**Mark bands**
	Band 1 (1–3 marks) The candidate provides a simple statement of the aim and hypotheses of the investigation. The candidate provides a brief, general description of the work in the field. Spatial context is limited. Basic theoretical context present.
	Band 2 (4–7 marks) The candidate outlines the aim and hypotheses of the investigation and provides a spatial context for the study. The candidate describes the methods they use in the field with a fair degree of accuracy. Spatial context is evident using GIS. Adequate theoretical context present.
	Band 3 (8–10 marks) The candidate states the aim and hypotheses clearly and concisely. The candidate clearly provides the spatial context of the investigation. The candidate describes the methodology used precisely. The candidate demonstrates a clear understanding of the methodology. Spatial context is detailed with a regional and local map present using GIS. Detailed theoretical context present.

What should be included in this section?

The Introduction is a very important section as it allows you to set the scene for the written report. Detail within each section of the Introduction is essential. You should include the following:

1. Title Page

Although you do not get any marks for including an attractive title page for your coursework, it is good practice to have a nice title page to start your write up. You should aim to include:

- The title for the Controlled Assessment task.
- Your name and your exam candidate number.
- A relevant photograph/picture to illustrate the task.

2. Introduction to the coursework question (including hypotheses)

You do not need to write a vast amount of information in this section, but you should aim to set the scene for the coursework and explain a little about what you are intending to address in the report. You should aim to include:

- An explanation of why you chose your particular task title over the other options.
- An explanation of any theory/geographical context relating to your study. What patterns and trends has your geographical background led you to expect? Try to quote specific theories, ideas and expectations from textbooks and other sources. You might even use some diagrams to help to explain the theory.
- How you divided your study into a set of further questions (called hypotheses) to answer the main title question.
- List your two or three hypotheses.

3. Introduction to the study area/spatial context

You need to spend a little time introducing the reader to the area under study. You should aim to include:

- As many relevant facts about the area as possible to explain why it is a good place to study and more suitable than another location.
- A description, in words, of the precise location and situation of all sites used. Use six figure grid references where appropriate.

4. Maps and GIS

Your introduction should use a variety of maps, at different scales, which use aspects of GIS to demonstrate features of the location under study. You should aim to include:

- At least one map showing the location on a regional scale (Northern Ireland) with clear annotations.
- At least one map showing the location on a local scale (the precise location) with clear and detailed annotations.
- At least one map that was developed using GIS software. If you are analysing aspects of human geography you might find this a little easier to do than students completing physical geography coursework.
- You could also use some annotated photographs to show different aspects of the location under study.
- Do not forget that if you get your maps or images from books or the Internet you must reference the source of your information carefully.
- Any map presented should ensure that the usual conventions for basic cartography are followed and have:
  - *a:* a detailed title
  - *b*: clear and appropriate annotation
  - *c*: a scale and North point.

5. Methodology/Methods

The final section of your Introduction should include a detailed, precise statement which helps to describe the different methods that you used within your data collection. The length and detail for this section will depend on the particular task that you are addressing.

Imagine that the person who is reading your work knows nothing about geography and that they will be using your instructions to try and replicate the fieldwork that you had to carry out. You need to ensure that you explain the equipment that is required and how it should be used out in the field. You should also identify any health and safety issues, noting how they are to be overcome. You should aim to include:

- An accurate description of how and why you collected this data.
- A list of the equipment used to take the results.
- A description of how each measurement/observation was taken. You should explain clearly how any equipment was used, the order of the measurements and why each was taken.
- Some information about any health and safety measures that were taken to minimise risks in the field.
- Some diagrams of the equipment used or annotated photographs of you taking the measurements/observations.

Figure 4
An example of using an annotated photograph for rivers fieldwork

(B) Data Presentation (10 marks)

Assessment Criteria

(B) DATA PRESENTATION	
Candidates should be assessed on their ability to complete a written report: • Appropriate tables, graphs, annotated maps, photographs, overlays, etc.	**Mark bands**
	Band 1 (1–3 marks) The candidate makes some effort to present the data collected using simple, graphical and mapping techniques.
	Band 2 (4–7 marks) The candidate demonstrates some ability to sort the data collected and identify it for inclusion in each presentation technique. The candidate's techniques of presentation are generally appropriate and some use is made of ICT.
	Band 3 (8–10 marks) The candidate demonstrates the ability to sort the data collected and to use presentation techniques to illustrate relevant sets of data. The candidate demonstrates a high level of competence using ICT in the construction of the appropriate presentation techniques.

What should be included in this section?

The Data Presentation is one of the most important sections in the whole fieldwork report. It is essential that you get this section right and present enough information so that further writing in the Analysis and Interpretation sections can be completed. However, it is also very easy to present too much information in your report. Anything you include should be 'appropriate' – you do not need to have three different ways of presenting the same piece of information. Therefore, to ensure that you use as many appropriate graphical techniques as possible, you need to think carefully about how to organise and plan your presentation of results.

The best way to do this is to take each hypothesis separately and present the information which is relevant to it. Some hypotheses will encourage more information and data than others so do not expect each hypothesis to have the same amount of information.

You should remember the following:

1. You must make sure that some of your presentation is done using ICT.
2. The usual conventions for the production of graphs, maps and diagrams should be followed carefully. Each should have a title, key, scale, North point, label on axes and annotation as appropriate.
3. It is also good practice to give each graph, map or resource used a figure number for easy reference in the later stages.
4. Keep your graphs in a logical order, with all graphs relating to Hypothesis 1 first, followed by those for Hypothesis 2, and so on.
5. Although the method of presentation will change depending on the task titles each year, the following table gives some examples of how you can present your data effectively.

Maps	There are a variety of maps which can be used. Location maps, sketch maps, transect maps and distribution maps can all be used to illustrate and adapt collected information. OS maps should have significant input such as annotation or overlays. All maps should be relevant to the study area.
Tables and Diagrams	Results summary tables, annotated sketches, transect diagrams, systems diagrams, flow charts, cross sections and kite diagrams can all be used to illustrate key points.
Graphs	Line, Bar, Histograms, Pie, Divided/Proportional Bar, Scatter, Triangular, Pictogram, Gain/ Loss Bar Chart, Flow lines.
Photos	Photographs should be titled and annotated. Overlays may also be used.

6. All material presented must support the argument in the text and be used in the Analysis and Interpretation sections which follow.

3C: ANALYSIS AND EVALUATION OF FINDINGS STAGE

ANALYSIS AND EVALUATION OF FINDINGS	
	Maximum mark (out of 70)
(c) Data Analysis	10
(d) Interpretation	15
(e) Conclusions	15
(f) Evaluation	10

In the final stages of the Controlled Assessment, the level of control is **high**. This means that all work must be completed under the direct supervision of a teacher or exam invigilator. This is the most important section of the Controlled Assessment and will be completed under examination conditions.

You have a maximum of 6 hours to complete this stage. This does not have to be done in one sitting and often teachers will plan a schedule for completing each stage of the report. The security of your work is essential during this stage and all notes will be collected, stored securely and re-distributed as necessary until the completion of the task. You must work independently and are not allowed to communicate with other students or bring new materials into the classroom once the phase has started. Similarly, teachers are not permitted to offer advice in the analysis and evaluation of your findings or during the writing of the final report.

In order to work through the final sections of your report you can either:

Method 1: Use the individual sections as the chapters for your final report, with one chapter for Data Analysis, one for Data Interpretation, and one for Conclusions and Evaluation. You can also look at each hypothesis within each chapter in order. This is the methodology which is advised by the exam board.

or

Method 2: Look at each of your hypotheses independently, with one chapter for Hypothesis 1 (where you look at the Data Analysis, Data Interpretation and Conclusion for this hypothesis), one chapter for Hypothesis 2, and one chapter for Hypothesis 3. You could finish with a final round up Conclusion and your Evaluation.

There are merits in both methods and you need to seek advice from your teacher as to how you should organise these sections. Different coursework titles which might have overlapping data used in the hypothesis might favour Method 1. However, coursework titles where there is very discrete and separate data within each hypothesis might find Method 2 more useful.

(C) Data Analysis (10 marks)

Assessment Criteria

(C) DATA ANALYSIS	
Candidates should be assessed on their ability to complete a written report: • A description of the patterns and/or relationships in the data presented.	**Mark bands**
	Band 1 (1–3 marks) The candidate provides a simple, straightforward description of the patterns apparent in the presented data.
	Band 2 (4–7 marks) The candidate provides a reasonable description of the patterns and relationships apparent in the presented data.
	Band 3 (8–10 marks) The candidate demonstrates the ability to describe clearly and concisely the patterns and relationships apparent in the presented data.

What should be included in this section?

Most students find this the most difficult section of the Fieldwork Report. A little preparation and research before you go into the high control session can make a big difference to your marks.

Simple errors can be made if the information in the Data Presentation section is not analysed fully. Data Analysis is when you **describe** your results. You do not explain them here. You should look at each resource in turn and write down what it shows. Aim to write a paragraph about **every** graph or resource by carrying out the following:

1. Describe what data was used for the graph – where did the information come from?
2. Describe what your graph shows using figures – the highs, lows and any averages that there might be.
3. Describe any patterns/trends you notice in your graph.
4. Describe any relationships you notice between graphs.
5. Describe any unusual results you notice.
6. Do not forget to comment on any patterns or relationships that you notice inside or between the graphs.

Any graph or figure which is not described in your report should be removed from the Data Presentation section as it is not useful or relevant to your work. Remember to ensure that you talk about each resource in a logical order. Group your discussion so that you describe Hypothesis 1 first, followed by Hypothesis 2 and finally Hypothesis 3.

(D) Data Interpretation (including QWC) (15 marks)

Assessment Criteria

(D) DATA INTERPRETATION	
Candidates should be assessed on their ability to complete a written report: • Concise and valid explanations of the information collected in the fieldwork, supported by evidence and theory.	**Mark bands**
	Band 1 (1–5 marks) The candidate demonstrates their application of knowledge and understanding through simple, straightforward explanations of the patterns they have identified. There is limited use of evidence to support these explanations, and there are limited links to theory. Candidates present some relevant information in a form and using a style of writing which suits its purpose. The text is reasonably legible. Spelling, punctuation and the rules of grammar are used with some accuracy so that meaning is reasonably clear. A limited range of specialist terms is used appropriately.
	Band 2 (6–10 marks) The candidate demonstrates their application of knowledge and understanding through reasonable explanations of the patterns they have identified. The candidate supports these explanations through the use of evidence. There are reasonable links to relevant theory. Candidates present relevant information in a form and using a style of writing which suits its purpose. The text is legible. Spelling, punctuation and the rules of grammar are used with considerable accuracy so that meaning is clear. A good range of specialist terms is used appropriately.
	Band 3 (11–15 marks) The candidate demonstrates their application of knowledge and understanding through concise and valid explanations. They support these explanations by evidence and link to theory where appropriate. Candidates present and organise effectively relevant information in a form and using a style of writing which suits its purpose. The text is fluent and legible. Spelling, punctuation and the rules of grammar are used with almost faultless accuracy so that meaning is clear. A wide range of specialist terms is used skillfully and with precision.

What should be included in this section?

Data Interpretation is when you **explain** the reasons behind your results. In this case you need to explain the results that you have just Presented and Analysed and the reasons why they are that way. You should:

1. Restate the hypothesis that you are investigating.
2. Explain the results that you have found to support, prove or disprove the hypothesis.
3. With reference to your Introduction and Methodology section, compare your results with geographical theory and information. You can use photos, data and details from the Data Presentation section to support your answer.
4. Explain any results that are unusual or unexpected and try to put forward an explanation for these anomalies.

(E) Conclusions (15 marks)

Assessment Criteria

(E) CONCLUSIONS	
Candidates should be assessed on their ability to complete a written report: • Summary statements about the outcomes of testing the hypotheses.	**Mark bands**
	Band 1 (1–5 marks) The candidate gives some limited statements relating to the stated hypotheses..
	Band 2 (6–10 marks) The candidate's conclusions are generally logical and relate to the evidence collected for the hypotheses..
	Band 3 (11–15 marks) The candidate demonstrates the ability to state clearly logical conclusions in relation to their hypotheses. which they support by the evidence they have collected.

What should be included in this section?

In this section you refer back to your hypothesis and see if you agree or disagree with it. What evidence do you have to support this? You need to look at each hypothesis separately and conclude what you have learnt about this. You should:

1. Copy out the hypothesis one last time – only this time you are going to make a statement as to whether you found the hypothesis to be True or False.
2. How strongly did you find the hypothesis to be True or False?
3. Are you surprised by your results? If so, what outcome did you expect?
4. What piece of geographical theory did your investigation prove?
   - What particular part of the theory best matches with what you found?
   - What particular part of the theory does not fit with your investigation?
5. Can/have you explained all of the unusual results that you have found?
6. What have your learned about this investigation that you did not know before you started?

(F) Evaluation (10 marks)

Assessment Criteria

(F) EVALUATION	
Candidates should be assessed on their ability to complete a written report: • Advantages and disadvantages of the methods used to carry out the investigation. • Further improvements to the investigation. • Evaluation of the conclusions.	**Mark bands**
	Band 1 (1–3 marks) The candidate is able to identify a few of the strengths of the methods used or conclusions drawn in the investigation.
	Band 2 (4–7 marks) The candidate offers some evaluation of the methods and conclusions and suggests some improvements.
	Band 3 (8–10 marks) The candidate offers a thorough evaluation of the methods and conclusions and suggests improvements to the investigation.

What should be included in this section?

Often students fail to leave enough time from their 6 hour limit to deal with this appropriately. Ensure that you leave enough time to complete your evaluation as it can be very easy to get high marks in this section.

An **evaluation** is when you reflect on the work that you have carried out, in this case over the last few weeks. There are two aspects that need to be evaluated within this section:

1. Evaluation of the methods

- You need to explain any problems that you found with your data collection techniques and any possible improvements.
- You need to explain any problems that you encountered with the methods used for data collection. What were the advantages and disadvantages of each technique used? How reliable and how accurate do you think your results were?
- Could your methods, accuracy, equipment and data collection techniques be improved?

2. Evaluation of the conclusions

- In this section you decide if the conclusions you made were accurate and reliable, explain why and suggest evidence which supports your claims.
- Think about the title of the coursework – was this a good question to be looking at in the first instance?
- Was the location chosen for the study appropriate?
- Think about any problems with your conclusions. Apart from issues with your methodology, what reasons might you have for inaccuracies? Was the location of your survey sites good enough? Did you questionnaire enough people?
- Was the quality of your initial hypotheses good enough? Were they relevant to the Controlled Assessment task? Were they clear and easy to use?
- How might you improve your investigation so that you would be able to trust your conclusions more?

3D: TIPS AND GUIDANCE FOR WRITING UP YOUR FIELDWORK REPORT

Using ICT for your write up: ICT can help streamline the Controlled Assessment process but there are some strict rules on the storage of information and the access that you can have to your materials in the High Control sections of the coursework. Your teacher will give you further guidance on this and advise you on best way to use ICT within your own school.

Time limits: You need to be very aware of the time limits that are set by CCEA on how long you have to plan, prepare and write up various sections of the coursework. You do not want to run out of time at the end.

Avoiding plagiarism: Using information from outside sources within your Controlled Assessment is a good way to show your knowledge of the topic. However, you must not claim that someone else's words or ideas are your own, as this will be considered as cheating. Exam boards set out very clear guidance on how to show what information is your own and what comes from another source. Any time that you use another source of information (including the Internet) for maps, ideas or quotes you must make sure that you acknowledge and reference the material carefully. If you use the exact wording as an outside source, you must also place quotation marks around the extract. Even though your fieldwork can be carried out with other people in your Geography class, you must write your own report, ensure that all the work is your own and you cannot copy from another student. Be careful as plagiarism is taken very seriously and the penalties are harsh.

Health and Safety: It is very important that together with your teacher, you carefully consider aspects of Health and Safety at all stages through your fieldwork.

Final preparations

It is really important that, towards the end of your write up, you organise your completed report in a logical manner. Do not use plastic poly pockets to present your coursework but make sure that your work is securely bound in a folder.

Student checklist

The following list will help you check that you have completed all the sections of your coursework.

Title Page	✓
Index / Contents list	✓
Introduction and methodology	✓
Aims and Hypotheses	✓
Maps (relevant to your study area and at least ONE use of GIS)	✓
Methods of data collection and equipment used	✓
Data Presentation: maps and graphs (labelled and given a heading)	✓
Annotated photographs (where appropriate)	✓
Analysis for each hypothesis	✓
Interpretation for each hypothesis	✓
Conclusion	✓
Evaluation	✓
Bibliography (showing all sources used)	✓
Appendices: data collection sheets	✓

3E: FOCUS ON CONTROLLED ASSESSMENT FOR A RIVER

It is impossible in a textbook to cover every possible example of a Controlled Assessment for Geography in Northern Ireland. Instead, this book focuses on the example of a Controlled Assessment for a river, as this is always a popular option with both students and teachers.

This section is designed to give you some ideas, tips and advice about things that you might need to consider in the early stages of planning for this particular Controlled Assessment piece.

Recent Rivers Titles

YEAR OF SUBMISSION	TITLE
2012	An investigation of river erosion along a chosen stretch of river
2013	An investigation of river features/characteristics along the long profile of a chosen river
2014	A study of how river features change downstream
2015	An investigation of changing characteristics along a chosen stretch of river

Source: CCEA GCSE Geography Controlled Assessment Tasks 2013–2015, © CCEA

Figure 5

Students taking measurements in the Glenarm river, Co Antrim

Most of the titles focusing on rivers to date have allowed students to look at the 'big picture' changes that occur along the course of a river. This allows students to address the specific issues that occur at different places along the river. Students can investigate the depth and width (to help show processes of erosion and deposition), velocity in the river (which can help with wetted perimeter and discharge calculation), and the size and shape of bedload found in the river (which again indicates erosion, transportation and deposition at work within the river). The river title for 2013 allows students to combine their knowledge of the processes that shape the land to investigate the different features that occur along the path of the river.

Working on a piece of controlled assessment that involves rivers is exciting. However, it also has the potential to be dangerous if great care is not taken during planning and implementation.

The following information is designed to help you understand some of the techniques that you might need to collect some of the basic data that will allow you to complete your Controlled Assessment.

Planning for Rivers Fieldwork

1. Planning Issues

Most river studies require you to consider the river channel variables upstream and how the different processes depend on each other. In most cases you will need to find an accessible river that allows you to study the upper, middle and lower courses to allow a comparison between all three.

For a study which is looking at changes along the long profile of a river, it would be advisable to visit the upper, middle and lower course of the river. CCEA recommends that 5 or 6 sites are needed for a sound report. The following considerations might affect your planning for this:

- How many people are working with you in a group to collect results?
- How quickly will you be able to accurately collect the necessary information at each site?
- Are the places that you want to collect data accessible? Is it always safe to enter a river in the lower course?
- Do you need to consider taking results at straight sections or meanders in the river?
- Do you need to employ some type of sampling technique? Are your choices for sites pragmatic (because they are easily accessible), systematic (because you are visiting sites at the

same interval from the last site) or random (generated through the use of a random number table and not influenced in any way by you)?

Figure 6

Taking river measurements in the upper course of Wolf Water, a tributary of the Glenarm river, Co Antrim

2. Health and Safety issues

Rivers can be dangerous environments and before entering the water you need to check that it is safe and take note of the water levels and velocities. You must be able to stand comfortably within the river, which usually means that the deepest part of the river should be no more than 1 m deep. If the river is higher than usual following a flood, you need to take extra care and may even need to postpone your field trip. The velocity (speed) of the water should not be so fast that you cannot work easily in the river. Check with your teacher before entering any river environment that you think is unsafe. In some cases you might also need to do a chemical test to check the pollution levels in the water. You might obtain further information from the Rivers Agency or the Met Office.

You should never take risks to collect river data. What do you think are the key indicators that you should NOT go into a river? Think about the different things that make rivers unsafe.

Health and Safety equipment

Incidents in the river can be minimised if you plan carefully and ensure that everyone understands their role before going into the water.

a. Equipment to help you cross the river safely

- **Waders**
  These are long thigh high or chest high boots that will allow you to cross the river safely by keeping your feet and legs dry. However, if water gets into the waders this can make walking in the river very difficult.

- **Ranging Poles**
  Although these are designed for measuring slope angles, they can also useful for steadying yourself when crossing the river. The Ranging Pole can also be used to help find suitable places for people to put their feet as they cross the river and to test depths.

- **Safety Rope**
  A strong rope can be tied about 1 m above the surface of the water to provide something for you to hold on to if you feel that you are falling. The rope should be pulled taught and either held by other team members or tied around a tree.

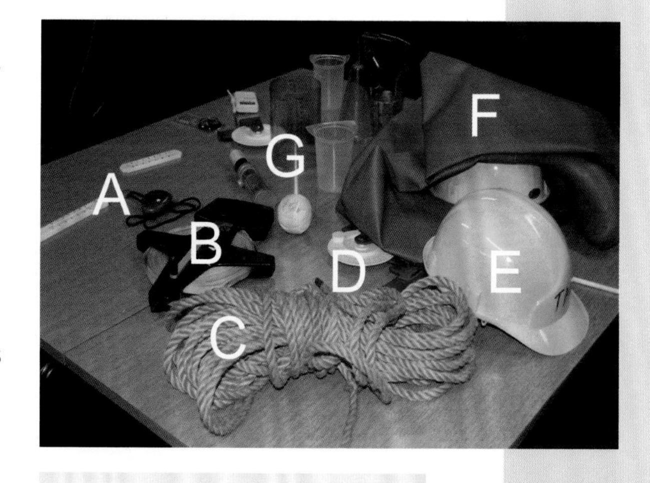

Key:

A Stopwatch
B Tape measure
C Safety rope
D Clinometer (for slope angle)
E Safety helmet
F Waders
G Polystyrene float

Figure 7

Some typical Geography fieldwork equipment

Figure 8

Some of the safety equipment that can be used when crossing a river

b. First Aid equipment

As with all Geography field trips, it is essential to carry a fully stocked first aid kit.

c. Other essential emergency equipment

- **Emergency whistle**
  A plastic whistle should be issued to each group so that in the event of an emergency they can call for assistance quickly. Do you know the international distress signal when using a whistle?
- **Emergency throwrope**
  It can sometimes be useful for a teacher to carry a special emergency throwrope or life belt/ring that can be thrown out to students if they are struggling in the water.
- **Bivvy bag**
  This is a simple plastic bag that can help to carry one person in emergency situations.
- **Mobile Phone**
  It is an essential piece of kit for any expedition to call for help and the emergency services. However, you need to see if the phone can be used in every location, for example, you might find that the upper course site of the river might be without mobile phone coverage. How could you plan for this?

Can you think of any other guidance that people might need to help them work and collect information in a river? How should people work together as a team? What is the best number of people in a group working together for a study like this? How should people move in the water to minimise splash and make sure they do not lose their footing in the river?

3. Hypotheses and Objectives

There are a wide range of potential objectives and hypotheses that are available to a river study. You need to think carefully about which ones you want to use, as it is better to choose two or three hypotheses and do these well rather than to write ten hypotheses that are only half completed.

The following are some examples of hypotheses that you could consider:

RIVER INFORMATION	HYPOTHESIS	OBJECTIVE/DETAILS
Stream Channel Characteristics	The shape of the river channel will increase downstream.	In these hypotheses you are going to look at the basic aspects of change within the stream. Use the data collection methods opposite to help you work out what is being recorded and why this is important. • What are the changes you expect to see in the river? • Does the river you are studying do what rivers are expected to do?
	The velocity of the river will increase downstream and is linked to depth.	
	There is a significant, positive correlation between width and depth.	
	There is an increase in discharge with increasing distance from the source.	

RIVER INFORMATION	HYPOTHESIS	OBJECTIVE/DETAILS
River Load	The bedload in the river will become more rounded and smaller as you move downstream.	Collecting and testing the bedload/sediment in the river can be a biased way of collecting information about the river. Often students are more likely to select large, odd shaped stones across the river channel, rather than taking a more systematic approach to selection. However, it does help to gain information about the amount of erosion, transportation and deposition taking place along the river course.
	There is a decrease in sediment size as you move downstream.	
	The sediment gets more rounded with increasing distance from the source.	
Hydraulic Radius	There is a significant, positive correlation between velocity and hydraulic radius.	This helps to identify any further changes that may take place in geology along the river.

4. Data Collection methods

a. River cross sectional area

Equipment: metre ruler, long tape measure

One of the most visual and effective demonstrations of graphical skills in river fieldwork is drawing the cross section of a river channel and comparing the size and shape at different sites along the river.

Figure 9
Taking depth and width measurements in a river

- **Channel width (m)**
  Use the tape measure and stretch it across the full width of the river above the water by about 1 m. The tape measure needs to be kept clear of the water and should be held as tightly as possible. The measurement is usually taken from the inside river bank edge to the far bank edge.

- **Channel depth (m)**
  Use the metre ruler to record the depth of the water from the river bed to the surface of the water. The depth measurement should be taken at 30 cm intervals from the inside river bank to the far bank edge.

- **Cross section profile**
  When you have completed these measurements you will be able to use a suitable scale on graph paper to plot the shape of the river bed beneath the water level. This is the cross section profile for the river at the site that you have visited.

- **Cross sectional area (m^2)**
  This is a very simple but useful calculation that you can use to compare the area of water at each site.

  CSA (m^2) = width (m) × depth (m)

b. River velocity

Equipment (Method 1): Tape measure, Float, two Ranging Poles, Stop Watch
Equipment (Method 2): Flowmeter, Stop Watch

The velocity is the speed of the water at a particular point in the river. The velocity will change along the course of a river and the factors that cause the velocity to change are the gradient, the volume, the shape of the river channel and the amount of friction which might be created by other things in the river such as the river bed, its load and plants. Velocity is measured in metres per second (m/sec).

There are two main ways that you can measure velocity:

- **Method 1: Using a float**

  The first thing you need to do is measure a 10 m stretch of river downstream along the near bank. You can use the ranging poles to identify the start and finish line. Have one member of your group stand upstream ready to place a float (for example, a tennis ball or an orange) carefully into the river and another to catch the float when it has travelled the 10 m downstream.

  Usually, you will need to take nine measurements at different points across the river. Three should be taken towards the near bank, three should be taken midstream and another three should be taken towards the far bank.

  You then can calculate the average time taken at each of the three points across the river and work out the differences of velocity across the river, as well as the differences in velocity from one site to the next. Do not forget that your measurement is for a 10 m stretch of river and you need to do a final calculation to work out what the surface speed/velocity is for a 1 m stretch of river:

$$\textbf{Surface speed/Velocity} = \frac{\textbf{Distance (m)}}{\textbf{Time (secs)}}$$

- **Method 2: Using a flowmeter**

  In some ways using a specialist flowmeter can be easier to use out in the river but can be more complicated when trying to work out the final velocity measurement.

  You use a flowmeter by placing the propellor device facing upstream. Usually the device should be held so that it is just clear of the river bedload. Measurements should be taken to count the number of propellor rotations that take place within a timed minute.

  Generally, you will need to take nine measurements at different points across the river. Three should be taken towards the near bank, three should be taken midstream and another three should be taken towards the far bank.

  You then can calculate the average time taken at each of the three points across the river and work out the differences of velocity across the river, as well as the differences in velocity from one site to the next. Do not forget that your measurement is the number of rotations that the propellor has made within 60 seconds so you will need to use the guidance that comes with your flowmeter to calculate how the rotations will give you a reading for the velocity per second.

  For your evaluation section, you may wish to consider which of the two velocity methods you found to be more effective, more efficient or more accurate.

c. Friction

Rivers tend not to be very efficient systems for transporting water from one place to another. They use and lose a lot of energy through the process of transportation downstream. As water flows over the river bed and alongside the river banks it experiences friction in the form of drag. Sometimes further friction is caused on the surface of the water as air moves across the water (wind).

Internal Friction in a river is called turbulence. It can be very difficult to estimate the amount of turbulence within a river, especially in calm conditions. However, as a general rule, calm conditions usually indicate little friction, whilst fast flowing water will have a lot more friction.

External Friction is caused when the water in the river touches the river bed and the river banks. The amount of friction depends on the proportion of the river that is contact in this way. The shape of the river is important in this regard – a shallow, wide river will experience a lot more friction than a deep, narrow river channel.

You can use something similar to the Friction Table in Figure 10 to give a score to a particular piece of river.

	Friction Score	Internal Friction (Turbulence) Calm water	External Friction (Banks and bed)
Low Friction	1	Usually lower course.	River bed and banks usually quite smooth.
	2	Water velocities different across the river channel.	River bed fairly smooth, quite sandy.
	3	Some water is choppy across the river channel.	River bed is deeper on one side than the other – some build up of deposited material on one side.
	4	White water evident in the river, moving over larger pieces of bedload. More usual in upper course.	The river bed is irregular and coarse.
High Friction	5	Very fast moving water, severe white water, rapids. Usually only found in upper course or flood conditions.	River bed is very irregular and is strewn with large boulders.

Figure 10
Friction Table

d. Wetted Perimeter and Hydraulic Radius

- **Wetted Perimeter**

  This is the total length of the river channel cross section which is in contact with the water. It can be measured in the field using a tape measure or a chain which is laid across the river in a straight line. It is very difficult to measure this accurately, especially if the velocity of the river is elevated. However, it can also be calculated by measuring the amount of river bed that is under water on the cross section profiles that you drew earlier. The wetted perimeter is measured in metres.

- **Hydraulic Radius**

  This compares the wetted perimeter (the amount of river bank and bed that is in contact with the water) with the cross sectional area. It is usually expressed as a ratio. A river that has a result close to 1.0 will be more efficient and will have less friction (and have less water in contact with the river bed and banks). A river less than 1.0 will be less efficient, have more friction (and have more water in contact with the river bed and banks). Geographers generally agree that the greater the efficiency of the channel, the less likely the river is to flood.

$$\textbf{Hydraulic Radius} = \frac{\textbf{Cross sectional Area (m}^2\textbf{)}}{\textbf{Wetted Perimeter (m)}}$$

e. River Gradient

Equipment: Clinometer, Ranging Poles, Tape Measure

In some cases you will want to measure the gradient of the river channel. This allows you to see how steep the river is.

You should use the tape measure to measure a 10 m stretch of river and use the ranging poles to again indicate the distance. You should make sure that the bottom of the ranging pole is resting on the river bed in each case. Line up your clinometer and work out the angle from the downstream to the upstream pole.

Sometimes, measuring the gradient of the river channel can help to explain some of the velocity and discharge readings that you get along the course of the river.

f. Discharge

The discharge of a river is the volume of water which flows through a particular cross section at a particular time.

You can calculate the discharge (in metres cubed per second) by multiplying the cross sectional area by the average velocity of the channel.

g. Bedload

Equipment: Metre ruler or callipers

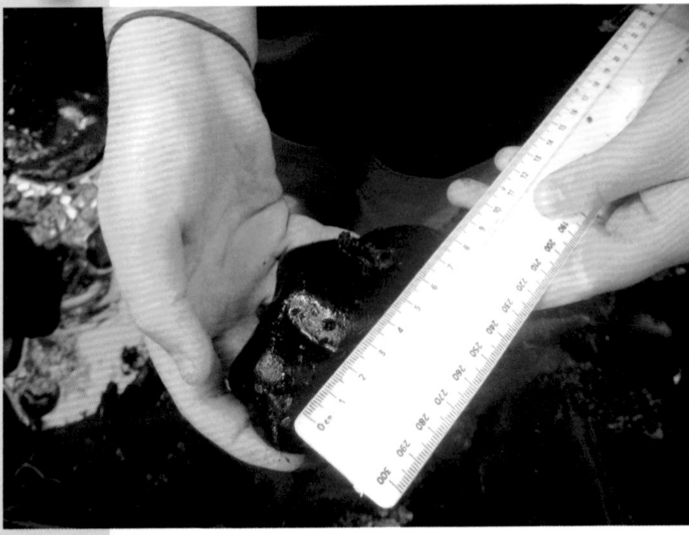

Figure 11

Using a meter ruler to measure the size and shape of bedload

The collection of the river bedload is often something that students find difficult to do accurately in the field. Being able to measure the bedload in the river is very important, as the size and shape of the stones in the river can help to explain the processes that are taking place at a particular place.

The aim of this measurement is to try to sample the size and shape of the pebbles and boulders that you find across the river bed, from one side of the river to the other. You should use a random method of collection. Insert a meter ruler into the river at random places across its width and lift whatever stone the stick is touching. Aim to collect 20 stones from across the river. Measure the long axes of each stone (the longest two points) and assess the stone's shape using the Powers index of roundness.

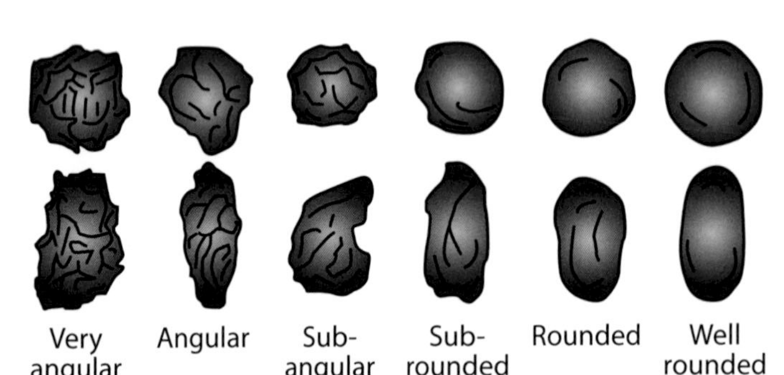

Figure 12

The Powers index of roundness (used to observe stone shape)

All you need to do now is design some recording sheets to help you to collect the information.

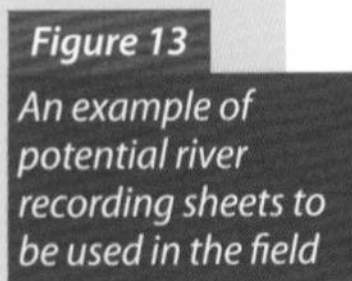

Figure 13

An example of potential river recording sheets to be used in the field

It is important to have a safe, businesslike attitude when working in the field, but fieldwork in Geography can be fun too. Take time to look around the physical landscape where you are working, as this often gives clues to the different processes that are at work within the river environment.

Once you get back into the classroom it is then over to you to put together some different data presentation techniques which will allow further discussion of the hypotheses that you have selected.

Copyright Information

continued overleaf

OECD/IEA: World Energy Outlook 2011 © OECD/IEA 2011, fig 10.1, p354: 237 (middle)

Office for National Statistics (licensed under the Open Government Licence v1.0): 156

Ordnance Survey (of Great Britain) ©Crown copyright 2013 License number 100054410: 47 (bottom); 48

Ordnance Survey of Northern Ireland (This material is Crown Copyright and is reproduced with the permission of Land & Property Services under Delegated Authority from the Controller of Her Majesty's Stationery Office, ©Crown Copyright and database right 2014 PMLPA No 100496): 20 (A); 22; 170 (bottom, underlying data only); 176 (roads & coast only); 178; 181

Seppo Leinonen, seppo.net: 121

Shutterstock: 251

Slemish College: 206 (bottom)

SSE and Dong Energy: 241 (both)

Titianic Quarter Limited: 188; 189

United Nations: 200 (Development Programme); 238 (UNEP Global Environment Outlook 5)

United States Army Corps of Engineers: 31 (middle)

United States Department of Energy: 88 (bottom, see notes beside graph for details)

United States Environmental Protection Agency: 88 (middle)

United States Geological Survey: 35 (both), 88 (top); 110 (top); 114 (photo by Scott Haefner); 129 (top left, photo by Walter Mooney); 130 (both)

United States Geological Survey Astrogeology Science Center http://astrogeology.usgs.gov: 87 (bottom)

United States Library of Congress: Library of Congress Prints and Photographs Division Washington, D.C. Reproduction number: LC-DIG-ppmsc-09840: 50 (middle); 56 (top); 204

United States Marine Corps photo by Lance Cpl. Garry Welch/Released: 123 (bottom)

United States Navy photo by Mass Communication Specialist 3rd Class Alexander Tidd/Released: 124 (bottom)

University Of Cambridge Collection Of Air Photographs/Science Photo Library: 47 (top)

Richard Vogel: 152

what'sthatpicture.com / James Morley: 31 (top left)

The following images are licensed under the Creative Commons Attribution-Share Alike 3.0 Unported license: 20 (B, Adfern); 32 (top left, Antoine Taveneaux); 32 (middle), US Army Corps of Engineers); 65 (middle, Alan Sim); 66 (top, Alan Sim); 96 (bottom, Sam Beebe, Ecotrust); 97 (left, Oxyman); 97 (map, base data only, OpenStreetMap); 97 (bottom, Nevilley); 125 (Morio); 199 (Alex Covarrubias); 213 (http://en.wikipedia.org/wiki/File:ActiveBlocs.PNG)

The following images are licensed under the Creative Commons Attribution 2.0 Generic license: 18 (bottom, NASA Goddard Space Flight Center); 20 (C, John Holm); 20 (E, Jim McDougall); 21 (Doc Searls); 31 (top right, David Dickenson); 31 (bottom, US Army Corps of Engineers); 33 (middle, Hugh Venables); 44 (top, Rev Stan); 44 (middle, Tom Maloney); 45 (top right, artq55); 46 (bottom right, Fearlesspunter); 55 (US Army Corps of Engineers); 63 (middle, Amanda Slater); 64 (top, Andres Rueda, http://www.flickr.com/photos/andresrueda/); 64 (bottom, Osa Mu); 65 (top, the aucitron); 66 (second from bottom, Arjan Richter); 67 (top right, National Oceanic and Atmospheric Administration (NOAA)); 67 (bottom right, Michael Pereckas); 68 (middle & bottom, National Oceanic and Atmospheric Administration (NOAA)); 90 (bottom, Department for International Development); 94 (middle, John LeGear); 94 (bottom, Peter Blanchard); 97 (top, Alex Graves); 98 (top, Elliot Brown); 129 (top left, Colin Crowley); 131 (Trees for the Future); 168 (middle, Richard Luney); 174 (top, Abdallah); 186 (right, Francisco Antunes); 206 (top, Paul Miller); 210 (Star5112); 224 (Rick Scavetta, U.S. Army Africa); 227 (Res A middle, Blake Patterson); 233 (left, Peter Kaad Andersen); 234 (top, Ansel Chen Lin); 234 (bottom, Metro Centric); 235 (Metro Centric)